Process Gas Chromatographs

Process Gas Chromatographs

Fundamentals, Design and Implementation

TONY WATERS

WILEY

This edition first published 2020

Registered Offices
John Wiley & Sons, Inc., 111 River Street, Hoboken, NJ 07030, USA
John Wiley & Sons Ltd, The Atrium, Southern Gate, Chichester, West Sussex, PO19 8SQ, UK

Editorial Office
John Wiley & Sons Ltd, The Atrium, Southern Gate, Chichester, West Sussex, PO19 8SQ, UK

For details of our global editorial offices, customer services, and more information about Wiley products visit us at www.wiley.com.

Wiley also publishes its books in a variety of electronic formats and by print-on-demand. Some content that appears in standard print versions of this book may not be available in other formats.

Library of Congress Cataloging-in-Publication Data

Names: Waters, Tony, author.
Title: Process gas chromatographs : fundamentals, design and implementation
  / by Tony Waters.
Description: Hoboken, NJ, USA : Wiley, 2020. | Includes index.
Identifiers: LCCN 2020001270 (print) | LCCN 2020001271 (ebook) | ISBN
  9781119633044 (cloth) | ISBN 9781119633006 (adobe pdf) | ISBN
  9781119633013 (epub)
Subjects: LCSH: Gas chromatography–Equipment and supplies.
Classification: LCC QD79.C45 W38 2020 (print) | LCC QD79.C45 (ebook) |
  DDC 543/.850284–dc23
LC record available at https://lccn.loc.gov/2020001270
LC ebook record available at https://lccn.loc.gov/2020001271

Cover Design: Wiley
Cover Image: Process Gas Chromatographs at the INEOS Olefin Plant in Cologne, Germany. Photo © INEOS in Cologne, 2019.

Set in 11/13pt STIXTwoText by SPi Global, Chennai, India

MIX
Paper from responsible sources
FSC
www.fsc.org
FSC® C013604

To Marilyn

Contents

Process Gas Chromatographs installed in a prefabricated air-conditioned analyzer shelter for an ethylene plant in Texas. Image © Yokogawa Corporation of America, 2018. Reproduced with permission.

Preface

Welcome to the world of Process Gas Chromatography!

This book focuses on the **Process Gas Chromatograph** (PGC). There are dozens of fine books on the science of gas chromatography but few on the technology of the process instrument. I found only two previous books dedicated to online gas chromatographs (Huskins 1977; Annino and Villalobos 1992).

Process gas chromatographs are complex instruments, and the people that design and operate them need special knowledge and unique skills. With that in mind, I designed the book to serve the needs of the journeyman analyzer technician, the process instrument engineer, and the process analyzer specialist.

PGC is a practical technology, and this is a practical book. It's an effective classroom training manual for those currently learning the art and a handy reference manual for those already practicing it.

Chapters are deliberately compact, suitable for a weekly reading program or as focused lessons in an educational course. Each chapter ends with a summary of knowledge gained and a self-assessment quiz with answers provided. In addition, there are nine optional test questions for students; three easy, three moderate, and three challenging.

Why is such a book necessary?

Anyone working in the fluid processing industries knows that their knowledge base is in full flight. Due to staffing reductions and mass retirements our industry is losing decades of hard-won experience.

Walter Jennings and Colin Poole recently expressed this situation rather well (Jennings and Poole 2012, 72):

> *This [automation of gas chromatographs] has led to a continuing decline in the expertise of the average practicing chromatographer from the mid-1980s to the present time. This can be perilous, because everything from column selection to trouble-shooting skills is based on a fundamental knowledge of chromatographic principles, the absence of which degrades the quality and usefulness of the information acquired by these instruments. To address these problems requires a massive educational effort before the knowledge is lost and the usefulness of gas chromatography to decision makers is called into question.*

There can be no clearer call to justify this book. While the authors were writing to laboratory chemists, those working on process gas chromatographs also need a *fundamental knowledge of chromatographic principles* presented in a way that facilitates a *massive educational effort*. This textbook sets out to satisfy those needs. It's primarily written for process analyzer engineers and technicians but should be helpful to anyone using or maintaining a process gas chromatograph.

To succeed in its mission, a book needs to so excite readers that they want to read more. It should be so useful that they immediately return to it when they need information. Yet the average book on gas chromatography is abysmally boring and poses an intellectual challenge even to post-doctoral scientists, let alone the lonely guy faced with fixing a broken process chromatograph at midnight.

This text teaches the fundamental knowledge of process gas chromatography by encouraging the reader to think critically about what is happening in the instrument, mostly without recourse to analogy or math. It also describes some practical procedures for design or troubleshooting.

So, here you have it. A clear yet detailed book that is ideal for classroom instruction, private study, or distance learning. Focused chapters unfold the technology of a process gas chromatograph to an engineer or technician who may have no previous experience of the technique. The content is basic, yet thorough, so it should meet the needs of many readers.

I'm glad that you're here. I hope you enjoy the book!

Tony Waters
Atascadero, California
January 2020

References

Cited

Annino, R. and Villalobos, R. (1992). *Process Gas Chromatography*. Research Triangle Park, NC: Instrument Society of America.

Huskins, D.J. (1977). *Gas Chromatographs as Industrial Process Analyzers*. New York, NY: Pergamon Press.

Jennings, W.G. and Poole, C.F. (2012). Milestones in the development of gas chromatography. In: *Gas Chromatography* (ed. C.F. Poole), 1–18. Oxford, UK: Elsevier.

Contributors

An international team of expert chromatographers has peer-reviewed the technical content of this text. This **Editorial Advisory Board** comprised the experienced analyzer engineers listed below. We gratefully acknowledge their contributions. Culpability for remaining errors or omissions rests entirely on the author.

Jerry Clemons, PhD
Process Gas Chromatograph Consultant

Formerly, General Manager
ABB Process Analytics
Ronceverte, West Virginia, USA

- Jerry has worked with gas chromatographs during his entire career starting at Virginia Polytechnic University where he earned his PhD with Dr. Harold McNair.
- He has held many engineering and management positions at ABB Process Analytics and its predecessors, always focused on their process gas chromatographs. Now retired from active duty, he continues to provide his technical expertise as a consultant to that company.
- Jerry has 50 years of experience working with process gas chromatographs.

R. Aaron Eidt, BSc
Process Analyzer Consultant
PEAK PERFORMANCE Analytical Consulting Ltd.
Delta, British Columbia, Canada

Formerly, Analyzer and PGC Manager
Dow Chemical Canada
Fort Saskatchewan, AB, Canada

- Aaron is a chemist with 25 years of experience developing new GC methods for research and industrial chromatographs at Dow Chemical Canada. Aaron specialized in process analyzer validation, troubleshooting and performance improvement. For several years, he led the Dow Global Process Chromatography Technology Network.
- Since retiring from Dow, Aaron has had process analyzer consulting engagements with the Sadara Chemical Company in Jubail, Saudi Arabia, and with MEGlobal.
- Aaron has developed and instructs both introductory and advanced troubleshooting training courses in process gas chromatography for analyzer maintenance technicians.
- Aaron has 30 years of experience practicing industrial gas chromatography.

Zoltán Hajdú, RNDr
Marketing Manager
Analyzer System Integration
Yokogawa Europe

Formerly, Analyzer Systems Consultant for Yokogawa
Central and East Europe

- Responsible for analyzer system design and analyzer selection, including on site start up, and trouble-shooting of Yokogawa process chromatographs and analyzer systems throughout Central and East Europe.
- Previously, Supervisor of Process Analyzers at Slovnaft Refinery in Bratislava. Now responsible for analyzer system sales for Yokogawa in Europe.
- Zoli has 12 years' experience working with various process gas chromatographs.

Phil Harris, BSc MSc
Process Analyzer Consultant
President
Insight Analytical Ltd.
Calgary, Alberta, Canada

Formerly, Engineering Manager AMETEK Western Research
Calgary, Alberta, Canada

- Phil has a BSc in Physics and a Master's in Chemistry. His career began in the Research Chemistry branch of Atomic Energy of Canada, where he designed spectroscopic analyzers and built algorithms for numerical analysis of spectral and chemical data.
- Phil has been an independent consultant since 1998, primarily on the development of process analytical solutions in the oil, gas, and petrochemical industry. He provided services to AMETEK for a number of years and developed most of the numerical analysis algorithms used on the 900 series of Analyzers.
- He has published over 25 papers and has given training courses on spectroscopy and process analyzer sample systems all over the world.
- Phil has 35 years of experience with industrial process analyzers, mainly with process spectrometers.

Michael Hoffman
Business Development Manager
Siemens Industry, Inc.
Analytical Products & Solutions
Houston, Texas USA

- Michael started in industry at Phillips 66, and continued the journey with Standard Oil Chemicals, BP, Innovene, and INEOS.
- His initial work was with laboratory chromatographs. After transitioning to process chromatographs, he focused on online analyzer reliability, advanced control support, materials handling, and analyzer data management technologies.

- Michael joined Siemens in 2007. He now provides marketing and technical support for analytical solutions, communications, PGC applications, and sample handling system designs.
- Michael has 37 years of experience working with laboratory and process gas chromatographs.

Dirk Horst
Process Analyzer Consultant
Heerhugowaard, Netherlands

Formerly Global QMI Consultant
Shell Global Solutions Team
Amsterdam, Netherlands

- Dirk has long experience with process analyzers, including startup assignments at Shell jobsites in Germany, India, Nigeria, and Russia. He is also well known for his many classroom and practical training programs for analyzer maintenance technicians.
- Dirk has 34 years of experience working with process gas chromatographs.

Dr. Daniel Kuehne
Process Gas Chromatograph Consultant
Siemens AG
Analytical Products and Solutions
Manufacturing Karlsruhe, Germany

- Daniel studied Chemistry at the University of Bremen and did his diploma and doctorate thesis in Analytical Chemistry.
- He joined Siemens in 2005 as method developer for process GCs. He stayed in method development for 11 years, whereof the last five years being the head of the PGC method developer team.
- Since 2016 he has been making technical evaluations of PGC inquiries and working as a technical consultant for sales and customers and additionally as technical advisor for the GC method development team.
- Daniel has 14 years of experience working with process gas chromatographs.

James Leonard, PhD
Process Analyzer Specialist
Eastman Chemical Company
Kingsport, Tennessee, USA

- James received his PhD in Analytical Chemistry from The Ohio State University. He has 20 years of experience working in the field of Process Analytics at Eastman Chemical. During this time, James has designed, installed, and commissioned analyzer systems incorporating modern on-line techniques throughout the world.

- He has presented lectures on process analytics at universities and other organizations to promote the use of on-line technologies to improve process control and reduce waste.
- James has 17 years of experience working with process gas chromatographs.

Harald Mahler
Process Analyzer Engineer
Siemens AG
Analytical Products and Solutions
Karlsruhe, Germany

- Harald studied chemistry at the University of Applied Science in Reutlingen. Since 1989 he has gained experience in process analytics in various engineering and management positions within Siemens AG. He has authored and presented many technical papers within the process analytical community.
- Harald has held engineering and management roles in application and method development, project management, industry marketing, and product management. Currently he is Global Sales and Business Development Manager for process analytics within the Process Automation Division, serving mainly the petrochemical, oil and gas, and renewable energy markets.
- Harald has 29 years of experience working with process gas chromatographs.

Gen Matsuno, ME
Product Manager
Quality Analyzer Systems

General Manager
IA-PS Analyzer Center
Yokogawa Electric Corporation
Mitaka, Tokyo, Japan

- Matsuno-san was leader of the Yokogawa GC8000 PGC development team. In addition to his experience of designing process gas chromatographs, he has five years of experience as a laboratory GC user.
- Gen-san has 12 years of experience working with process gas chromatographs.

Takashi Matsuura, BE
Senior Field Engineer
Nippon Swagelok FST, Inc.
Yokohama, Japan

Formerly, Manager of Process GC Development
Yokogawa Electric Corporation

- Taka designed the Yokogawa GC1000 PGC oven and was leader of the engineering team that developed the Yokogawa GC1000 Mk2 PGC. He also wrote the specifications for the GC8000 PGC.
- Taka has over 25 years of experience working with process gas chromatographs.

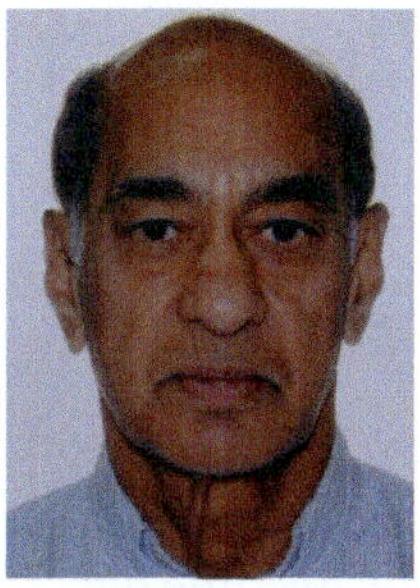

Suru Patel, PhD
Process Analyzer Consultant
Patex Controls Ltd.
Calgary, Alberta, Canada

Formerly, Distinguished Engineering Associate for Process Analyzers
Exxon Chemical Company, Sarnia, Canada, and Singapore

- In addition to his process analyzer engineering work, Suru developed PGC training courses for process analyzer technicians and PGC data users.
- Previously, for several years, Suru was a PGC Applications Engineer at Servomex Company in the UK and was the Lead Analyzer Engineer in Houston for Exxon's Singapore Chemical Complex project. He was also the development engineer for a new flame ionization detector.
- Suru has 40 years of experience working with process gas chromatographs.

Ivan Rybár, PhD
Head of Process Analyzer Group
Slovnaft MaO, a.s.
Bratislava, Slovakia

Formerly Research and Teaching
Assistant, Department of Analytical Chemistry, Comenius University
Bratislava, Slovakia

- For 10 years now, Ivan has been responsible for the reliability of all process analyzers at the refinery, including the maintenance of existing systems and the design of new installations. As the supervisor of 26 analyzers, he creates work procedures and provides training and support for his team.
- In 2013 and 2015, he was twice awarded the accolade "Slovnaft Star."
- Previously he worked as an analyzer engineer for several companies providing engineering services to industrial plants, including the selection of analyzers and the design of complete sampling systems.
- During his time at university, Ivan developed new methods and taught several graduate courses in liquid chromatography. He has recently published four scientific papers on this work.
- Ivan has 15 years of experience working with process gas chromatographs.

Eric Schmidt, PhD
Principal Research Scientist
The Dow Chemical Company
Analytical Sciences
Freeport, Texas, USA

- Eric Schmidt received his PhD in Analytical Chemistry from The University of Texas at Austin. He has worked at the Dow Chemical Company in Freeport, Texas, for over 20 years as a Research Scientist where he spends his time developing new on-line process measurements for R&D and manufacturing. He is currently leading the On-line Chromatography Strategic Capability Team at Dow.
- Eric has 20 years of experience working with process gas chromatographs.

Acknowledgments

Hearty thanks all friends and associates who contributed material to this text. Many of those listed below audited the beta-test of an online tutorial based on this textbook. Their contributions of time and knowledge are much appreciated.

Minh Anh	Vietnam
Brian Aplin	South Carolina
Dale Arstein	Ohio
Masafumi Awano	Japan
Ken Backus	Texas
Hesham El Banna	Saudi Arabia
Eddie Beezemer	Netherlands
Linda Bonnette	Texas
Mark Booth	Scotland
Danny van den Burg	Netherlands
Bruno Chaurand	France
Alice Chin	Malaysia
Alex Chu	England
Scott Cookson	Australia
Marcus Creaven	Ireland
Dave Demsey Sr.	Pennsylvania
Matt Dixon	Ohio
Ana Dominguez	Switzerland
Kevin Fajri	Indonesia
Mikhail Fedorets	Russia
Mike Frost	Australia
Victor Alberto Fuentes	Spain
Keisuke Fukada	Japan
Yves Gamache	Quebec
Udo Gellert	Germany
Roger Glass	England
Urich Gokeler	Texas
Matt Hasenohr	Oregon
Darryl Hazlett	Texas
Jack Holland	England
Tom Huddle	England
Damian Huff	West Virginia
Humberto Serrato Hurtado	Colombia
Joe Iveljic	Ohio
Nick Iverson	Minnesota
Samson Jacob	Abu Dhabi
Jayson Zhang Ji	Singapore
Kyle Juist	Ohio
Eric Kayla	California
Eric Kvarda	Ohio
Bert Laan	England
André Lamontagne	Quebec
Wilco Landkroon	Netherlands
Rudi Lehnig	Germany
Tim Lenior	Netherlands
Hank Liu	Singapore
Aldemar Figueroa Loza	Colombia
Karim Mahraz	Ohio
Rogério Matos	Brazil
Bill Menz	Ohio
John Meyer	Germany
Thomas Neuhauser	Oklahoma
Kentaro Nomura	Japan
Doug Nordstrom	Ohio
Tatsuya Ohkoshi	Japan
Bob Perusek	Ohio
Stacey Phillips	Alberta
Wouter Pronk	Netherlands
Venkat Rao	Dubai, UAE
Syed Jawwad Raza	Qatar
Reino van Rensburg	South Africa
Joe Rodriguez	Ohio
Zaffar Shariff	Singapore
Joel Siallagan	Indonesia
Trey Sinkfield	Texas
Charlie Smith	Louisiana
Sharon Sng	Singapore
Mike Strobel	Pennsylvania

Max Sukuma	Australia	Mark Welch	England
Asad Tahir	Texas	Henk van Well	Germany
Jorge Trillos	Colombia	Martin Wieser	Germany
Steve Trimble	Oklahoma	Kenta Yamada	Japan
Kunawat Wattanakij	Thailand	Norbert Zeug	Germany

Part One

PGC fundamentals

1975 Beckman Model 6800 Air Quality Chromatograph.

Figure 1.1 A Classic Process Gas Chromatograph.
Source: Beckman Historical Collection, Box 58, Folder 28. Science History Institute, Philadelphia. https://digital.sciencehistory.org/works/474299142
Reproduced with permission of Rosemount, Inc.

"***We cannot teach people anything; we can only help them discover it within themselves.***"

Attributed to Galileo Galilei
1564–1642

Process Gas Chromatographs: Fundamentals, Design and Implementation, First Edition. Tony Waters.

Why study this?

Part One introduces the art and science of gas chromatography (GC) as applied to the industrial process instrument.

These four chapters explain how a GC column works, why the compounds in the injected sample form the characteristic peak shape, how one peak becomes separate from another peak, and how we can predict the position and shape of peaks on a chromatogram from known patterns of peak timing and width.

The text presents this information in an easy-to-read and mostly non-mathematical manner. Yet it shuns simplistic analogies of what happens inside a GC column because they tend to mislead rather than to inform. Instead, it offers a challenging insight into real chromatographic behavior.

The knowledge gained here is a necessary preparation for understanding the function of the hardware devices and software techniques introduced in later chapters of the book. For those who aspire to be proficient in the application or troubleshooting of process gas chromatographs, mastery of these concepts is not optional.

1

An introduction

> *"Books on gas chromatography, of which there are many, usually start by reviewing the historical development of the science, so we won't do that here. Instead, we'll start by understanding the basic technique: what a chromatograph does and how it does it. To read the fascinating history of chromatographic science, see the beautiful book by Ettre (2008)".*

Chromatographic separation

Let's start by looking briefly at the various forms of **chromatography**.

Chromatography by itself is not a complete analytical technique. It's just a way to separate one kind of **molecule** from another kind of molecule. Of course, for those reading this book, the reason for separating those molecules is to measure them alone, without interference from other molecules. This is the analytical use of chromatography.

While analytical measurement is the main use of chromatography, it is not the only one. Some laboratory-scale and industrial-scale processes use a chromatographic separation to isolate extremely pure batches of valuable chemicals. This usage is known as preparative chromatography, and it works with much larger quantities of material than analytical chromatography does. This textbook focuses on analysis, so it doesn't further discuss the preparative use of chromatography.

When used as part of an analytical technique, chromatography is a very effective way to separate the measured compounds from each other and from all the other chemical compounds present in the analyzed material. After all desired compounds have been isolated, another device measures each one independently.

Keep in mind, then, that chromatographic analysis is always a two-stage process: first separation, then measurement.

There are many ways to produce a chromatographic separation, and they involve all possible combinations of gases, liquids, and solids. While

Process Gas Chromatographs: Fundamentals, Design and Implementation, First Edition. Tony Waters.

quite different in practice, these various forms of chromatography share some common features. All practical chromatographic separations involve a fluid material moving across the surface of a stationary material.

In the formal terminology of chromatography, the moving material is the **mobile phase**, and the immobile material is the **stationary phase**.

The mobile phase may be a gas or a liquid, from which we derive the terms:

- **gas chromatography**, in which the mobile phase is a gas.
- **liquid chromatography**, in which the mobile phase is a liquid.

A few applications have used a **supercritical fluid** as the mobile phase.

This book is about the analytical use of gas chromatography for the online measurement of industrial processes. We won't be discussing liquid chromatography.

In gas chromatography, the mobile phase is always a gas, and it's common to call it the **carrier gas**.

The carrier gas flows through a long narrow tube called a chromatographic **column**, which contains the stationary phase. The stationary phase may be an adsorbent solid or a non-volatile liquid. More about that later.

The gas chromatograph

The basic instrument

A **gas chromatograph** is an analytical instrument that uses the techniques of gas chromatography to measure the concentration of selected chemical compounds in a small sample containing a mixture of compounds.

In a gas chromatograph, the mobile phase is a gas carefully selected for the application. It's usually hydrogen, helium, or nitrogen, but any gas will do the job, as long as is doesn't react with the sample components or the column materials. The gas should not contain oxygen or water vapor, as these substances might damage the columns.

The pressure of the carrier gas is closely controlled, after which it flows continuously through the column.

The analyzed fluid can be a gas or a volatile liquid. A special valve injects a small volume of that fluid into the flowing carrier gas. A liquid sample usually vaporizes instantly upon injection, so it's all vapor by the time it reaches the column.[1]

> *Note that when a gas chromatograph accepts a liquid sample, it doesn't become a liquid chromatograph. A liquid chromatograph is an entirely different instrument that employs a liquid*

[1]There are some exceptions to the principle of instant vaporization that are beyond the scope of this introductory text.

mobile phase and separates components in the liquid phase. Liquid chromatographs are rarely employed as industrial online analyzers and are not considered here.

After injection, the carrier gas carries the gas or vapor sample into the column, where it contacts the stationary phase. It's the contact with the stationary phase that accomplishes the desired **separation**.

It's common to use the word **component** for a chemical substance or a group of chemical substances that are present in the sample. The gas chromatograph may not measure every component, but each component measured is an **analyte**.

A gas chromatograph can separate and measure one, several, or all the components in a gas or liquid sample.

After separation, the carrier gas carries the components into a **detector** that provides a measurable signal to the data-processing circuits.

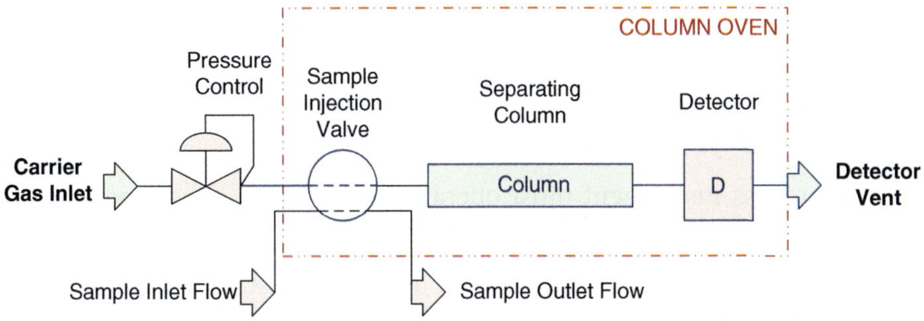

When actuated, the sample injection valve transfers a minute aliquot of the sample fluid into the flowing carrier gas. Later chapters provide full details of the many varieties of sample injector valve used in process gas chromatographs.

Figure 1.2 Basic Gas Chromatograph.

Figure 1.2 introduces the essential hardware devices found in any gas chromatograph:

- A column oven with one or more controlled temperature zones.
- A carrier gas supply and pressure control system.
- A sample injector to inject a repeatable volume of sample into the flowing carrier gas.
- One or more separating columns.
- One or more detectors.

All gas chromatographs have these basic functions, yet we see a large variation in their design and fabrication.

The process instrument

The early development of the gas chromatograph was unusual. After the invention of the technique in 1952, the oil and chemical companies soon recognized its potential for process control, and those industrial companies did much of the original development work. The contribution of the instrument manufacturers came later.

Consequently, gas chromatographs intended for process monitoring and control evolved differently from those intended for laboratory use. Although both types of instrument use the same core technology, their sphere of application is quite different.

For example, a process gas chromatograph performing a two-minute analysis receives 720 samples per day. The laboratory chromatograph might only receive three.

Thus, the design specifications for a gas chromatograph installed in an industrial processing plant are quite different than for a gas chromatograph sitting on a laboratory bench. The main reasons for these differences are:

- The process instrument operates in a potentially hot, cold, dusty, wet, windy, corrosive, or hazardous environment.
- The process instrument operates continuously twenty-four hours per day, seven days per week.
- The process instrument must operate reliably with almost no human intervention – perhaps only one calibration check each month.
- The process instrument can focus on measuring just a few of the components in a sample – the ones needed for process control.
- The process instrument suffers from a fanatical quest to reduce analysis time, so its measurements are valid for process control.

For all the above reasons, a process chromatograph (PGC) may include devices not shown in Figure 1.2. Later chapters will further discuss those devices. To whet your appetite, expect to see:

- Devices external to the instrument to condition the incoming process sample to make it compatible with the chromatograph; *i.e.* a **sample conditioning** system.
- Multiple columns with special valves to switch analyte molecules from one column to another, thus maximizing the rate that separated components arrive at the detector. This is an additional complexity rarely found in laboratory instruments.
- **Housekeeping columns** that allow strongly-retained components to quickly exit the column system. A laboratory instrument used only a few times each day has plenty of time to recover between sample injections.
- Robust column systems and stable devices, all designed to operate for a long time without adjustment. In contrast, the laboratory staff can frequently check and adjust their instruments, as necessary.

- Automatic validity checking and automatic calibration as necessary. Most laboratories analyze a quality control sample every day.
- Hardened electronic devices to capture and process the detector signal and to schedule timed events.
- An analyzer enclosure, shelter, or house to protect the analyzers and workers from the plant environment.

The following paragraphs introduce the basic function of the hardware devices. Later chapters detail their performance and technology.

The oven

Temperature control

The hardware devices used by a gas chromatograph and the separations that occur within its columns are sensitive to temperature change, so a gas chromatograph needs very fine temperature control.

In the first makeshift gas chromatographs the temperature-controlled enclosure was literally a laboratory oven, and the name stuck; the column compartment of a gas chromatograph is still the **column oven.**

Early PGCs had a single isothermal oven that housed the sample injection valve, column, and detector; and sometimes the pressure regulator too. The temperature setting was then a compromise that didn't always satisfy the needs of the individual devices. More recent instruments include several temperature-controlled zones for columns, valves, and detectors, thereby allowing individual temperature settings.

The chromatographic columns are very sensitive to temperature change. A change of column temperature will change the time that a component spends in that column, which might cause an error in analyte detection and measurement. Most columns today reside in a separate column oven often controlled to better than ±0.03 °C.

Temperature programming

A separate column oven may also support **temperature programming**, a sometimes-useful technique that gradually increases the temperature of a column during analysis. When temperature programming is employed, the analyzer needs a reproducible cooling system to rapidly lower the column temperature to its original starting point.

Temperature programming is common in laboratory gas chromatographs and allows them to separate a wide range of components, but it's rare in process gas chromatograph due to cost and analysis time issues. This may change with the introduction of less complex methods of heating and cooling, as discussed in Chapter 11.

The sample injection valve

Laboratory and online practice

To produce a chromatographic separation, the instrument needs a small **sample** of the gas or volatile liquid for analysis. The introduction of this sample into the carrier gas stream is known as sample injection. After injection, the carrier gas carries the sample into the column. As used here, a **volatile liquid** is one that will rapidly and completely vaporize at the injector temperature.

It used to be a standard laboratory practice to inject samples manually, using a glass syringe, but this routine procedure is now automatic. In the laboratory, an **autosampler** accepts an array of small vials containing the liquids for analysis. Then, according to a preloaded time program, it pulls a sample from each vial in turn and injects it into the chromatograph.

In contrast, an online gas chromatograph needs to periodically extract a minute sample from a continuously flowing process fluid and inject that sample into the carrier gas flow. To do this, most PGCs use a mechanical **sample injector valve** having a pneumatic actuator powered by an air signal from the chromatograph control unit. A few use electric power.

Figure 1.3 shows a typical valve configuration for injecting gas samples. For clarity, the diagram shows a rotary valve, but there are several other types of valve in use, including slide valves, diaphragm valves, and plunger valves. Chapter 8 details the function, design, and usage of these valves.

Plug injection

The injector valve must inject the measured sample volume all at once, in the form of a compact plug. If the injection is slow and the sample starts to mix with the carrier gas, the sample molecules will start to spread out in time even before they reach the column. This would not be good because it's more difficult to separate a wide band of molecules than it is to separate a narrow band. Separation is easier when the injected molecules tightly pack together.

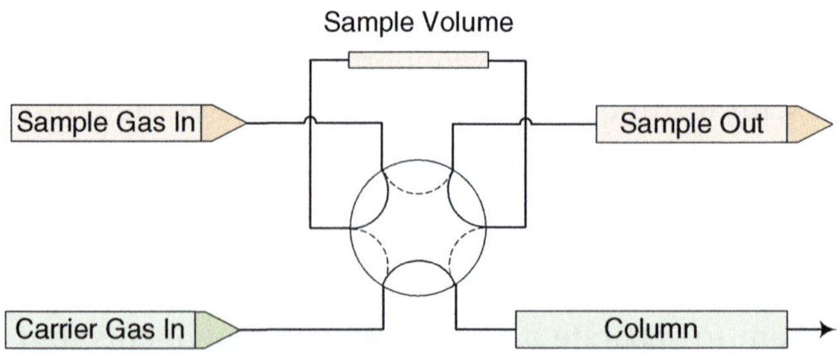

PGCs use several types of valve. As an example, this sketch shows a rotary valve. The rotor turns 60° to inject a sample.

Figure 1.3 Typical Gas Sample Injector Valve.

The sample volume is determined by the application. It's crucial to inject the same volume of sample each time because the detector output signal is proportional to the number of molecules it sees. Should the injected volume change, so would the output signal, even if the concentration of the analyte remained the same.

Gas sample injection

The injected volume of a gas sample is typically less than 0.25 mL.

The number of molecules in a fixed gas volume increases with pressure, so it's necessary to maintain constant sample pressure for each injection. Therefore, most PGCs block and bleed the sample line to allow the sample gas to come to atmospheric pressure, a technique known as **atmospheric referencing**. Chapter 7 discusses some valve systems to achieve this.

But that still leaves the normal variation of atmospheric pressure, which is quite small, as you can see by inspecting any barometer. Discounting stormy weather, the jobsite pressure variation should not be more than about ±2 %.

In practice, atmospheric referencing works well enough for most applications. If greater precision is desired, it's best to measure local barometric pressure and adjust the measurement values to compensate for any variation found.

Some PGCs have a sensor to measure the absolute pressure of the gas sample and use an algorithm to correct for detected changes.

Liquid sample injection

With liquid samples, the main challenge is to avoid gas bubbles in the injected sample as these will cause erratic measurements. To guard against bubbles, keep the pressure of a liquid sample as high as possible, consistent with the pressure rating of the sample injector valve.

A volume of liquid contains about 300 times as many molecules as an equal volume of vapor. Therefore, to inject the same number of molecules, a liquid sample volume needs to be very small, usually less than one microliter (1 μL). In such a small volume, even the smallest bubble will displace a significant amount of the sample volume and cause low measurement values.

> *It's easy to visualize a microliter since it's the same size as a one-millimeter cube (1 mm^3). A volume of one thousand microliters is equal to one milliliter (1 mL) and to one cubic centimeter (1 cm^3), commonly called a cc.*

Another challenge with liquid samples is getting a complete and instant vaporization without making the sample too hot, lest it start to react or decompose. The small volume is helpful, and most process liquids quickly vaporize without significant decay.

The column

The separating device

The chromatographic column is the heart of any gas chromatograph. It separates the analytes from the other sample components, and from each other, so the detector can measure them individually.

Figure 1.4 pictures some typical chromatographic columns.

The carrier gas carries the injected sample molecules into the column, where they touch the selected stationary phase. It's the contact with the stationary phase that causes separation. The stationary phase delays the sample molecules – some more than others – so different components end up with different transit times through the column. Each component emerges from the column after its own characteristic **retention time**.

It takes time

A chromatographic separation takes time. In most process applications, the **analysis time** is from one to ten minutes, depending on the complexity of the analyzed mixture. Some complex separations take longer.

It's important to realize that separation is just a prelude to analysis. The enormous power of the gas chromatograph comes from its ability to physically separate almost any chosen component from all other components, and then to measure it. Other analytical techniques attempt to measure the concentration of one substance in the presence of all the other substances,

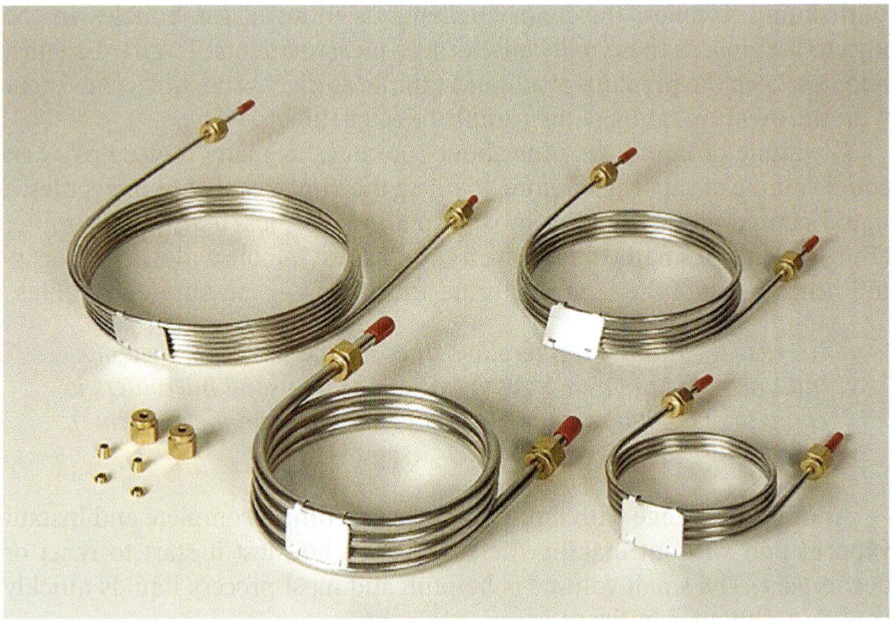

Figure 1.4 Typical Gas Chromatographic Columns. *Source: Ohio Valley Specialty Company, Inc. Reproduced with permission.*

a goal that few accomplish well. Gas chromatographs separate the analytes first, and then measure them individually. When properly designed, a process gas chromatograph may be the only process analyzer that doesn't suffer interference from other stuff in the sample.

Of course, it's possible for two or more components to have about the same retention time in a column, so a column might not separate every component from every other component present in the sample. The task of a PGC column system is to separate the measured components from all the others. It's neither necessary nor desirable to separate everything.

Multiple columns

The choice of separating column is always the key to a successful analysis. In practice, it's difficult to achieve the desired separation using just one column, so process gas chromatographs usually employ multiple columns to achieve the necessary separation in the shortest possible time.

With multiple columns, certain partially separated components from one column must flow into another column to achieve the desired separation. To divert flows between columns, an online gas chromatograph typically uses one or more **column valves** that are usually similar in design to its sample injection valve. Figure 1.5 shows an example of a simple column system.

Intercolumn valves must not leak. They must also have very low internal volume and smooth flow paths, lest separated components start to remix. For the same reason, a PGC typically employs $^{1}/_{16}$-inch o.d. tubing for all its internal plumbing.

PGCs are individually configured for a particular application. During this procedure, known as **application engineering**, the application engineer chooses a column system to perform the desired separation and decides on the stationary phase needed for each column. Refer to Chapter 9 for a review of some standard column configurations and the function of each column.

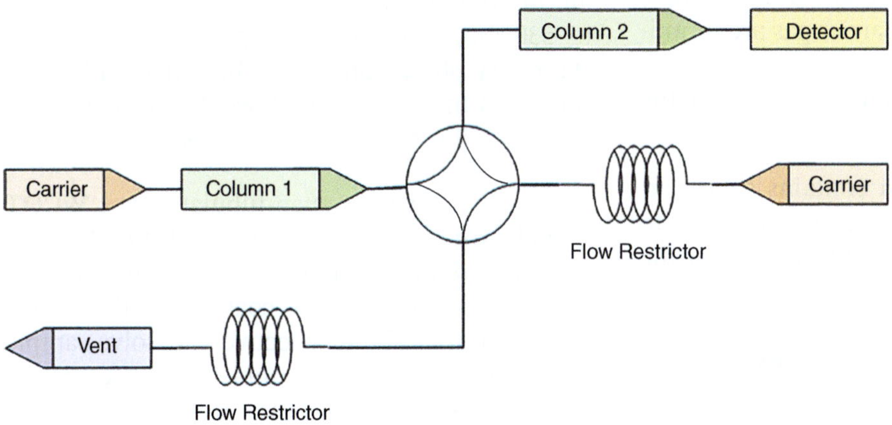

For simplicity, the figure shows a rotary valve that rotates 90° when actuated, thereby flushing later peaks to vent. Other valves have a similar function.

Figure 1.5 A Simple Column Switching System.

SCI-FILE: *On Column Types*

Introduction to SCI-FILEs

The more theoretical and mathematical content of the book resides in separate segments called SCI-FILEs. These contain optional reading that may or may not be part of a course of study.

Each SCI-FILE is a supplement to the main text that you can safely omit if not of immediate interest. Treat them as reference sources to consult when needed.

Two kinds of column

The stationary phase must be secure inside the column so it doesn't move. The packed column and the open-tubular column differ per the method they use to anchor the stationary phase in place.

Packed columns

A **packed column** most often uses several meters of 1/8-inch o.d. stainless steel tubing, although early PGCs used larger diameters, and some PGCs now employ 1/16-inch o.d. "micropacked" columns.

In the traditional packed column, the packing is a granular porous solid with particles about the same size as granulated sugar. These particles pack tightly together inside the tube so that any sample molecules moving with the carrier gas are in intimate contact with them.

The type of column so produced depends on the role of the solid particles:

- An **active-solid column** contains solid particles having a large activated surface area to selectively *adsorb* certain molecules from the sample gas.

  Since the stationary phase is solid, this technique is **gas-solid chromatography** (GSC).
- A **liquid-phase column** contains solid particles having a coating of non-volatile liquid to selectively *dissolve* certain molecules from the sample gas.

  Since the stationary phase is liquid, this technique is **gas-liquid chromatography** (GLC).

Many columns now use proprietary stationary phases, often made from specialized polymer material. These columns don't easily fit into the old classifications of GSC or GLC, so the terminology is becoming passé.

In a liquid-phase column, we call the granular solid an **inert support**. In real life, an inert support might not be completely inert; it sometimes affects the performance of a column.

The thickness of the liquid film coated on the support is an important variable. The **liquid loading** gives the percentage by weight of liquid on support.

The first gas chromatographs used packed columns, and they are commonly found in PGCs today.

Open-tubular columns

An **open-tubular** or **capillary column** uses several tens of meters of capillary tubing having an internal diameter ranging from about 100 to 530 μm. The mode of operation differs: the stationary phase adheres to the inner wall of the tube, and the carrier gas flows down the middle. Figure 1.6 illustrates three versions (Harvey 2017):

- A "wall-coated open-tubular" or **WCOT column** uses tubing made of fused silica. The stationary phase is a very thin layer of a non-volatile liquid coated on the inside wall of the tube to selectively dissolve sample molecules from the sample gas. PGCs rarely use these columns as they are fragile and tend to be unstable in use.

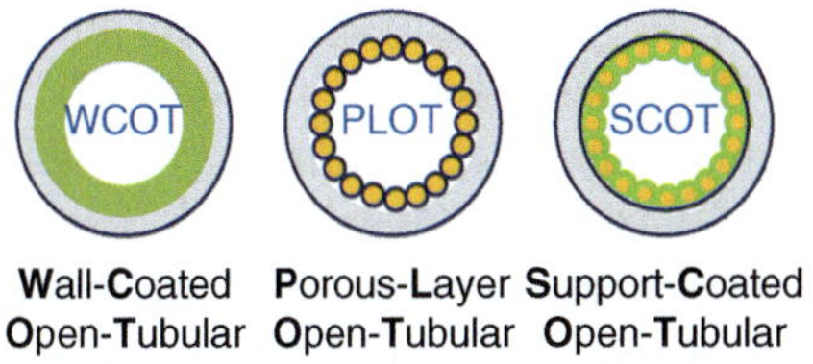

The WCOT columns typically use fused-silica tubing and tend to be too fragile for process use. The PLOT and SCOT columns mostly use steel capillary tubing.

Figure 1.6 Three Kinds of Capillary Column.

- A "porous-layer open-tubular" or **PLOT column** uses stainless steel capillary tubing. The stationary phase is a very thin layer of solid material coated on the inside wall of the tube to selectively adsorb sample molecules from the sample gas.
- A "support-coated open-tubular" or **SCOT column** typically uses stainless steel capillary tubing. The stationary phase is a coating on very fine support particles in a uniform layer on the inner wall of the tube. These rugged columns have become quite popular in process gas chromatographs.

Open-tubular columns have smaller diameters than packed columns and require special operating techniques. As in packed columns, the film thickness is an important variable, but we'll defer discussion on that. While they achieve better separations, the operating conditions of open-tubular columns can be difficult to sustain in the process environment.

For more information about column types and column liquid phases, refer to the excellent detailed review by Rahman et al. (2015).

The detector

Making the measurements

A chromatographic separation cannot produce a measurement. Chromatography is merely a separating technique; it doesn't measure anything. To measure the concentration of the analytes, the analytical instrument must estimate the quantity of selected molecules as they **elute** from the column. It follows that every gas chromatograph needs a device to generate a signal proportional to the number of sample molecules exiting the column. This is what a detector does.

> *In any gas chromatograph, two things are happening in series. First the column separates the analytes, and then the detector measures them. To improve your troubleshooting ability, keep that distinction in mind.*

Many detectors are available for gas chromatography, most developed for applications that require selective measurement or enhanced sensitivity.

The **thermal conductivity detector** (TCD) was the first gas chromatograph detector, and after much improvement it is still popular today. The TCD responds to the difference in thermal conductivity between pure carrier gas and carrier gas that contains sample molecules. So, when a TCD is used, the carrier gas is chosen to maximize the difference in thermal conductivity between the carrier gas and the analytes. The TCD is a general-purpose detector that will respond to any analyte.

Most other detectors are selective; they respond only to certain kinds of molecules and often do so with very high sensitivity. For instance, the **flame ionization detector** (FID) responds only to compounds containing both carbon and hydrogen, so it's very useful in the analysis of hydrocarbons. The **flame photometric detector** (FPD) is also very sensitive, but only to sulfur or phosphorus compounds. It's most used to measure sulfur compounds in fuels and stack emissions to ensure compliance with environmental regulations.

Generally, detectors operate in the differential mode. When pure carrier gas is passing through a detector, its output signal should be constant. The analytical instrument reads that signal and offsets it to a value close to zero. We call that the **baseline**. Then, when the detector responds to the presence of analyte molecules, the instrument outputs a change in signal level proportional to the concentration of that component.

Chapter 10 provides a detailed review of the three detectors most used in process gas chromatographs (TCD, FID, and FPD) and briefly mentions some other detectors that are common in laboratory instruments but only occasionally deployed for online process applications.

The chromatogram

The **chromatogram** is a graphical display of the detector signal plotted against elapsed time. The PGC may print the chromatogram on a chart or display it on a computer screen. Note that it's also common to refer to the raw signal from a detector as the **chromatogram signal**.

Be careful of the terminology: the **chromatograph** is an instrument, not a graphical plot.

The horizontal axis of the chart indicates elapsed time from the instant of sample injection. Chromatograms from an older PGC using a strip-chart recorder may look like the one in Figure 1.7a, with the zero-time mark on the far right and elapsed time progressing from right to left. But modern PGCs now display and print chromatograms with time-zero on the left and elapsed time increasing from left to right, as in Figure 1.7b. You need to be comfortable with either mode of display.

> *Note: The book uses many simplified chromatograms like these to illustrate chromatographic principles: they do not represent real chromatograms.*

The vertical axis of the chart indicates detector response. Chromatograms in this book don't show a calibrated vertical axis because it's rarely significant. The chromatogram is just the raw detector response, and the PGC must heavily process that signal to yield the analytical measurements.

When working on a computer-controlled PGC, you can expand or compress the horizontal or vertical axes of the chromatogram to obtain a more detailed view of the area of current interest.

When only carrier gas is coming out of the column, the chromatogram should display a flat baseline with no change over time. Then, when some

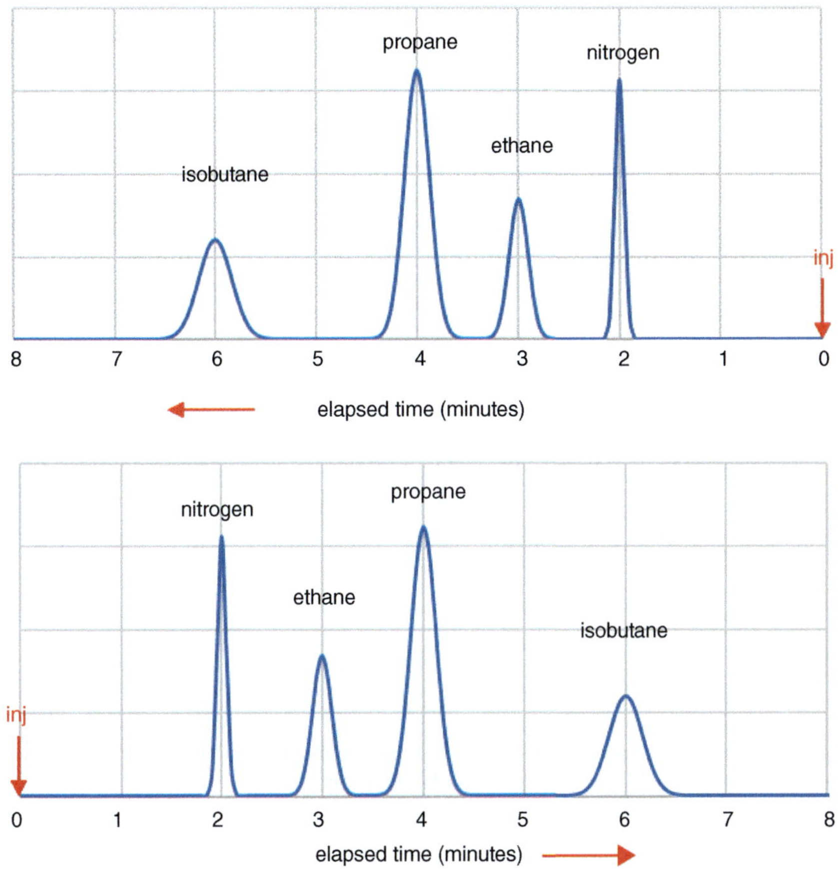

Illustrative graphics: (a) Upper. Classic strip-chart record; sample injection at right, time progresses right to left. (b) Lower. Modern on-screen display; sample injection at left, time progresses left to right.

Figure 1.7 Typical Chromatograms.

component molecules emerge from the column and enter the detector, the detector outputs a signal proportional to the rate of arrival of those molecules.

The number of component molecules arriving at any instant starts at zero, rises to a maximum, and then falls to zero, creating the characteristic chromatogram **peak**. The ideal peak shape is symmetrical per Figure 1.7, but many imperfections in the instrument conspire to distort that perfection. Figure 1.8 shows some peaks on a real chromatogram.

The output signal of the detector is proportional to the instantaneous number of component molecules it is seeing. Therefore, the area under a chromatogram peak is a true measure of the total number of component molecules passing through the detector. As a compromise, we shall see the occasional use of **peak height** as a surrogate for **peak area** as it may be easier to measure.

To relate the measurement of peak area or peak height to the concentration of the analytes, the analyzer uses a **calibration factor**. In an early

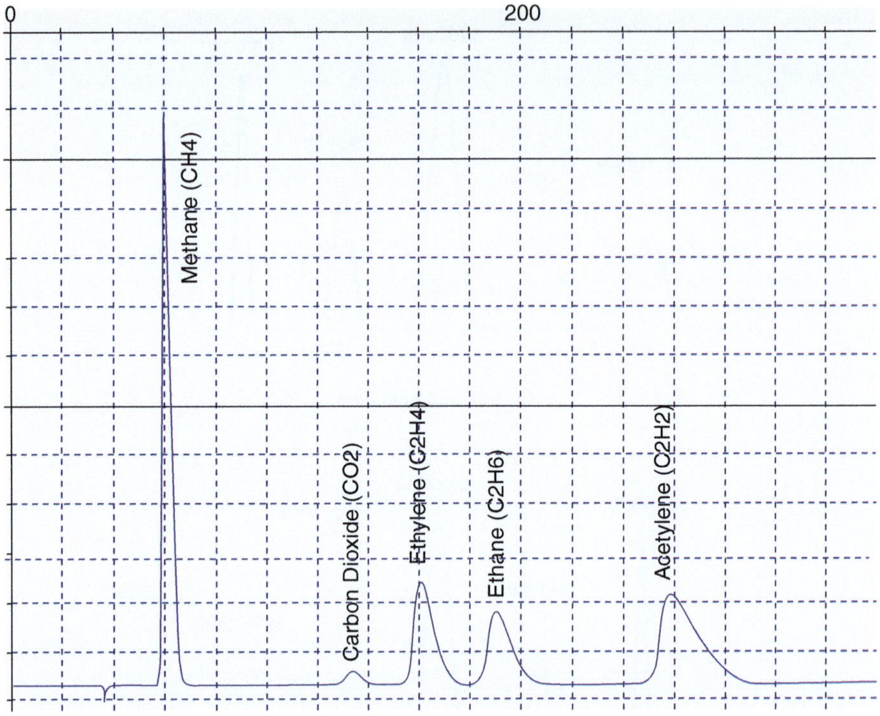

Real peaks are not perfect in shape. Note the gradual increase in peak width with elapsed time (left to right) and the asymmetric peak for acetylene.

Figure 1.8 A Real Chromatogram. *Source: Yokogawa Electric Corporation. Reproduced with permission.*

PGC, the calibration factor was simply the mechanical setting of an attenuator; usually a variable resistor. Modern instruments get the calibration factors from a calibration procedure that analyzes one or more standard samples containing a known concentration of each analyte. During calibration, the analyzer calculates and stores a calibration factor for each analyte based upon its peak area or height and its known concentration. Then, during process analysis, the analyzer measures the area or height of each analyte peak and multiplies that value by its stored calibration factor to deduce the analyte concentration.

Chromatographers are so used to seeing peaks on the chromatogram that they even use the word "peak" to describe the molecules of a component as it travels through the column system. In colloquial chromatography, each cluster of molecules in the column is a *peak* even if those molecules never reach the detector and never appear as a real peak on the chromatogram. This shorthand terminology makes it easier to describe the movement of sample molecules in the column: it's far easier to say:

"the column separates the propane peak from the ethane peak"

than:

> *"the column separates the cluster of propane molecules from the cluster of ethane molecules"*

By the way, we drew most chromatograms in this book with rather wide peaks so you can easily see their shape. Real chromatogram peaks are often much narrower than these, and it can be difficult to see small variations in their topography. You might need to expand the time scale on a computer screen or run at a higher chart speed on a recorder to measure the peak width or see the exact trajectory of the peak.

The chromatogram readout is a vital design and troubleshooting tool; so much so that it's difficult to overstate its usefulness. Discounting electrical failures that are easy to fix, all faults are visible on a chromatogram, either directly or by comparing the current chromatogram with a previous one.

> *All chromatographic faults are visible on the chromatogram. So, to become an expert troubleshooter, you must learn to read the chromatogram!*

Some faults are directly observable, such as a timing error or a missing peak. But a chromatogram holds a lot more information than that; the position and shape of the peaks tell us what they are, and whether the columns are working at full efficiency.

As an example of chromatogram reading skill, review the chromatograms in Figure 1.7. The peaks are nicely separate from each other, and it looks like a good analysis. But an expert observer would know that the column is not working well. By adjusting a few settings, an expert could improve the performance of the column and do the analysis in half the time!

The expert observer would also notice that the author has misidentified the last peak; it can't be isobutane on any column. It's probably n-butane instead. Expert users see information in the chromatogram that eludes novice users. That's why they are experts.

> *Keep on reading! The next three chapters will show you how to discern patterns in the shape and position of chromatogram peaks. From those patterns, you will make the same deductions about Figure* 1.7 *as our expert observer did.*

A chromatogram contains all the information needed to optimize column performance or to diagnose chromatographic problems. If you aspire to be an expert PGC applications engineer or troubleshooter, learn to read the chromatogram. Waters (2017) has summarized these diagnostic skills in a compact format.

Knowledge Gained

- *Chromatography is a general method of separation and includes many different practical techniques.*
- *Chromatography uses a fluid mobile phase passing over a liquid or solid stationary phase.*
- *In gas chromatography, the mobile phase is gas; in liquid chromatography, the mobile phase is liquid.*
- *Chromatographic analyzers separate the desired analytes and then measure them one by one.*
- *Other analyzers attempt to measure the analyte molecules in the presence of other molecules.*
- *Gas chromatographs inject a tiny volume of sample into the flowing carrier gas.*
- *The sample must be a gas or a volatile liquid that quickly vaporizes and enters the column as a vapor.*
- *In the laboratory, sample injection is by glass syringe, either manual or by an autosampler.*
- *The TCD, FID, and FPD are popular detectors in process gas chromatographs.*
- *The detector output signal forms a chromatogram display when plotted against elapsed time.*
- *Analyte molecules cluster together at different times to form separate chromatogram peaks.*
- *Process gas chromatographs (PGC) use an automatic injection valve to inject the sample.*
- *The carrier gas carries the sample into the column where it contacts the stationary phase.*
- *The stationary phase may be a solid adsorbent or an immobilized non-volatile liquid.*
- *Contact with the stationary phase delays some peaks more than others, so separation occurs.*
- *Special routing valves may direct the peaks into different columns to finish the desired separation.*
- *PGCs can now use either packed columns or capillary (open-tubular) columns.*
- *The separation process takes time; typically one to ten minutes, sometimes longer.*
- *The carrier gas elutes peaks from the column into a chosen detector for measurement.*
- *The detector responds to a property of analyte molecules that differs from carrier gas molecules.*
- *The PGC may measure peak area or peak height to compute the concentration of an analyte.*
- *The PGC uses stored calibration factors to calculate the concentration of each analyte.*
- *The chromatogram is a most valuable source of information, but one must learn to read it.*
- *To the expert troubleshooter, all chromatographic faults are visible on the chromatograms.*

Did you get it?

Self-assessment quiz: SAQ 01

Q1. In gas-liquid chromatography, what are the gas and liquid doing?
Select the one correct answer:

A. They are both moving.
B. The gas is moving, and the liquid is stationary.
C. The gas is stationary, and the liquid is moving.
D. They are both stationary.

Q2. In a gas chromatograph, which one of the gases listed below would not suitable as the carrier gas?
Select the one correct answer:

A. Oxygen
B. Nitrogen
C. Hydrogen
D. Helium

Q3. In a gas-liquid chromatograph, which one of the materials listed below might be the stationary phase?
Select the one correct answer:

A. An inert gas
B. A granular adsorbent solid
C. A volatile liquid
D. A non-volatile liquid

Q4. In a gas-liquid chromatograph, what really causes the separation?
Select the one best answer:

A. The carrier gas causes the separation.
B. The mobile phase causes the separation.
C. The stationary phase causes the separation.
D. The chromatogram causes the separation.

Q5. In a gas chromatograph, what do the columns do?
Select the one best answer:

A. They convert all the components into peaks.
B. They separate the analytes from all other components and from each other.
C. They separate all the components of the sample.
D. They allow only measured components to enter each detector.

Q6. In a gas chromatograph, what does a detector do?
Select the one best answer:

A. It provides a continuous flat baseline as a reference for measuring the peaks.
B. It generates a signal proportional to the instantaneous number of component molecules leaving the column.
C. It measures either the area or the height of each peak.
D. It converts each component of the sample to a concentration.

Q7. Why is the chromatogram so important?
Select the one best answer:

A. It shows the baseline used for measuring the peaks.
B. It shows the shape of each peak.
C. It shows the separation between peaks.
D. All of the above.

Check your SAQ answers with those given at the end of the book.

Student evaluation test: SET 01

Your instructor will provide the answers to these test questions.

S1. In gas-liquid chromatography, what is the physical state of the mobile phase?
Select the one best answer:

A. Gas
B. Gas or liquid
C. Liquid
D. Solid

S2. In gas-liquid chromatography, what is the physical state of the stationary phase?
Select the one best answer:

A. Gas
B. Gas or liquid
C. Liquid
D. Solid

S3. In gas-liquid chromatography, what is the physical state of the injected sample?
Select the one best answer:

A. Gas
B. Gas or liquid
C. Liquid
D. Solid

S4. According to the chapter text, what is a typical number of analyses done per day by a process gas chromatograph, relative to the number done by a laboratory gas chromatograph?
Select the one best answer:

A. About 80 times as many
B. About 120 times as many
C. About 240 times as many
D. About 720 times as many

S5. According to the chapter text, what is the main function of a housekeeping column?
Select the one best answer:

A. To separate all the measured components.
B. To separate only the unmeasured components.
C. To allow strongly-retained components to quickly exit the column system.
D. To permanently absorb one or more unmeasured components.

S6. Imagine you are changing a PGC from liquid sample injection to gas sample injection and the gas sample will be at atmospheric pressure. If the previous

liquid sample volume was 1 μL, what new gas sample volume would you install to get about the same peak heights?
Select the one best answer:

A. 0.001 mL
B. 0.3 mL
C. 0.9 mL
D. 3.0 mL

S7. PGC detectors may not respond to every analyte. Consider the statements below; which statements are correct?
Select all of the correct statements and none of the incorrect statements:

A. No detector can respond to hydrogen sulfide, H_2S.
B. The thermal conductivity detector does not respond to hydrocarbons.
C. The flame ionization detector is very sensitive and can measure low concentrations of carbon dioxide, CO_2.
D. The flame photometric detector does not respond to carbon monoxide, CO.

S8. What is a chromatogram, and why is it important?
From the list below, select all the correct statements and none of the incorrect statements:

A. The chromatogram is a graphical plot of the detector signal against elapsed time.
B. On a chromatogram, the horizontal axis represents elapsed time, which always increases from left to right, with the zero-time marker on the far left.
C. On the chromatogram from a modern process gas chromatograph, the vertical axis represents the concentration of the analytes and is scaled from zero to 100 %.
D. Chromatographic faults produce symptoms visible on the chromatogram that an expert user can diagnose.

S9. How is each analyte peak on the chromatogram measured?
From the list below, select all the correct statements and none of the incorrect statements:

A. Most modern PGCs measure the peak height, but a few measure the peak area.
B. To calculate the concentration of an analyte, the PGC multiplies the analyte peak height or peak area by a stored calibration factor.
C. The calibration factor depends on the detector in use but is the same for each analyte in the sample.
D. The calibration factors come from the analysis of a calibration sample containing known concentrations of the analytes.

References

Cited

Ettre, L.S. (2008). *Chapters in the Evolution of Chromatography* (ed. J.V. Hinshaw). London, UK: Imperial College Press.

Harvey, D. (2017). Gas chromatography. In: *LibreTexts, Section 2.4* (updated July 28, 2017), accessed October 23, 2018 at https://chem.libretexts.org

Rahman, M.M., El-Aty, A.A., Choi, J., Shin, H., Shin, S.C., and Shim, J. (2015). Basic overview on gas chromatography columns. In: *Analytical Separation Science* (eds. J.L. Anderson, A. Berthod, V. Pino, and A.M. Stalcup), 823–834. Verlag, Germany: Wiley-VCH. doi:10.1002/9783527678129.assep024

Waters, T. (2017). The fine art of chromatogram reading. *Proceedings of the 2017 Analysis Division Symposium*, Pasadena, California (April 24–26, 2017). Research Triangle Park, NC, USA: International Society of Automation.

Figures

New technical terms

When first introduced, these words and phrases were in bold type. You should now know the meaning of these technical terms. If still in doubt, consult the Glossary at the end of the book:

active-solid column
analysis time
analyte
application engineering
atmospheric referencing
autosampler
baseline
calibration factor
capillary column
carrier gas
chromatogram
chromatogram signal
chromatograph
chromatography
column
column oven
column valve
component
detector
elute
flame ionization detector
flame photometric detector
gas chromatograph
gas chromatography
gas-liquid chromatography
gas-solid chromatography
housekeeping column
inert support
liquid chromatography
liquid loading
liquid-phase column
mobile phase
molecule
open-tubular column
packed column
peak
peak area
peak height

PLOT column
retention time
sample
sample conditioning
sample injector valve
SCOT column
separation
stationary phase
supercritical fluid
temperature programming
thermal conductivity detector
volatile liquid
WCOT column

In addition, we introduced several chemical names and you need to know what they are. If you are not familiar with chemical names, refer to the SCI-FILE: *On Chemical Names* in Chapter 4. You can also look up individual chemical names in the Glossary.

2

Peak shape

> *"What happens inside a column? What makes the peaks form? Why are peaks that funny shape? Why are some peaks wider than other peaks? These are good questions, and now is the time for answers".*

How columns work

The secret to understanding process gas chromatographs is knowing how the columns separate the components of the sample. PGC training courses often omit this important knowledge, preferring to focus instead on the mechanics and electronics of the instrument itself. It's true that special skills are required to properly set up and maintain the equipment. And you must learn those skills. Yet, even if you gain perfect knowledge of the electromechanical systems, you won't be competent with process gas chromatographs until you clearly understand what the columns are doing.

You'll need to know how a column *really* separates molecules of one kind from molecules of another kind. It's not sufficient (nor true) to say that some kinds of molecule move faster than others do.

You'll also need to know what determines the shape of a chromatogram peak, particularly its width. So let's look a little closer at some typical peaks.

Looking back at the chromatogram in Figure 1.7, if you examine any individual peak, it is easy to see that even identical molecules don't reach the detector at the same time. Relative to the time of the peak apex, some molecules arrive earlier, and some arrive later. Take a look at the propane peak, for instance: its base width is about 40 s, which means propane molecules start arriving at least 20 seconds before their most frequent and average time (at peak apex) and continue for at least 20 seconds after that, gradually dropping back to zero. This variation in the elution time of

Process Gas Chromatographs: Fundamentals, Design and Implementation, First Edition. Tony Waters.

identical propane molecules *determines* the width of the propane peak and its characteristic shape.

At this point, you should be wondering *why* identical molecules don't spend an identical amount of time in the column. Whatever happens in there, surely identical molecules must experience identical delay and emerge from the column at the same time? No, they don't. Some emerge a little earlier, and some emerge a little later. Any useful explanation of chromatography must account for that inconvenient fact.

Of course, anything that makes a peak wider is a nuisance because it's more difficult to separate wide peaks from each other than to separate narrow peaks. So, as a practical matter, we need to know how to minimize the peak width, and that is one of the most important questions in gas chromatography! The answer will become apparent as you work through the book.

What happens inside the column

Inside the column, sample molecules that are traveling in the carrier gas touch the stationary phase. What happens next depends on what kind of stationary phase is present, a solid or a liquid.

Most PGC columns employ a liquid stationary phase, which works by selectively *dissolving* the component molecules. Since they are so common, it is reasonable to use liquid-phase columns as our example for explaining how the chromatographic process works. Therefore, the rest of this chapter will discuss only gas-liquid interaction.

Columns employing a solid stationary phase are used to separate simple gases like hydrogen, oxygen, nitrogen, or methane. Gas-solid columns work by selectively *adsorbing* the sample molecules. This is a different mechanism, but it has the same effect; the column retains one kind of molecule longer than it retains another kind of molecule.

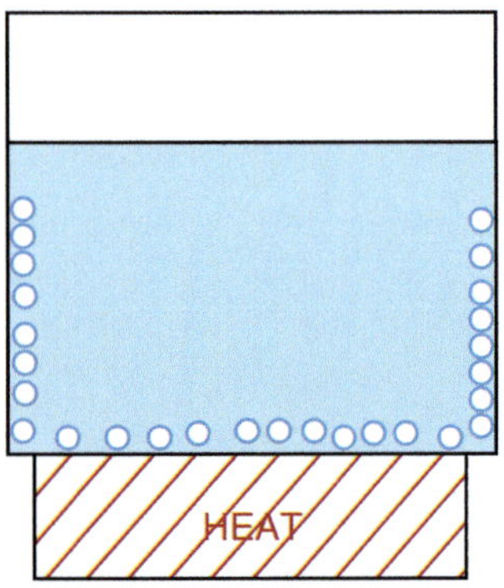

Gases dissolve less at higher temperature. Heating the water expels the dissolved gas, visible as a cluster of small bubbles.

Figure 2.1 Gases Dissolve in Liquids.

How gas and liquid interact

In a gas-liquid column, the component molecules touch a stationary liquid phase and obviously interact with it in a way that causes separation. So we need to start by discovering how a gas can interact with a liquid.

Several household examples illustrate the interaction between a gas and a liquid. For example, put some cold tap water in a saucepan and heat it gently on the stove. We have all done this and seen the result. As the water warms, thousands of tiny bubbles appear and cling to the side of the pan, as illustrated in Figure 2.1. This happens long before the water boils. The bubbles are obviously gas – but what gas is it, and where did it come from?

A moment's thought should reveal the answer. The gas can't be steam because steam bubbles can't exist under water unless the water is boiling. The gas can't be hydrogen because water doesn't decompose at this low temperature. No, the gas must be coming from something dissolved in the water, most likely oxygen and nitrogen from the air.

This first example demonstrates two important principles:

- Gases dissolve in liquids.
- Gases dissolve *less* in hot liquids than they do in cold liquids.

Another common example is the bottle of champagne illustrated in Figure 2.2. Closely inspecting the unopened bottle, we see very little gas; the bottle is nearly full of liquid. Yet, upon popping the cork, an enormous quantity of gas suddenly appears. This gas is carbon dioxide, a different gas than in our first example. The liquid is not quite the same either, but it's mainly water. And we can be sure that pure soda water would behave in exactly the same way.

Again, where did the gas come from? There is only one place that it could have been hiding; it was dissolved in the liquid. This confirms our theory that gases dissolve in liquids, but far more was dissolved this time; the **solubility** of carbon dioxide in water is greater than the solubility of air in water.

This second example demonstrates two additional principles:

- Gases dissolve *less* at low pressure than they do at high pressure.
- Some gases dissolve in a given liquid more than other gases do.

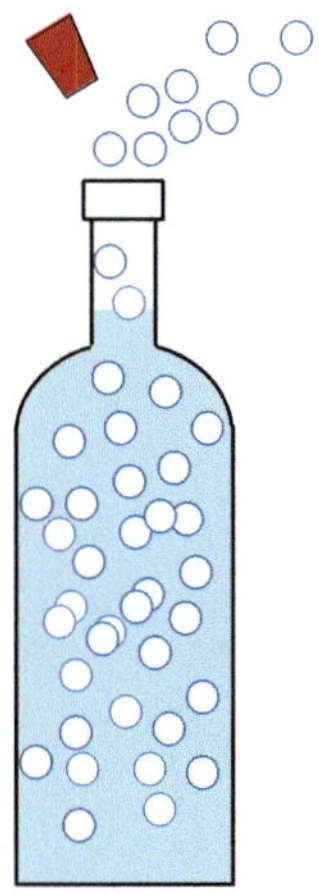

Gases dissolve less at lower pressure. Popping the cork lets the pressure drop, releasing lots of bubbles.

Figure 2.2 A Different Gas.

A final example (not illustrated) is what happens when you heat cooking oil in a pan. No bubbles appear. Apparently, the oil doesn't dissolve any air – and it won't dissolve much carbon dioxide either!

This last example gives a fifth principle:

- Some liquids dissolve more of a gas than other liquids do.

These five principles are all the science you will need to truly understand what happens in a column. They remind you that a gas can dissolve in a liquid and that the amount of gas that can dissolve depends on only four simple variables; the temperature and pressure, the type of gas, and the type of liquid.

That's it. Nothing else affects the solubility of a gas in a liquid.

The four variables are very easy to understand, yet they are the hidden foundation of all gas chromatography. Let's see how that can be …

In most process gas chromatographs, three of the four variables are closely controlled and do not vary:

- The column temperature is held constant.
- The carrier gas pressure is held constant.
- The liquid phase is predetermined and doesn't change.

The fourth and most important variable is due to the different gases in the sample. And this is the real cause of chromatographic separation:

- Different gases have different solubility in the liquid phase.

Gas chromatography works because each component to be separated has a different solubility in the liquid phase. We shall see that the less soluble peaks move quickly through the column while the more soluble peaks take longer to get through.

This is the process of separation. It's all about solubility.

Chemists call the liquid phase a **solvent** and each dissolved component a **solute**. But no real chemistry is involved. If a chemical reaction occurred, it might destroy some of the molecules that we are trying to measure and likely would cause a gradual and irreversible deterioration of the column itself.

Before moving on, a quick reminder. The discussion in this chapter focuses on the most common kind of column; one that has a liquid stationary phase. As noted earlier, another kind of column uses a solid stationary phase. The solute molecules can't dissolve in a solid, but they can and do adhere to its solid surface, and the final outcome is much the same.

Troubleshooting tips

The household examples used above may provide some valuable help with troubleshooting and are worth remembering:

- When a column works at higher temperature, gas solubility is *reduced*, and all the peaks come out earlier on the chromatogram, thereby reducing their separation. For an easy way to remember this, recall the heated water!
- When a column works at higher pressure, gas solubility is *increased* and all the peaks come out later on the chromatogram, thereby increasing their separation. For an easy way to remember this, recall the bubbly champagne!

These troubleshooting tips assume that the carrier gas flow rate is held constant. Later chapters discuss the effect of other variables.

The rest of this chapter explains how solubility causes the classic peak shape. The following chapter examines how a difference in solubility will cause the peaks to become separated from each other.

How peaks form

Forming an equilibrium

To examine the interaction between a gas and a liquid, consider a small enclosed space that's internally divided into a gas space and a liquid space, as in Figure 2.3a. For explanatory purposes, the diagrams in Figure 2.3 show the gas space and liquid space as deep layers that would not work in practice. In a real column, the gas and liquid layers are very shallow, so the sample molecules can move quickly between them.

Let's assume that the gas space in Figure 2.3a is full of helium. This is equivalent to the carrier gas in a column; it's always there. The helium molecules contact the liquid phase, and a few of them dissolve in it. Since helium is always there, the amount dissolved soon becomes constant and can be ignored.

Now consider what happens when a small sample of (say) propane is injected into the helium gas in the enclosed space. To keep it simple, let's say there are only 32 propane molecules. The same logic applies to 32 trillion molecules, or to any other number of them. This is shown in Figure 2.3b.

The propane molecules move randomly in the gas phase and soon encounter the liquid surface where some of them dissolve. Initially, all the propane molecules are in the gas phase, so they frequently collide with the liquid surface and their rate of entry into the liquid is high. Then, as more of the molecules dissolve in the liquid, there are less of them in the gas phase, and their rate of entry declines.

The dissolved propane molecules move slowly in the liquid phase and eventually encounter the gas-liquid surface, where some of them have enough energy to escape back into the gas phase. Initially, there are no propane molecules dissolved in the liquid phase, so none can escape; their rate of escape is zero. As more and more propane molecules dissolve, their rate of escape increases, as in Figure 2.3c.

With the rate of entry falling and the rate of escape rising, there must soon come a time when the two rates become equal. At this instant and beyond, every molecule that dissolves replaces one that escapes. The number of molecules in the gas phase is then constant, as is the number of molecules in the liquid phase. They will stay that way forever, as long as the operating conditions don't change.

This balancing act between two opposing and dependent processes is common in chemistry. Chemists call it a **dynamic equilibrium**.

There is nothing in our example that specifies the number of propane molecules in the gas phase and in the liquid phase once equilibrium has been achieved. That would depend on the solubility of propane in the selected liquid phase and would vary with different chemical compounds. To make it easy, though, let's assume that 50 % of the propane dissolves. Then, after reaching equilibrium, half of the molecules will be in the liquid phase, and the other half will be in the gas phase. This is the situation shown in Figure 2.3d.

Actually, it's reasonable to assume the propane solubility is 50 %, as that would generate a pretty good chromatogram. Yes, we can predict the position of peaks on the chromatogram from their solubility! You'll soon see how that works out.

In practice, it would not be difficult to set the propane solubility to exactly 50 %. We already know that the solubility of a given substance in a given liquid depends on temperature and pressure. So, to adjust the propane solubility simply change the temperature. It really is that simple. In fact, that's one way you can optimize the performance of a column.

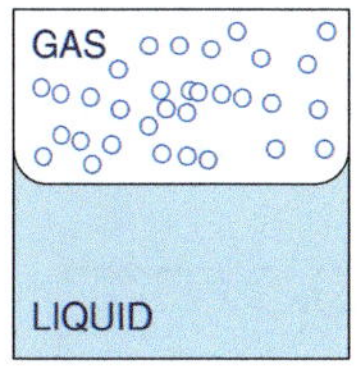

(a) Consider helium gas in contact with a liquid, then add propane.

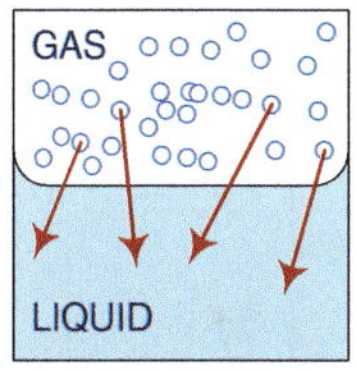

(b) The initial rate of entry into liquid is fast with zero rate of escape.

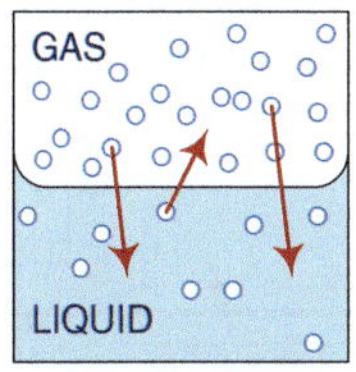

(c) Then, the rate of entry drops as the rate of escape increases.

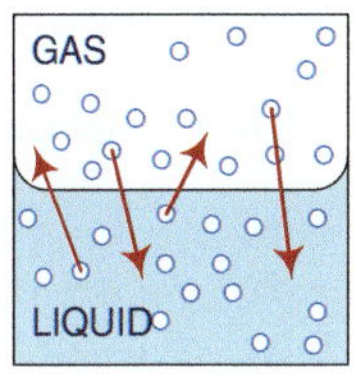

(d) Equilibrium occurs when rate of entry equals the rate of escape.

Figure 2.3 Forming an Equilibrium.

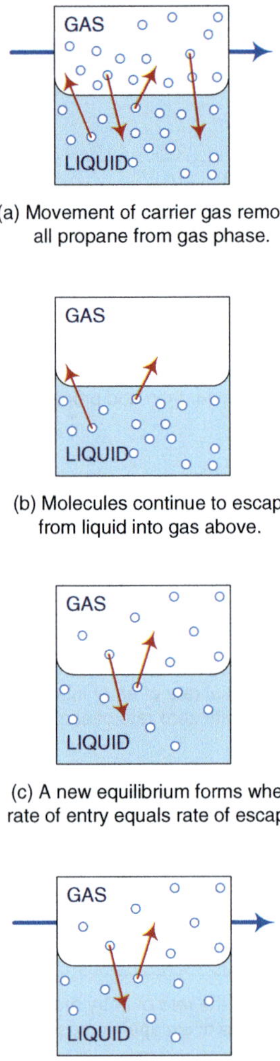

(a) Movement of carrier gas removes all propane from gas phase.

(b) Molecules continue to escape from liquid into gas above.

(c) A new equilibrium forms when rate of entry equals rate of escape.

(d) Movement of carrier gas again removes propane, cycle repeats.

Figure 2.4 The Carrier Gas Moves.

The effect of movement

So far, the discussion about equilibrium cannot explain chromatography. There is something missing from Figure 2.3, something that is essential for chromatography to occur. Figure 2.3 starts to explain what happens in a column, but it's not enough.

What is missing?

The gas phase is not moving! Recall that chromatography occurs when something moves across something that doesn't move. And in a gas chromatograph, it's the carrier gas that moves.

When the carrier gas moves, any propane molecules that happen to be in the gas phase are carried along with it, as illustrated in Figure 2.4a. In this figure, fresh carrier gas enters from the left and pushes the propane molecules out to the right replacing them with pure helium. In Figure 2.4b, the 50 % propane molecules are gone from the gas phase, and the other 50 % are stuck in the liquid phase. Pure helium now occupies the gas space, upsetting the original equilibrium.

Let's see what happens next. Imagine the small enclosed space is again sealed. The absence of propane molecules in the gas phase doesn't affect the behavior of the molecules trapped in the liquid. They continue to escape from the liquid into the clean helium above, as they did before. See Figure 2.4b. It should come as no surprise that as soon as some of the molecules reenter the gas phase, they start to dissolve in the liquid again, quickly forming the new equilibrium in Figure 2.4c.

Of course, it doesn't stop there. When the carrier gas again moves it disrupts the equilibrium of Figure 2.4c and the cycle starts again, as shown in Figure 2.4d – but with fewer molecules this time.

Pause for a moment to reflect. Figure 2.4 suggests that the carrier gas moves, then stops until a new equilibrium forms, then moves again. Clearly, this is not true. Chromatography is a smooth process, not a jerky one. But the jerky model is very useful for explaining what happens inside a column. It's a bit like taking a movie of the process and then looking at each frame in turn.

It's a long movie. The number of equilibria generated by a typical column ranges from about 5,000 to 50,000. Even a slow peak with a retention time of 1000 s would need to average one equilibrium every 50 ms to get 20,000 plates. That's equivalent to 20 movie frames per second – not a bad analogy!

With such a large number of data points, our jerky model is not so jerky after all. And it's a powerful way of evaluating column efficiency. We shall soon discover that having more equilibria in the column causes better separations.

There is a theoretical connection between the shape of a peak on the chromatogram and the number of times that equilibrium has occurred. Yes, we can figure the effective number of equilibria by measuring the resulting peak shape. This is yet another glimpse of the information buried in a chromatogram. We'll exhume it later.

A peak appears

An important thing just happened.

In Figure 2.4c, notice that the same *percentage* of the molecules dissolved in the liquid phase even when there were fewer molecules available; the solubility ratio is constant. In our example, for instance, we always end up with exactly half of the propane molecules dissolved in the liquid phase. Constant solubility is a very useful property of liquid phases because it generates symmetrical peak shapes. You are about to see how that happens.

There are some rare exceptions to the rule of constant solubility. When solubility varies with solute concentration, some adverse distortion of peak shape occurs that you will need to recognize when troubleshooting. This isn't the time to discuss the problem, so let's leave it for later.

In Figure 2.4 there's only one small enclosed space where an equilibrium forms and is quickly disrupted by the movement of the carrier gas. Now imagine that a column has lots of these small enclosed spaces arranged in series, as depicted in Figure 2.5, so the gas leaving one of the spaces enters the next one, where it encounters fresh clean liquid. It then quickly forms a new equilibrium.

The lower section of Figure 2.5 shows the first two of these equilibria side by side. So far, it's not very interesting because not much has happened; the original molecules have divided into four equal parts.

But they don't stay that way for long; again, the carrier gas moves. Figure 2.6 shows this movement and the formation of the third equilibrium. Look closely at the third equilibrium. The distribution of the propane molecules in the gas phase is now **4:8:4**. Already, a peak has appeared!

In Figure 2.7, the process continues and forms four new equilibria. Notice that each time the carrier gas moves, the number of molecules in the original *small enclosed space* is reduced by a factor of two. Clearly, if this same

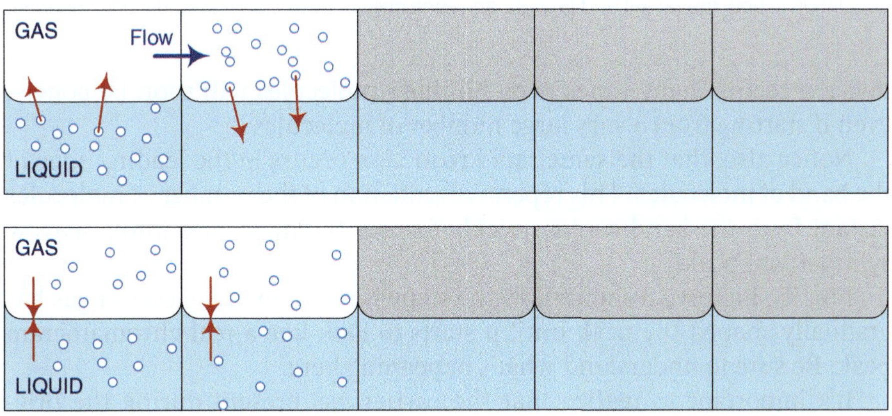

The upper diagram shows how the carrier gas movement carries the propane molecules into the next part of the column, where it encounters fresh liquid phase. The lower diagram imagines that the carrier gas stops for a moment to allow two equilibria to form, each one involving only half of the original molecules.

Figure 2.5 The Second Equilibrium.

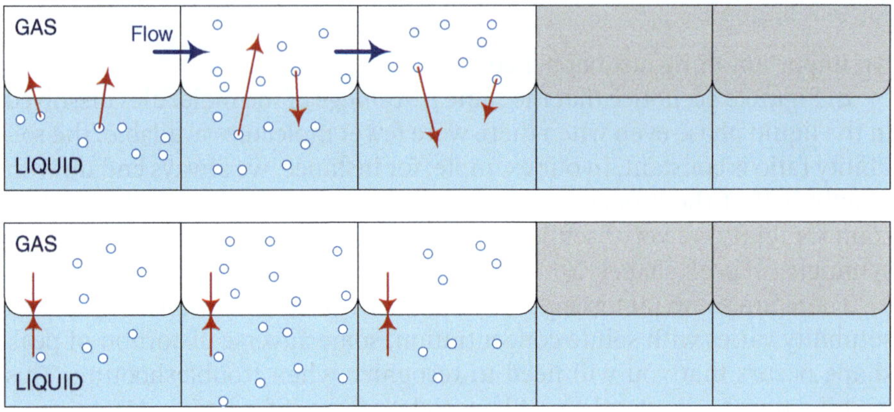

Figure 2.6 The Third Equilibrium.

Again, the upper diagram shows the movement of the carrier gas, which carries all the propane molecules to the next part of the column. The lower diagram then shows how three new equilibria form, but the center one contains half of the original molecules.

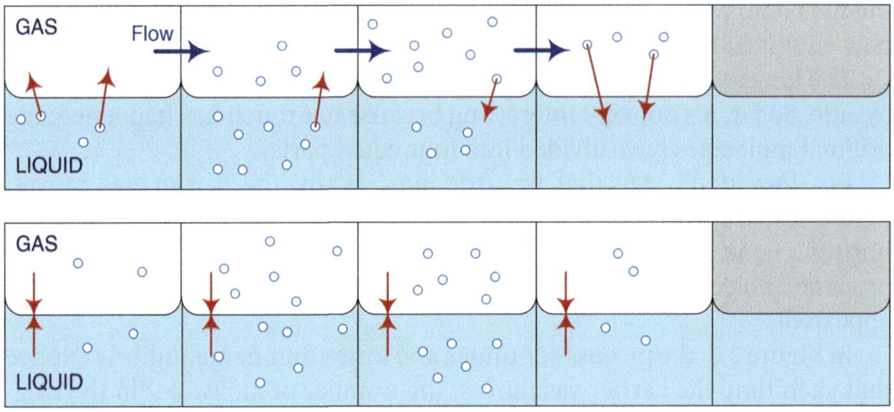

Figure 2.7 The Fourth Equilibrium.

Again, the upper diagram shows the movement of the carrier gas. The lower diagram then shows how four new equilibria form. Notice that each time the carrier gas moves, the molecule population in both the far left and far right equilibria divides by two and is rapidly disappearing.

division recurs many times over, all those molecules will soon be gone – even if starting from a very large number of molecules.

Notice also, that the same rapid reduction occurs in the leading edge of the band of molecules. This repetitive reduction of the number of molecules distant from the band center quickly focuses the molecules into a narrow symmetrical peak.

Finally, Figure 2.8 shows how the stepwise motion of the carrier gas has gradually shaped the peak until it starts to look like a real chromatogram peak. Be sure to understand what's happening here.

It's important to realize that the carrier gas present during the original sample injection has traveled down the column and is now in the fifth equilibrium. This progression of the carrier gas through the five equilibria is indicated by the white background color in Figures 2.5 through 2.8.

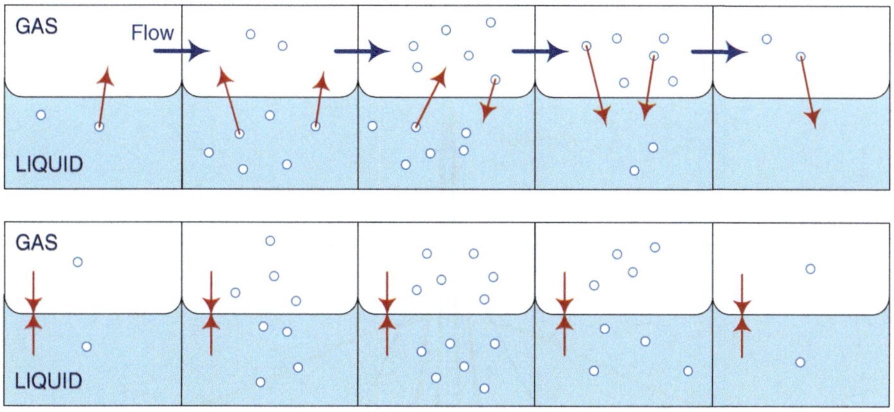

Again, the upper diagram shows the movement of the carrier gas. The lower diagram then shows how five new equilibria form. The molecule distribution is now in the shape of a peak with the highest concentration of molecules at the center, and the lowest concentration at the edges. This is how real peaks form in columns.

Figure 2.8 The Fifth Equilibrium.

By the time the carrier gas reaches the fifth equilibrium, most of the propane molecules have been left behind; only one stayed with the carrier gas during the five moves considered so far. The distribution of propane molecules in the gas phase is now **1:4:6:4:1**, and the peak concentration is right at the middle. Notice that the average rate of movement of a propane molecule is half the speed of the carrier gas. If you are following closely, you'll realize why. The propane solubility is 50 %, so at any instant of time only half of the molecules are in the gas phase, moving. The other half are stationary.

Effect of more equilibria

Figure 2.9 is a graphic rendering of the equilibrium model that tracks the movement of the propane molecules relative to the movement of the carrier gas. The solubility of the molecules is again assumed to be 50 %. The *carrier movement* represents the distance moved by carrier gas molecules that were present at the instant of sample injection.

The gold curve in Figure 2.9 is a smoothed version of the **1:4:6:4:1** distribution we obtained with five equilibria. This embryonic peak would appear in the first two millimeters of a regular packed column and would take less than one fourth of a second to form.

Continuing the jerky mechanism for more equilibria would be tedious, but luckily it can be done mathematically. Figure 2.9 also displays the curves for 50, 500, and 5,000 equilibria, plotted as if they occurred along the same length of column. They look just like chromatogram peaks! And that's exactly what they are. You have just witnessed how the standard peak shape forms.

In chromatography theory, the **plate number** (N) is the number of equilibria achieved by a column. Of course, the notion of separate equilibria occurring in the column is just a theoretical concept. One might argue that the gas and liquid never quite achieve equilibrium because the gas is

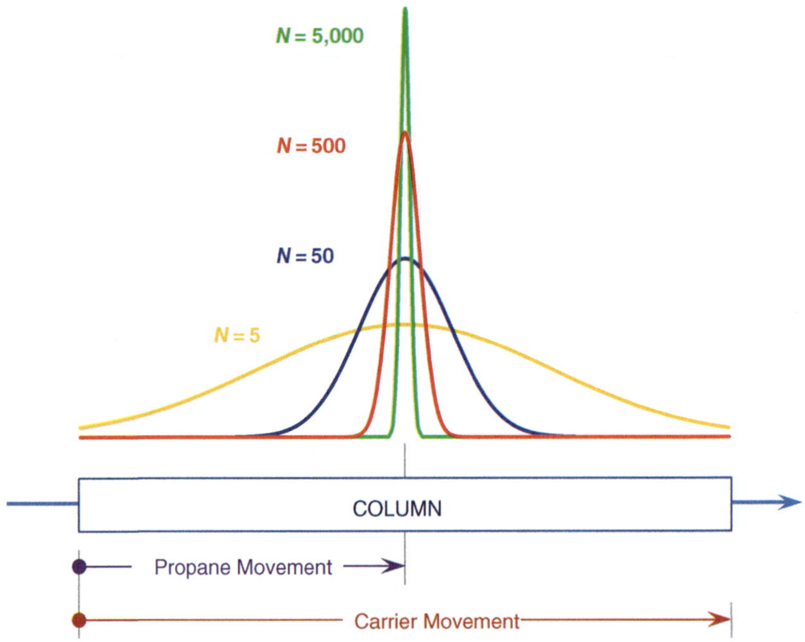

Each colored trace shows the distribution of molecules and the corresponding peak shape obtained for a different number of equilibria (*N*). In practice, the narrower peaks would be much higher: the scale of the vertical axis is not the same for each trace. When a column operates under optimum conditions, a peak experiences more equilibria as it passes through the same length of column, resulting in narrower peaks that are easier to separate.

Figure 2.9 Effect of Having More Equilibria.

constantly moving. True; and that leads to another theory of chromatography, which assumes equilibrium never happens. Nevertheless, even that theory ends up with a *plate number* to express the efficiency of a column, its ability to produce narrow peaks.

The plate number is a useful parameter for evaluating column performance. As is evident from Figure 2.9, anything that increases the plate number must be a good thing, because it reduces peak width and narrower peaks provide better separations.

Some conclusions

Figure 2.9 contains a lot of information. Let's pause for a moment and see what it can tell us.

Identical molecules – different behavior

Notice that the propane molecules do not all spend the same time in the column, even though they are identical to each other. They enter the detector at different times.

> *The variable arrival time of identical molecules at the detector is the root cause of chromatogram peak shape.*

Clearly, some of the propane molecules are delayed longer in the liquid phase than others are. It's a crap shoot in there, a totally random process where some get left behind and others don't. As a result, the peak shape follows the standard *bell-shaped curve* that is found in many random processes. You saw this happen in Figures 2.5–2.8.

All peaks are symmetrical

The statistical bell-shaped curve is named for Carl Friedrich Gauss, the acclaimed German mathematician who derived its mathematical equation. The two major theories of chromatography both deduce that all the peaks in the column should be **Gaussian** in shape.

Thus, our ideal peaks in Figure 2.9 are perfectly symmetrical. Unfortunately, real peaks are not perfect, and they may be slightly or grossly asymmetric due to other mechanisms occurring in the columns or in the instrument itself.

More equilibria – narrower peaks

On a given column, the peaks always get narrower as the number of equilibria increases. This is a very important observation because narrow peaks are highly desirable. It's so much easier to separate narrow peaks from each other than to separate wide peaks from each other. Therefore, to ensure that the peak widths are as narrow as possible, we adjust the column operating conditions – particularly the carrier gas flow rate – to generate the maximum number of equilibria.

In practice, each peak exhibits a different plate number. We can calculate the plate number of a selected peak from chromatogram measurements of its width and retention time. This calculation is used to evaluate column performance and to optimize column operating conditions. We'll look at it later.

It follows that a certain length of column must be necessary to generate each equilibrium. You can calculate it if you wish: just divide the column length by the plate number. In most columns the answer will be about 0.5 mm. We call this the **plate height** (H). It's a very important parameter for optimizing the performance of a column.

More equilibria – taller peaks

The apex of the peak gets higher when the plate number increases. It actually gets very much higher because the peak area remains constant (the vertical axis in Figure 2.9 is not drawn to scale). A higher peak in the graph means there are more molecules in the detector and a larger detector output signal, making the peak easier to measure. The chromatogram baseline always suffers from a little detector **signal noise**, so a higher peak has a better signal-to-noise ratio.

Retention at the apex

The apex of the peak is where the most propane molecules are. It's the highest concentration of propane molecules in the carrier gas. With symmetrical peaks like these, it's also the position of the average molecule. Since the recorded peak retention time should represent an average molecule, the elapsed time at peak apex is our best estimate of its retention time.

More equilibria – same retention time

In Figure 2.9, the retention time of the propane peak is always the same, regardless of the plate number. The plate number of a column affects only the peak width; it has no effect on the peak retention time.

In Figure 2.9, the peak apex has traveled along the column exactly half the distance that the carrier gas has traveled. This remains true during all of its journey through the column. When the carrier gas reaches the end of the column, as depicted, the propane peak has reached the mid-point. At that time, it's only halfway through the column.

> *That's a lot of information to absorb, and it's all important for understanding how columns separate components. Before we look into what causes separation, you might want to read it again.*

The SCI-FILE that follows is optional reading, but is not difficult to understand. It gives a more technical account of solubility, so you may find it interesting.

SCI-FILE: *On Solubility*

Solubility

The main text uses the concept of "solubility" because it's intuitive and easy to understand. The solubility of a component depends on how easily its molecules can interact with the molecules of the stationary phase.

In that text, we make the simplifying assumption that "75 % solubility" means 75 % of the molecules dissolve in the liquid phase and 25 % remain in the gas phase.

While convenient and applicable to a given column, this representation fails to account for the volume of each phase inside the column, an important variable. Instead of using the number of molecules, the formal expressions of solubility use the *concentration* of the solute in each phase, thus incorporating the volume of each phase.

Partition

In the gas chromatography literature, the process and outcome of equilibrium is often called **partition**. The *partition coefficient* is the ratio of solute concentrations in each phase.

Chemists now prefer to reserve *partition* for liquid-liquid systems. For gas-liquid systems, the equivalent term is **distribution**, and the ratio of the solute concentrations in the gas and liquid phases is called the **distribution constant**.

Distribution

The distribution of molecules between gas and liquid phases in a chromatograph column is a special case of **Henry's Law**: the concentration of a gas dissolved in a liquid is directly proportional to the partial pressure of that gas in contact with that liquid.

The distribution constant (K_c) is defined for a specified solute A as the equilibrium concentration of A in the stationary phase divided by the equilibrium concentration of A in the mobile phase:

$$K_c = \frac{[A]_S}{[A]_M} \tag{2.1}$$

Where the square brackets are a standard chemical shorthand for [*the concentration of*].

Limitations

The theory assumes that the distribution constant will remain *constant*. It is assumed, for example, that the distribution of propane molecules at equilibrium is constant regardless of the number of propane molecules present, and that the distribution is unaffected by the presence of other solute molecules in the stationary phase.

While generally true at the low concentrations of solute normally reached in a liquid phase as a peak passes through, either of these assumptions may fail at higher concentrations.

The first assumption is that the distribution is constant for varying amounts of solute. At low concentrations of solute in the liquid phase, the distribution constant is indeed constant for most solute-solvent pairs. But high concentrations of solute may exceed the linear range of Henry's Law, causing a large peak to become wider and distorted in shape.

The second assumption is that solute molecules are at such low concentration in the liquid phase that they do not interact with each other; i.e., the presence of one kind of molecule dissolved in the liquid phase will not affect the solubility of another kind of molecule.

This independence of solubility is generally found to be true in practice. A rapidly moving peak will pass right through a slower one, usually with no effect on the retention time of either of them. Even so, you should be aware that a very large peak can saturate the liquid phase and displace a low-concentration peak from solution, thereby reducing the distribution constant and the retention time of the smaller peak.

Knowledge Gained

- *It's not sufficient to say that some kinds of molecule travel faster in the column than others do.*
- *Identical molecules don't spend the same time in the column; some elute earlier than others do.*
- *Variation in the retention time of identical molecules is the root cause of the peak width and shape.*
- *Wide peaks are more difficult to separate (resolve) than narrow peaks are.*
- *Solid columns work by a different mechanism than liquid columns, but the end result is similar.*
- *Gases dissolve in liquids and rapidly reach dynamic equilibrium.*
- *Component solubility is affected by temperature, pressure, and the kind of liquid; nothing else.*

- *Component solubility in a liquid phase decreases with temperature increase or pressure decrease.*
- *At equilibrium, component molecules enter the liquid phase at the same rate as they are escaping.*
- *Although chromatography is continuous, it can be modeled as a succession of discrete equilibria.*
- *A succession of equilibria forms a symmetrical peak shape which narrows with more equilibria.*
- *Ideally, peaks would have perfect symmetry, but real peaks are often somewhat asymmetric.*
- *For every injected molecule there are only two speeds along the column; stop or go.*
- *All component molecules in the gas phase must move along the column at full carrier gas speed.*
- *The molecules that by random chance are stuck in the liquid phase cannot move with the carrier gas.*
- *Even identical molecules suffer random time in the liquid, which is the main cause of peak width.*
- *Narrower peaks are taller, and more easily separated from adjacent peaks.*
- *Plate number is the effective number of equilibria and is estimated from the chromatogram.*
- *Peaks become narrower as the plate number increases, making separation easier.*
- *Peaks become taller as the plate number increases, thereby increasing signal-to-noise ratio.*
- *The plate number affects only the width of a peak, not its position on the chromatogram.*
- *The peak apex is at the elution time for an average molecule, so is the best measure of retention time.*

Did you get it?

Self-assessment quiz: SAQ 02

Q1. The way a column works depends on whether the stationary phase is a solid or a liquid. What is the mechanism involved when the stationary phase is liquid?
Q2. What is the mechanism involved when the stationary phase is solid?
Q3. What are the four variables that determine the solubility of a gas in a liquid?
Q4. When a gas and a liquid are in equilibrium with each other, what is happening?
Q5. Why do peaks get wider as they travel through a column?
Q6. Predict the effect of increased column temperature. If all else were constant, what effect would it have on the peaks?
Q7. Predict the effect of increased column pressure. If all else were constant, what effect would it have on the peaks?

Check your SAQ answers with those given at the end of the book.

Student evaluation test: SET 02

Your instructor will provide answers to these questions.

S1. What are the two main mechanisms at work in the different kinds of gas chromatographic column?
Select the two correct statements:

A. Columns containing a solid stationary phase work by selectively adsorbing components of the sample.
B. Columns containing a solid stationary phase work by selectively dissolving components of the sample.
C. Columns containing a liquid stationary phase work by selectively adsorbing components of the sample.
D. Columns containing a liquid stationary phase work by selectively dissolving components of the sample.

S2. Consider the identical molecules within a certain component peak, say, propane. Which of the statements listed here are false?
Select the two false statements:

A. If all variables are held constant, identical molecules take the same time to pass through a column.
B. When identical molecules travel through a column, some randomly stay in the liquid phase longer than others do.
C. When identical molecules travel through a column, some randomly stay in the gas phase longer than others do.
D. The width of a peak is due to identical molecules randomly taking different times to travel through a column.

S3. What is the effect of temperature and pressure on the solubility of sample gases in the liquid phase?
Select the two correct statements:

A. The solubility of a gas in a liquid increases when the temperature is increased.
B. The solubility of a gas in a liquid decreases when the temperature is increased.
C. The solubility of a gas in a liquid increases when the pressure is increased.
D. The solubility of a gas in a liquid increases when the pressure is decreased.

S4. When applied to the discontinuous model of chromatography presented in the text, which two of these statements are true?
Select the two true statements:

A. When the carrier gas moves, all the sample molecules move with it.
B. Each time the carrier gas moves, the same number of sample molecules move with it.
C. Each time an equilibrium forms, it contains the same percentage of the original sample molecules.
D. Each time an equilibrium forms, the same percentage of the available molecules is dissolved in the liquid phase.

S5. Consider a gas-liquid equilibrium. Which of the statements listed here is true?
Select the one true statement:

A. Whenever a gas is in equilibrium with a liquid, 50 % of the gas molecules are dissolved in the liquid.
B. At equilibrium, the liquid phase cannot dissolve any of the sample molecules that are still in the gas phase.
C. At equilibrium, the rate of molecules dissolving in the liquid has become zero.

D. At equilibrium, the rate of molecules dissolving in the liquid is exactly equal to the rate of molecules leaving the liquid phase.

S6. For a 10,000-plate peak having a retention time of 500 s, what is the average time for the column to generate one equilibrium?
Your answer must be an integer number of milliseconds.

S7. In the example given in Figure 2.9, why is the propane peak exactly halfway along the column when the carrier gas gets all the way through?
Select all the correct answers and none of the incorrect explanations:

A. Because, in the example, the propane solubility was adjusted to 50 %.
B. Because, when in the gas phase, the average speed of a propane molecule is half the speed of the carrier gas.
C. Because, on average, the propane molecules stop in the liquid phase for half the time, so they only travel in the gas phase for half the time.
D. Because, in the example, the plate number of the column was $N = 5$.

S8. What change in operating variables might influence the height of a peak on the chromatogram?
Select all the correct answers and none of the incorrect answers:

A. A change in the number of molecules of that component present in the injected sample.
B. A change in the width of the peak.
C. A change in the plate number of the peak.
D. A change in the volume of sample injected.

S9. This question requires you to make deductions from what you have learned so far.

For a particular sample gas, what properties of the gas-liquid equilibrium are necessary to produce a symmetrical peak?
Select all the correct answers and none of the incorrect answers:

A. The proportion of molecules in the gas and liquid phases must remain constant when the component concentration changes.
B. The equilibrium must form very rapidly so more equilibria occur while the peak is in the column.
C. The overall time the molecules spend in the gas phase must be equal to the time they spend in the liquid phase.
D. The solubility of the sample gas must not change with its concentration.

References

Figures

2.8 The Fifth Equilibrium
2.9 Effect of Having More Equilibria

Equation

2.1 $K_c = \frac{[A]_S}{[A]_M}$ Distribution constant

Symbols

$[A]_M$	Equilibrium concentration of substance A in mobile phase
$[A]_S$	Equilibrium concentration of substance A in stationary phase
H	Plate height
K_c	Distribution constant
N	Plate number

New technical terms

When first introduced, these words and phrases were in bold type. You should now know the meaning of these technical terms. If still in doubt, consult the Glossary at the end of the book:

distribution
distribution constant
dynamic equilibrium
Gaussian
Henry's law
partition
plate height
plate number
signal noise
solubility
solute
solvent

3

Separation

> *"It's a paradox worth repeating. The different kinds of molecules that are injected into a column all travel along the column at the same speed, yet emerge from the column at different times, thus becoming separated from each other. Separation is what chromatographs do. You certainly need to know how".*

How peaks get separated

Looking back at Figure 2.9, why is the propane peak exactly halfway along the column? Think about it.

Taking the question one stage further: What would have to happen for another peak to be at a different location in the column, separated from the propane peak?

Simple explanations of gas chromatography often say that different peaks move at different speeds and come out of the column at different times. While this is empirically true, it's a circular argument and explains nothing. It's plainly true that if the peaks come out of the column at different times they must be traveling at different average speeds, but it's illogical to then conclude that the peaks are separated *because* they move at different speeds. It doesn't explain anything.

A more realistic explanation

We need a better explanation. Let's start from the obvious truism that when the sample molecules are in the column, they must always be in the gas phase or the liquid phase. There's nowhere else for them to be.

And we know that the liquid phase doesn't move.

Therefore:

- When the sample molecules are dissolved in the liquid phase, they are held in place and are not moving along the column.
- When the sample molecules are in the gas phase, they move along the column at the same speed as the carrier gas.

Process Gas Chromatographs: Fundamentals, Design and Implementation, First Edition. Tony Waters.

Note that we refer to movement along the column. Molecules are always moving randomly in every direction, and this contributes to peak broadening, but random motion contributes nothing to positive movement along the column.

There are only two speeds along a chromatograph column; stop or go!

This is true for every peak. Every injected molecule must spend enough time in the gas phase to transit through the column. Therefore, each molecule travels at the same speed and spends the same time traveling as the other molecules do. It's not true to say that different molecules travel at different speeds.

Separation is not caused by motion at all. It's caused by stopping; the time that different molecules stay motionless in the liquid phase. The less soluble molecules don't stop for long and move quickly through the column, while the more soluble ones hang around in the liquid, slowing them down. It's that simple.

Notwithstanding the stop-go mechanism, one might successfully argue that the peaks really are travelling at different average speeds through the column. Yes, of course, this is the overall result. But talking about the average speed of a peak obscures what is really happening in the column. Instead, chromatographers may refer to the **migration rate** of the peak along the column. The word "migration" infers a gradual movement by alternately stopping and going, to remind us what is really going on in there.

It should now be clear why the propane peak in Figure 2.9 is exactly in the middle of the column. We assumed that 50 % of the propane molecules dissolve in the liquid phase and 50 % remain in the gas phase. Remember that those molecules are rapidly moving between the two phases. So the average propane molecule spends 50 % of its time in the gas phase moving at carrier speed and 50 % of its time in the liquid phase going nowhere. Therefore, when the carrier gas that was present during sample injection reaches the end of the column, the propane peak is exactly half-way.

Figure 3.1 applies this logic to two additional peaks, one more soluble and one less soluble than propane. The 1-butene peak is more soluble, so we would expect it to spend more time in the liquid phase and elute from the column later than the propane. And so it does. For instance, if the solubility of 1-butene is 75 %, only 25 % of the molecules move each time the carrier gas moves, and the peak stays in the column twice as long as the propane peak does. The carbon dioxide peak is less soluble: If its solubility is 25 %, then 75 % of its molecules move with the carrier gas.

> *This is the true cause of separation. When in the gas phase, all sample molecules move along the column at the same speed as the carrier gas. But when in the liquid phase, the molecules stop moving and the more soluble ones stop longer than the less soluble ones do.*

Figure 3.1 illustrates the net effect of solubility difference. It shows the location of four components peaks at the exact moment the air peak reaches the end of the column. A small white-and-blue equilibrium diagram indicates the solubility of each peak.

Figure 3.1 assumes that:

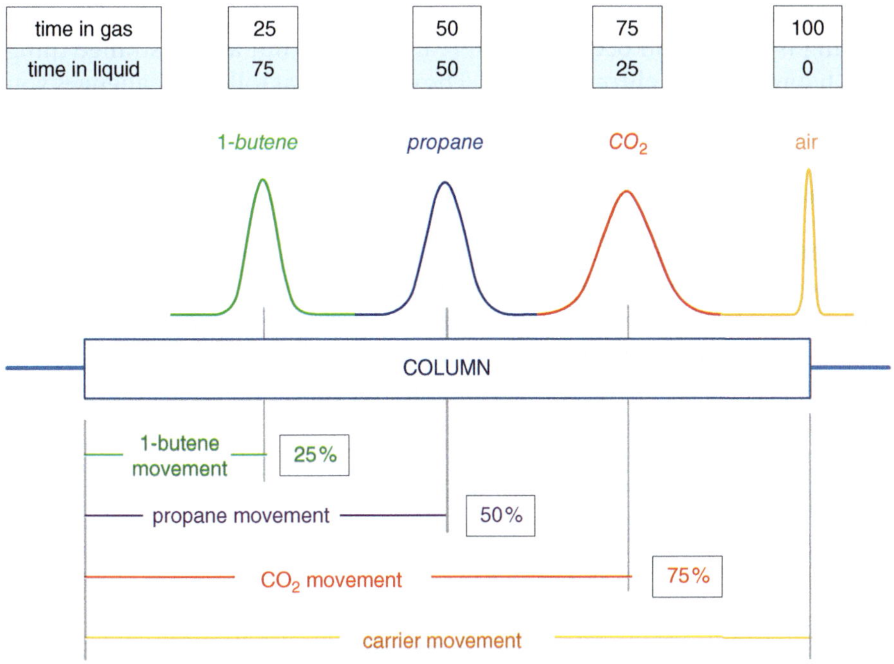

Drawing peaks above a column is a common way to show the position of component molecules within the column at a certain instant of time. Here, an air peak (which doesn't dissolve in the liquid phase) has moved along with the carrier gas and has just arrived at the column end. At that instant, the three colored peak drawings indicate the position within the column of the other component molecules, predictable from their solubility in the liquid phase.

Figure 3.1 Effect of Component Solubility.

- The stationary phase is a liquid, and the air peak is not soluble at all. The air peak therefore moves with the carrier gas, and its retention time is a good indicator of average carrier gas velocity.
- The carbon dioxide (CO_2) solubility is 25 %, so an average CO_2 molecule spends 75 % of the time traveling and only 25 % of the time stopped. Therefore, the CO_2 peak moves 75 % of the distance that the carrier gas moves.
- The propane solubility is 50 %, so an average propane molecule spends half the time traveling and half the time stopped. Therefore, the propane peak moves 50 % of the distance that the carrier gas moves.
- The 1-butene solubility is 75 %, so an average molecule spends only 25 % of the time traveling and 75 % of the time stopped. It follows that the 1-butene peak moves only 25 % of the distance that the carrier gas moves.

These convenient values for the component solubilities are just simple examples assumed for discussion purposes, but every substance has a real solubility in a given liquid that depends only on the temperature and pressure.

Figure 3.1 shows the location of each peak when the air peak has reached the end of the column and is about to flow into the detector. At that moment, the four peaks in our example are equally spaced along the column. Similar spacing is a common occurrence in real columns, but a very strange thing then happens to the chromatogram. The exercise that follows challenges you to discover what comes next.

A challenge question

Imagine that the effluent from the column in Figure 3.1 flows into a detector and the detector signal is recorded in the form of a chromatogram. For this exercise, assume Figure 3.1 shows the correct position of the four peaks in the column at three minutes after sample injection.

Now predict what the chromatogram will look like after all the peaks have exited the column.

> *It's worth your time to stop for a moment and try to do this; you will learn a lot from the exercise. Be careful; it's more difficult than it looks!*

You can draw your chromatogram on Figure 3.2. Sketch the chromatogram you would expect to get, showing your chosen time base in minutes and the exact positions of the four peaks.

How confident are you of your chromatogram? People rarely get it right at first attempt. The most common mistakes are:

- Inserting all four peaks between 0 and 3 mins on the chromatogram
- Assuming all four peaks are equally spaced
- Assuming three of the peaks are equally spaced
- Assuming any peak has a fractional retention time (none do)

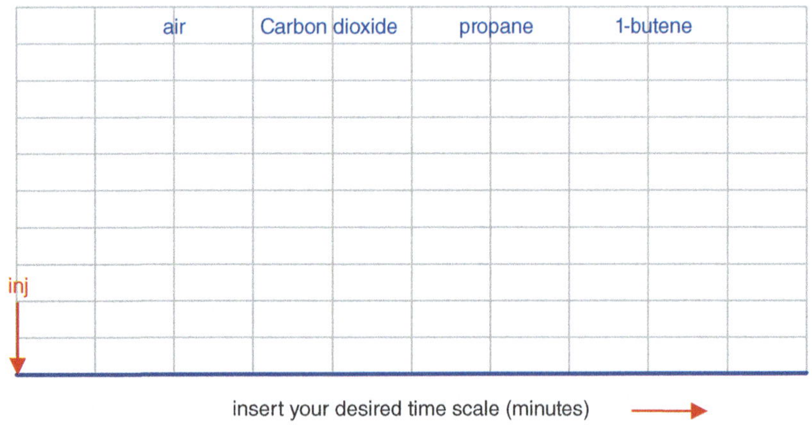

Based on the information given in Figure 3.1, draw the chromatogram you would expect to see after all four peaks have passed through the detector. Your chromatogram should show the four peaks at their correct retention times. The injection marker is time-zero on your chromatogram, but you must decide the time scale. In this exercise, don't worry about peak widths.

Figure 3.2 Draw Your Own Chromatogram.

Remember to measure retention times from sample injection to the apex of each peak; *i.e.*, the retention of the average molecule.

Want to try again? We'll reveal the correct answer later.

Significance of the air peak

The **air peak** is a valuable indicator of column performance. Since air doesn't dissolve in column liquids to any significant extent, the air peak remains in the gas phase all the way through the column – traveling at full gas velocity. We call it an **unretained peak**. Its position on the chromatogram indicates the elapsed time for the carrier gas to travel from one end of the column to the other end.

If using a detector that doesn't respond to air, another unretained peak can act as a surrogate. For instance, methane often serves as an adequate "air peak" on a flame ionization detector.

Since we know the column length (L), the air peak retention time (t_M) allows us to calculate the average carrier gas velocity (u_M) in m/s:

$$u_M = \frac{L}{t_M} \tag{3.1}$$

We don't need this information now, but Equation (3.1) is used in more advanced work to optimize the carrier gas flow rate.

The air peak location is good to know when troubleshooting because component molecules can't travel faster than the carrier gas, so no peaks from the current sample injection can elute before the air peak.

But the air peak has a much more significant role because it shows how long the carrier gas took to pass through the column. Recall that component molecules can move along the column only when they are in the carrier gas. So, to get through the column, all component molecules must spend the same time in the carrier gas as the air peak does.

As a way of emphasizing the point, imagine that the column had no liquid inside; where would the peaks be on the chromatogram?

Yes, of course, all the peaks would elute together with the air peak. With no liquid present, no separation can occur. The air peak reveals the time that component molecules spend in the gas phase, and it's the same for *every* component.

> *As an aside, realize from this imagined experiment that the time a peak spends in the liquid phase is proportional to the amount of liquid phase present in the column. In packed columns, the liquid film can slowly evaporate into the carrier gas, thereby, over time, reducing the observed peak retention times.*

On a more typical chromatogram such as the one in Figure 3.3, the component peaks come out some time after the air peak. This additional retention time, different for each component and plainly visible on the chromatogram, is the time that each component spends in the liquid phase.

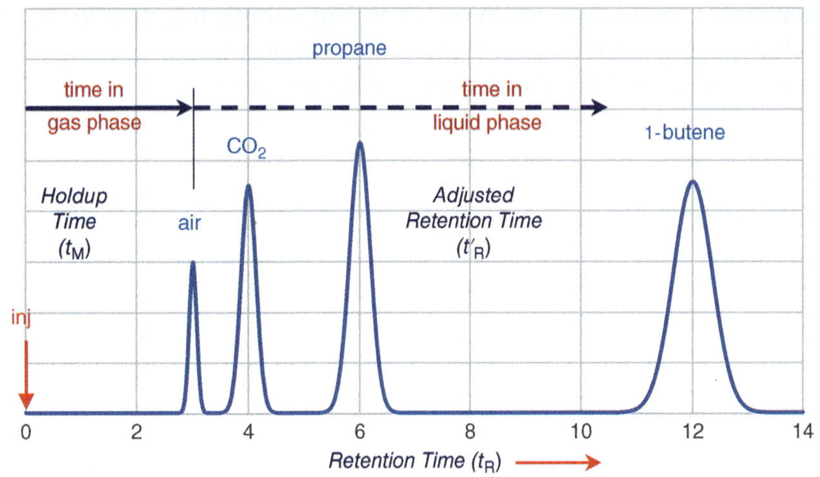

Figure 3.3 Significance of an Air Peak.

An air peak tells us how long the carrier gas takes to get through the column, which is also the time that each component spends in the gas phase. The additional retention time for each peak is the time that peak spends in the liquid phase.

Figure 3.3 illustrates the point: the peak retention time (t_R) is the sum of two times; the time that the peak spends in the gas phase (t_M) plus the time it spends in the liquid phase (t'_R):

$$t_R = t_M + t'_R \tag{3.2}$$

Notice that the air peak neatly divides the chromatogram into two time zones and makes them easy to measure:

- From injection to the air peak is the time that all components spend in the *gas phase*. This time is often called the "dead time" because it doesn't contribute to separation. The formal name for the time in the gas phase is the **holdup time**.
- From the air peak to the apex of each peak is the average time that each component spends in the *liquid phase*. The formal name for the time in the liquid phase is the **adjusted retention time**.

These chromatogram measurements are an essential prerequisite for optimizing or troubleshooting column performance. For example; they allow the optimum setting of column temperature. We plan a second book to explore these more advanced techniques.

The answer

Figure 3.3 is also the correct answer for the chromatogram you were invited to draw on Figure 3.2. The four peaks are centered at 3, 4, 6, and 12 minutes from sample injection. You may find this surprising; the peaks are no longer equally spaced! Let's see how that happened.

In Figure 3.1, none of the peaks have left the column yet, so your chromatogram needs to show a flat baseline for the first three minutes after injection – the so-called "dead time." It would be a mistake to draw any peaks in this zone.

The air peak is just leaving the column after three minutes of traveling. Since it doesn't dissolve in the liquid phase, its time in the liquid phase is zero. One way to indicate this is to show its gas:liquid residence time ratio as 3:0. Since the air molecules are not delayed in the liquid phase, the air peak should be very narrow, and it's reasonable to draw it centered at three minutes.

Most people find it very difficult to decide where the other peaks come out on the chromatogram, perhaps because the correct answer defies all expectations.

Actually, it's easy to get it right. Starting from the position of each peak in Figure 3.1, there are two ways of reaching the correct conclusion about its final position on the chromatogram. Both ways are noted below.

To locate the remaining peaks, it's easier to think about them in the following order, starting with the propane peak.

The propane peak took three minutes to reach the center of the column, so it should take another three minutes to reach the detector, for a retention time of six minutes overall. A better argument comes from its solubility ratio. Recall that retention time is the sum of the time a component spends in the gas phase and the time it spends in the liquid phase (Equation 3.2). To reach the detector, the propane peak must spend three minutes traveling in the gas phase. But, because of its 50:50 solubility ratio, it also spends three minutes stopped in the liquid – so its residence time ratio is 3:3 for a total retention time of six minutes.

The 1-butene peak took three minutes to move 25 % of the column length, so it should take four times as long to reach the detector, for a retention time of twelve minutes overall. A better argument invokes its solubility ratio. To reach the detector, the 1-butene peak must spend three minutes traveling in the gas phase. But, because of its 25:75 solubility ratio, it also spends nine minutes stopped in the liquid – so its residence time ratio is 3:9 for a total retention time of twelve minutes.

The carbon dioxide peak took three minutes to move 75 % of the column length, so it should take one additional minute to travel the last 25 % of the column, for a retention time of four minutes overall. A better argument comes from its solubility ratio. To reach the detector, the carbon dioxide peak must spend three minutes traveling in the gas phase. But, because of its 75:25 solubility ratio, it also spends one minute stopped in the liquid – so its residence time ratio is 3:1 for a total retention time of four minutes.

Notice that the different molecules all spend exactly three minutes in the gas phase traveling.

Notice also that the peaks were equally spaced along the column, but not on the chromatogram. This may seem strange, but it's the normal behavior of columns and explains why early peaks tend to cluster close to the air peak location, while later ones have much longer retention times. In the next chapter, we'll look at these patterns of retention in more detail.

> *If you closely followed what is going on here, you may be thinking that the pressure drop along the column must distort the predictions made in the above arguments. That's somewhat true, but way too complex to consider at this early stage.*

Measurements from chromatograms

A practical task

To evaluate various measures of column performance, we'll need to measure the peak retention time and the peak width. Some computer programs will make these measurements for you, but you won't get the significance of them until you learn to do it on a chromatogram.

Figure 3.4 illustrates the most important measurements. Measure the variables in seconds or in millimeters. Many of our optimization parameters are ratios, so the units cancel out.

It's not possible to measure the actual peak width at the baseline because the peak gradually fades away. Instead, draw tangent lines along the flat sides of the peak and extend the baseline across the width of the peak. Chromatographers call this procedure **triangulating** the peak.

Some chromatographers prefer to measure the width of the peak at half its height. This is often easier to do than triangulating the base width, and perhaps more accurate, since no triangulation errors occur. The two width measurements are related, and either of them can be used to evaluate column performance.

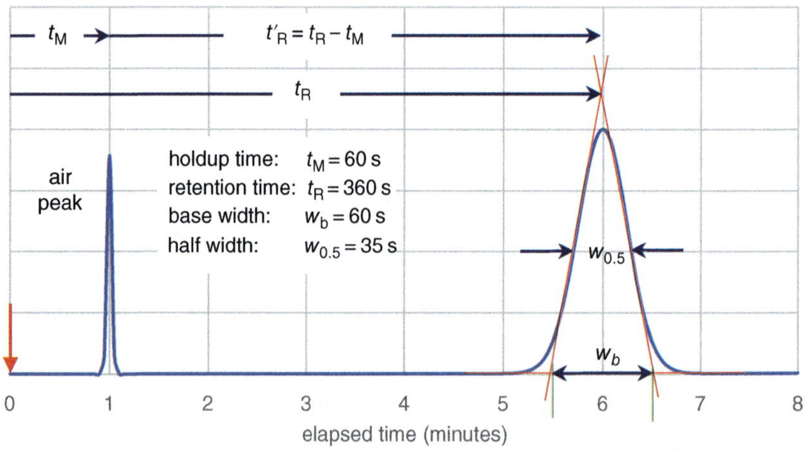

You can make simple measurements from a chromatogram to discover how well the columns are performing. The data most often collected are shown in this illustrative chromatogram above, and are further discussed in the text.

Figure 3.4 Typical Chromatogram Measurements.

Make the following measurements:

- Measure the holdup time (t_M) from the injection time mark to the apex of the air peak (you can triangulate the air peak if you wish).
- Measure the retention time (t_R) of each component peak from the injection time mark to the intersection of its tangent lines.
- Measure the **base width** (w_b) of each component peak between the intersections of the tangent lines with the extended baseline.
- Alternatively, measure the **width at half height** ($w_{0.5}$) of each component peak.

For clarity of display, Figure 3.4 shows a single wide peak. You may have many peaks, and most likely they will be narrower than that. It's difficult to measure the width of a narrow peak. To obtain a good measurement, you may have to expand the time base on a computer display, or increase the chart speed on a recorder.

Typical calculations

These few measurements are enough to evaluate the performance of a single column. By way of example, let's calculate the plate number (N) for the peak in Figure 3.4. The math is not difficult:

$$N = 16\left(\frac{t_R}{w_b}\right)^2 \tag{3.3}$$

Inserting the data from Figure 3.4 gives:

$$N = 16\left(\frac{360}{60}\right)^2 = 576$$

A plate number of 576 is very low and would indicate a very inefficient column, since most columns generate about 2000 plates per meter. Of course, this peak is intentionally drawn wide so the measurements are clearly seen. A real peak would be much narrower than this: a more typical peak might have one third of the width and nine times the plate number.

The alternative equation using the peak width at half height ($w_{0.5}$) is:

$$N = 5.54\left(\frac{t_R}{w_{0.5}}\right)^2 \tag{3.4}$$

Chromatographers calculate plate number as a way to evaluate the performance of a column. Any change of operating parameters that increases plate number automatically improves the separating power of a column.

Another performance indicator is the plate height (H), which is the length of a column required to generate one plate. It's usually reported in mm. Knowing the column length (L), you can easily calculate the plate height:

$$H = \frac{L}{N} \tag{3.5}$$

Plate height gives a measure of column performance that is independent of the column length and is the primary variable used for optimizing performance.

Knowledge Gained

- *It's obviously true to say that peaks come out of the column at different times.*
- *It's not true to say that molecules move at different speeds inside the column.*
- *For all injected molecules, there are only two speeds possible inside the column; stop or go.*
- *When molecules are in the liquid phase, they cannot move along the column.*
- *When molecules are in the gas phase, they move at full carrier gas speed.*
- *The "air peak" doesn't dissolve in the liquid phase so it travels at the same speed as the carrier gas.*
- *It's impossible for any peak to travel faster than the carrier gas.*
- *No peak from the same injection can appear on the chromatogram earlier than the air peak.*
- *To get through the column, all injected molecules must travel for the same time as the air peak does.*
- *If the column had no liquid in it, all the peaks would elute together with the air peak.*
- *Any additional retention time beyond the air peak time is the time a peak stopped in the liquid phase.*
- *Separation is not caused by motion, it's due to peaks stopping for different times in the liquid.*
- *The time that a peak stops in the liquid phase is directly proportional to its solubility in that liquid.*
- *The spacing of peaks on the chromatogram is different than their spacing inside the column.*
- *Peaks that migrate slowly along the column come out very much later on the chromatogram.*
- *Retention time is the sum of time traveling in the gas phase and time stopped in the liquid phase.*
- *Holdup time is the time in the gas phase; adjusted retention time is the time in the liquid phase.*
- *Column operating performance can be evaluated from chromatogram measurements.*
- *Chromatogram measurements may be made in seconds or millimeters.*
- *You can measure the holdup time, peak retention time, and peak width on the chromatogram.*
- *When measuring base width, triangulate the peak and extend the baseline under the peak.*
- *Alternatively, measure the peak width at half the peak height.*
- *Plate number is calculated from measurements of retention time and peak width.*
- *Any change in operating conditions that increases plate number increases column separating power.*

Did you get it?

Self-assessment quiz: SAQ 03

These questions relate to gas-liquid chromatography:

Q1. At what speed do component molecules move along the column when they are in the gas phase?
Q2. At what speed do component molecules move along the column when they are in the liquid phase?
Q3. What causes the separation between components?
Q4. What information may be derived from the retention time of an unretained peak, such as air?
Q5. If all the liquid phase was washed out of the column, where would all the peaks be on the chromatogram?
Q6. What causes the additional retention time beyond the holdup time?
Q7. What is the purpose of triangulating a peak?

Check your SAQ answers with those given at the end of the book.

Student evaluation test: SET 03

Your instructor will provide answers to these questions.
All of the questions relate to gas-liquid chromatography.

S1. Which one of the statements below best defines the retention time of a component peak? Select one:

- **A.** It's the average time that the component molecules spend in the column.
- **B.** It's the average time that the component molecules spend in the gas phase.
- **C.** It's the average time that the component molecules spend in the liquid phase.
- **D.** It's the average time that the component molecules spend in the detector.

S2. Which one of the statements below best explains the separation of component peaks on the chromatogram? Select one:

- **A.** The molecules of some components move at different speeds than others do.
- **B.** The molecules of some components spend less time moving than others do.
- **C.** The molecules of some components spend more time moving than others do.
- **D.** The molecules of some components spend more time not moving than others do.

S3. Which one of the statements below best explains the significance of the air peak? Select one:

- **A.** The air peak marks the time when the component molecules start to dissolve in the liquid phase.
- **B.** The air peak indicates the time that every component molecule spends in the gas phase.
- **C.** The air peak indicates the time that every component molecule spends in the liquid phase.
- **D.** Any peaks traveling faster than the air peak cannot be separated on that liquid phase.

S4. On the chromatogram shown in Figure 3.3, what percentage of the 1-butene dissolves in the liquid phase at equilibrium?

S5. If the holdup time on a certain column is sixty seconds, what would be the adjusted retention time of a peak having a liquid phase solubility of 60 %?
Express your answer in integer seconds.

S6. If the propane peak has a retention time of sixty seconds on a three-meter column and propane has a solubility in the column liquid phase of 50 %, what is the average velocity of the carrier gas?
Express your answer in m/s.

S7. Why do peaks that are evenly spaced in the column turn out to be unevenly spaced on the chromatogram? Combine two of the arguments given below to get the best overall explanation.
Select two:

- **A.** All of the peaks spend the same time in the carrier gas.
- **B.** The later peaks are more soluble in the liquid phase than the earlier peaks are.
- **C.** After the air peak reaches the end of the column, the later peaks have farther to go than the earlier peaks do.
- **D.** The later peaks migrate through the column more slowly than the earlier peaks do.

S8. This question asks you to make logical deductions beyond what you learned in the chapter.

Assuming that all other variables can be held constant, evaluate whether each statement listed below is true or false.
Select all of the true statements and none of the false statements:

- **A.** The holdup time would be proportional to the column length.
- **B.** The retention time of a component peak would be proportional to the column length.
- **C.** If the carrier gas flow increased, the holdup time would increase.
- **D.** If the column temperature increased, the adjusted retention time of a component peak would decrease.
- **E.** None of the statements A–D above are true.

S9. This question asks you to make logical deductions beyond what you learned in the chapter.

Consider a column that loses some of its stationary phase due to slow evaporation, but all other variables are held constant. Then, evaluate whether each statement listed below is true or false.
Select all of the true statements and none of the false statements:

- **A.** The holdup time would be reduced.
- **B.** The retention time of all peaks would be reduced.
- **C.** The adjusted retention time of all peaks would be reduced.
- **D.** The separation of component peaks would be increased.
- **E.** None of the statements A–D above are true.

References

Figures

Equations

3.1	$u_M = \frac{L}{t_M}$	Average carrier gas velocity
3.2	$t_R = t_M + t'_R$	Retention time as sum of times in gas and liquid phases
3.3	$N = 16\left(\frac{t_R}{w_b}\right)^2$	Plate number from base width
3.4	$N = 5.54\left(\frac{t_R}{w_{0.5}}\right)^2$	Plate number from width at half height
3.5	$H = \frac{L}{N}$	Plate height (usually expressed in mm)

Symbols

H	Plate height
L	Column length
N	Plate number
t_R	Peak retention time
t'_R	Peak adjusted retention time
t_M	Holdup time (air peak)
w_b	Peak base width
$w_{0.5}$	Peak width at half height
u_M	Average carrier gas velocity

New technical terms

When first introduced, these words and phrases were shown in bold type. You should now know the meaning of these technical terms. If in doubt, consult the Glossary at the end of the book:

adjusted retention time
air peak
base width
holdup time
migration rate
triangulating
unretained peak
width at half height

4

Peak patterns

"Newcomers to chromatograms see a row of peaks on a baseline. But experienced chromatographers see patterns, patterns that reassure or warn. Start to notice the position and shape of your peaks relative to other nearby peaks on the chromatogram".

Migration rate

In this chapter we discuss the rate of movement of peaks along a single column, sometimes called the peak migration rate. Up to now, we have been careful to note that when individual peak molecules move, they travel at full carrier gas speed. This is true, but the movement of whole peaks is entirely a different matter. It's quite obvious that whole peaks can have different migration rates and reach the end of the column at different times.

The following discussion is about the migration rate of whole peaks, so it talks about their *speed of movement.*

Predictable patterns in peak position

Space or time

It would be reasonable to ask why the chromatogram peaks in Figure 3.3 are no longer equally spaced, like they were inside the column. It's a frequent question in our training classes and causes a lot of confusion. It does seem rather odd.

Why are peaks that were equally spaced inside the column unequally spaced on the chromatogram?

It's not easy to answer that question. To do so, you'll need to distinguish between two kinds of separation; a separation in space (spatial) or a separation in time (temporal). Looking back at the illustrations in Chapter 3:

Process Gas Chromatographs: Fundamentals, Design and Implementation, First Edition. Tony Waters.

- Figure 3.1 shows a **spatial separation**; it indicates the *location* of the components along the column at a specific instant of time. Note that none of the peaks has yet reached the detector. Since all peaks have traveled for the same time, *this diagram shows no time separation at all.*
- Figure 3.3 shows a **temporal separation**; it records the *elapsed time* as each component arrives at the detector. When the chromatogram is complete, all of the peaks have left the column. Since all peaks have now traveled the same distance, *this diagram shows no space separation at all.*

For a group of components, their separation-in-space along the column is very different from their separation-in-time on the chromatogram. The distinction is important for designing, maintaining, or troubleshooting a **column system**.

Nearly all PGCs use a custom-designed column system, an arrangement of two or more columns in series. The first column partially separates the components. Then, selected groups of components flow into other columns for final separation, or are discharged to an atmospheric vent. Refer to Chapter 9 for a detailed review of column systems, including several examples.

When you are designing a column system or adjusting the column valve timing, you are dealing with the spatial separation of peaks along the column, not with their time of arrival at the detector. You need to know where the peaks now are in the column, not where they will be on the chromatogram.

Spatial or temporal separation

It may be helpful to explain the difference between spatial and temporal separation in a little more detail.

When the air peak reaches the end of the column, the other peaks are at various places along the column. They are separated in space, but not yet separated in time. In fact, that's when the time separation begins. From the instant that the air peak appears on the chromatogram, each peak takes its own time to reach the detector. Recall that this time, the time between the air peak and a component peak, is also the exact time that the peak spends in the liquid phase.

For the challenge question in Chapter 3, we imagined that Figure 3.1 shows the positions of the four peaks at three minutes after injection. The air peak is just leaving the column after three minutes of traveling in the gas phase. The propane peak is halfway through the column and will need another three minutes to reach the detector; six minutes in all. But at that instant, the 1-butene peak faces a longer journey to the detector than the propane does, and is migrating at only half the speed of the propane. The 1-butene peak therefore takes another nine minutes to reach the detector; twelve minutes in all.

When the air peak reaches the detector and the time separation begins, the late components get a double handicap: the slower they are moving, the farther they have to go. The net effect is that when the component peaks have equal spacing inside the column, their spacing is exponential on the chromatogram.

This is a very common pattern on chromatograms. The early peaks are clustered closely together near the air peak, and the later peaks move farther apart as their retention times increase. In a moment, we will use this pattern to predict the retention time of a peak, or perhaps to identify an unknown peak. Before getting into that, though, there is another consequence of the spatial separation within the column: it also affects the peak width.

Predictable patterns in peak width

Distance or duration

Another oddity of a spatial separation, like the one shown in Figure 3.1, is that the faster peaks are wider than the slower peaks. This is contradictory to our experience with chromatograms; the later peaks on a chromatogram are always predictably wider than the earlier peaks are.

In a spatial separation, all the peaks move for the same amount of time, so the duration of the separation cannot be responsible for the different peak widths. Instead, peak width is a function of the distance moved. The peaks that don't move very much don't get any wider, whereas peaks that travel farther tend to spread.

When working on a PGC, you should be aware of the reduced width of the heavy component peaks while they are in the column. This peculiar behavior of peaks might be important to know when setting up a multiple column system.

In a temporal separation, all peaks travel the same distance – the length of the column – so their different peak widths can't be caused by distance traveled. Only the *time of travel* is different. Therefore, the width of a chromatogram peak is a function of how long the peak has been inside the column. Later peaks are always wider than earlier peaks. If one of your peaks doesn't follow this predictable pattern of continuously increasing width, it's an interloper, not a component present in the injected sample. The alien peak may be left over from a previous injection or simply an artefact of valve switching. Don't accept it as part of your analysis.

Notwithstanding the possibility of a freak peak, be aware that the chromatogram obtained from a multicolumn PGC may sometimes show peaks that have passed through different columns. If so, the peaks from one column may be wider than the peaks from another column. To evaluate this possibility, you would have to know your column system design.

Before we discuss the predictable patterns of retention, the following SCI-FILE: *On Chemical Names* reviews the meaning of simple chemical names, and some of their shorthand notations used in gas chromatography. This knowledge is helpful for understanding how the structure of a molecule affects its retention time. The SCI-FILE is optional reading, and you can skip it if you already know some basic chemistry.

SCI-FILE: *On Chemical Names*

Hydrocarbons

Paraffins

When working with a process gas chromatograph, you will need to know what components are present in the sample and how they are related to each other. Since the majority of PGC applications are in the oil and chemical industry, this book uses **hydrocarbons** for most examples of separations.

Hydrocarbons are chemical compounds containing only hydrogen and carbon, nothing else. For predicting their behavior in columns, it's useful to understand their structure.

The carbon atom has the ability to form a chemical bond with four other atoms, but in a hydrocarbon, it can bond only with a hydrogen atom or another carbon atom. The smallest and lightest hydrocarbon molecule is methane, comprising one carbon atom and four hydrogen atoms. We represent it as CH_4.

When carbon bonds to carbon, it can form chains of carbon atoms of unlimited length, resulting in a myriad of chemical compounds. The smallest is ethane, which contains two carbon atoms bonded together, plus three hydrogen atoms attached to each carbon. Ethane can be represented as $H_3C{-}CH_3$, although we usually show its formula more simply as C_2H_6.

Hydrocarbons like ethane or propane containing only single bonds between carbons are called **paraffins**, a large family of stable chemical compounds. The first four paraffins have arbitrary names, but the others are named after the Greek numeral for the number of carbon atoms they contain. Here are the first eight:

• Methane	CH_4	Pentane	C_5H_{12}
• Ethane	C_2H_6	Hexane	C_6H_{14}
• Propane	C_3H_8	Heptane	C_7H_{16}
• Butane	C_4H_{10}	Octane	C_8H_{18}

Notice that paraffin names end in the suffix "–ane". This indicates a hydrocarbon containing only single bonds between carbon atoms. Chemists call a sequence of similar molecules like this a **homologous series**. Each member of the series differs from the previous one only by the addition of a methylene group ($-CH_2-$).

Longer carbon chains can be straight or branched, which may result in multiple **isomers**, compounds that have the same chemical formula but different structure. Straight-chain paraffins are considered "normal," and their name may be prefixed with "n-," as in n-butane. Crude oil is mostly n-paraffins.

Olefins

Carbon can also form a double bond with another carbon, the simplest example being ethene (or "ethylene") having the structure $H_2C{=}CH_2$ and the formula C_2H_4.

A hydrocarbon whose molecular structure includes one double bond is an **olefin**. When two double bonds are present, it's a **diolefin**.

Olefin names follow the same rules as paraffin names, but take the suffix "–ene", as in propene or 1-butene.

A numeral in a hydrocarbon name is not a quantity; it indicates the position of a feature along the carbon chain – in this case, the position of the double bond.

Olefins are chemically reactive and are often present in the process streams analyzed by our process gas chromatographs.

Acetylenes and more

Carbon can form a triple bond with another carbon. The simplest example is ethyne (or "acetylene") having the structure $HC{\equiv}CH$ and the formula C_2H_2. Hydrocarbons containing triple bonds are called **acetylenes**.

Acetylene names follow the same rules as paraffin names, except with the suffix "–yne," as in propyne or 2-butyne.

Acetylenes are very reactive and usually undesirable in process streams, so PGCs are often required to measure them.

Hydrocarbons can also form ring structures. If you are working with any of these, refer to the Glossary.

Shorthand notation

Chromatographers use a shorthand notation to identify peaks on a chromatogram. This is particularly useful for hydrocarbons because there are so many of them. A numeral indicates the number of carbon atoms in the molecule:

- methane CH_4 C1
- ethane C_2H_6 C2
- propane C_3H_8 C3
- isobutane C_4H_{10} iC4

We sometimes use the above shorthand symbols to mean all hydrocarbons with the same carbon number, like all the C_4s, for instance. Append a single prime to the above symbols when it's important to distinguish paraffins from the other hydrocarbons. For example, C_3' is specific to propane.

Multiple primes or dashes are used to identify olefins and acetylenes:

- ethene C_2H_4 C_2'' C2=
- ethyne C_2H_2 C_2''' C2≡
- propene C_3H_6 C_3'' C3=
- propyne C_3H_4 C_3''' C3≡
- 1-butene C_4H_8 1-C_4'' 1-C4=
- 1,3-butadiene C_4H_6 1;3-C_4'''' 1:3-C4==

The glossary at the end of the book lists the names of many of the chemical compounds that appear as peaks on chromatograms, together with their chemical structure and their shorthand notations. It's worth your time to learn the names of common chemicals so you know what peaks are possible. That knowledge will help you to identify anonymous peaks on your chromatograms.

Predictable patterns in retention

The doubling rule

In certain well-defined cases, the stretching of retention times seen in Figure 3.3 is very predictable. At constant temperature, for instance, the solubility of similar hydrocarbon molecules just about doubles for each additional carbon atom they have. The higher solubility has the effect of doubling the adjusted retention time; the time dissolved in the liquid phase. It happens on all liquid phases and under all operating conditions.

Figure 4.1 shows the **doubling rule** for propane and n-butane. Notice that the adjusted retention time of the n-butane peak is twice the adjusted retention time of the propane peak, and the other peaks are clustered near the air peak. From this chromatogram, it is easy to predict that the retention time of the n-pentane peak will be 18 minutes. Check that you can do it.

For predictability, the series of peaks must have similar molecular structures that differ only by the successive addition of a **methylene** group (CH_2). The normal paraffins are a good example.

Predictability is very useful for troubleshooting, as it makes it easier to identify unknown peaks. You can predict the retention time of all the

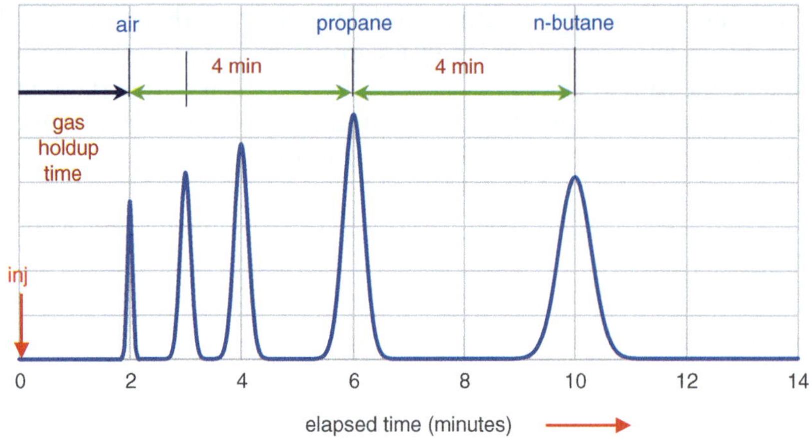

Figure 4.1 The Doubling Rule.

Starting from the air peak time (t_M), the adjusted retention time (t'_R) of n-butane (C_4) on this figurative chromatogram is eight minutes, double the adjusted retention time of propane (C_3). These times are just to illustrate the doubling rule; real PGC chromatograms are faster than this.

n-paraffins if you know the retention time of two of them. Or just one of them and the air peak time.

Challenge question

Try this; what is the likely identity of the two unmarked peaks in Figure 4.1?

You should be proud of your answer, but don't ever make the mistake of thinking that prediction can absolutely identify an anonymous peak. It is probable *but not certain* that the two anonymous peaks are methane and ethane.

This prediction follows from the doubling rule. By that lore, the ethane peak will elute exactly half-way between the air peak and the propane peak with a retention time of four minutes. That's the location of the second unidentified peak, so that peak might be ethane. In practice there are only a few components with such low retention times, so it's probable that the observed peak is ethane.

The first unidentified peak has a retention time of three minutes, exactly half-way between the air peak and the ethane peak. This location is where we expect to see methane, and there are few other components that would turn up there, so it's probable that the first unidentified peak is methane.

A process of elimination

Prediction is much better at saying what a peak is not. You may remember the likely error of peak identification (Figure 1.7) that we left unsolved in Chapter 1. Let's look at it again; the chromatogram in Figure 4.2 recalls the issue, showing four peaks:

- A peak with a retention time of 2 minutes, labeled nitrogen
- A peak with a retention time of 3 minutes, labeled ethane

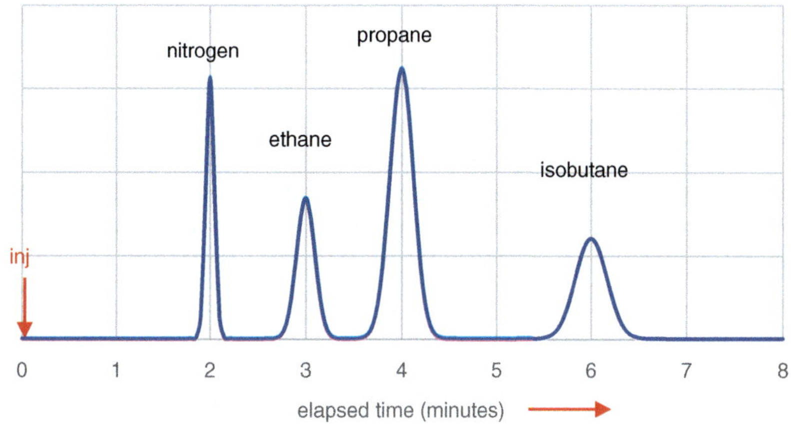

The doubling rule indicates that the peak mistakenly labeled isobutane is probably n-butane. On liquid columns, we know from experience that more-compact molecules elute before less-compact molecules, so the isobutane peak would appear between the propane and n-butane peaks. We conclude that the six-minute peak is definitely not isobutane.

Figure 4.2 Diagnosing Peak Identities.

- A peak with a retention time of 4 minutes, labeled propane
- A peak with a retention time of 6 minutes, labeled isobutane

We note that the adjusted retention time of the propane peak (two minutes) is double the adjusted retention time of the ethane peak (one minute). This is as expected, so the identification of the first three peaks is probably correct.

But the given identification of the last peak is suspicious. Actually, it's impossible for the six-minute peak to be isobutane, it's probably n-butane.

> *Using the doubling rule, you can predict that the six-minute peak is probably n-butane. Be sure that you know how to do that.*

We know about isobutane from experience, and because the **isoparaffins** are also a predictable homologous series. They always appear at about two-thirds of the time between the corresponding n-paraffin and the previous one. On that basis, the isobutane peak should be near the 5.3-minute mark:

$$t_R(iC_4) = t_R(C_3) + \frac{2}{3}(t_R(C_4) - t_R(C_3))$$

$$t_R(iC_4) = 4 + \frac{2}{3}(6 - 4) = 5.3$$

There's no way the isobutane peak can turn up at six minutes.

The doubling rule tells us that an n-butane peak would come out in six minutes, but that doesn't guarantee that the observed six-minute peak is n-butane! It could be another peak with a different molecular structure, such as propene, that just happens to elute at the six-minute mark.

You should start to see that the power of chromatographic separation is limited, but there are some helpful principles. It's impossible for ethane to elute after propane on a single liquid column. Likewise, no isothermal column will generate equal spacing between the ethane, propane, and n-butane peaks. These general principles will help you to correctly identify peaks on the chromatogram, and they are useful aids during troubleshooting.

We have been discussing the behavior of similar molecules – those that belong to a homologous series. When the analyte molecules have different molecular structures, the choice of liquid phase can have an enormous effect on separation. That's when chromatography gets interesting!

Temperature programming

The doubling rule is sometimes inconvenient as it results in very long retention times for larger molecules. For instance, a naphtha analysis might range from C_4 to C_{12}, which means the adjusted retention time of nC_4 in Figure 4.1 would be doubled eight times by the time it gets to nC_{12}. The retention time for nC_{12} would then be about 2,000 minutes and far too long.

Temperature programming is a way to overcome this excessive retention and to reduce analysis time. Continuously ramping up the column temperature during analysis accelerates the later peaks and brings them forward on the chromatogram. In effect, each peak elutes at a different average temperature. This allows the early peaks to separate at low temperature and the later peaks to separate at a higher temperature.

It turns out that increasing the temperature at a constant rate eliminates the doubling rule and allows a homologous series to elute with equal spacing, as is evident in Figure 4.3. The rate of change is normally set between two and ten kelvins per minute.

The kelvin (K) is the international SI unit of absolute temperature. A temperature change of one kelvin is identical to one Celsius degree.

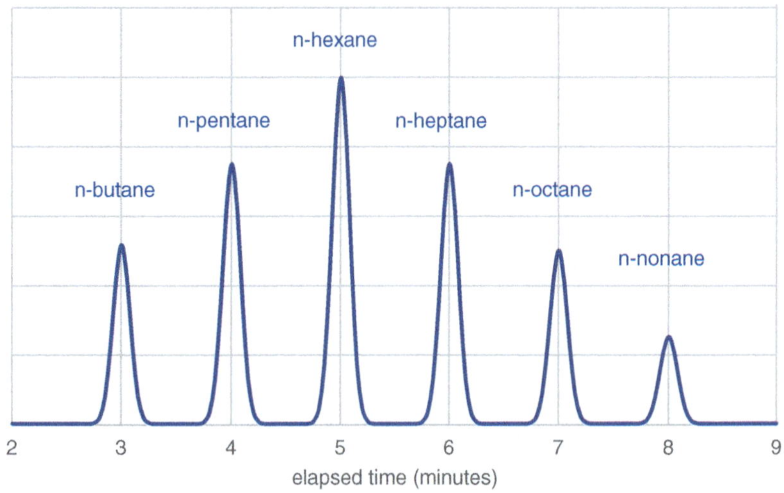

Figure 4.3 Effect of Temperature Ramp.

A gradual increase of column temperature eliminates the doubling rule, so these n-paraffins are now equally spaced on the chromatogram. Temperature programming is very effective for analyzing samples containing a wide range of components.

Surprisingly, the doubling rule is eliminated whatever ramping rate is chosen. The column temperature usually starts much lower than it would for an isothermal analysis, thereby allowing better separation of the lighter components. The heavier components don't move much until the temperature gets high enough for them to generate a significant vapor pressure.

Note that the peaks are all about the same width: another effect of temperature ramping. The later peaks are narrower and therefore taller than they would be in isothermal operation, and that makes them easier to measure.

Temperature programming is employed in most laboratory chromatographs as it can separate a wide range of possible components. It's now being used more often in process chromatographs for the fast analysis of multiple components. As an example, Figure 4.4 is a real chromatogram for the fast analysis of natural gas using temperature programming. Another common usage is the simulated distillation of gasoline to evaluate its boiling point behavior.

It's still true, though, that the additional complexity of heating and cooling isn't warranted for most process applications. Process measurements tend to be well defined and usually can be accomplished isothermally by column switching, which may be less expensive and is often more reliable.

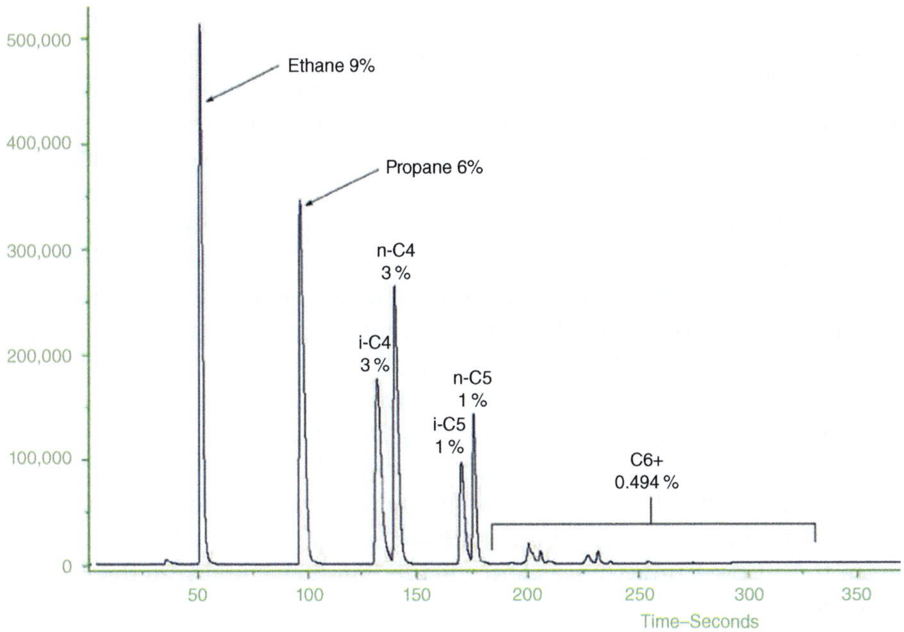

This real chromatogram shows the advantage of a fast temperature ramp. The light hydrocarbons are separated at low temperatures, providing an excellent separation of ethane and propane, while the many C6+ components are accelerated by higher temperatures into a compact band for totalization. The entire analysis is completed in less than six minutes.

Figure 4.4 Fast Analysis of Natural Gas.
Source: Teledyne Analytical Instruments, a business unit of Teledyne Instruments, Inc. Reproduced with permission.

Relative retention

The doubling rule shows that a fixed ratio exists between the adjusted retention times of any two n-paraffins on any liquid phase column; for adjacent n-paraffins that ratio is always close to 2.0. This is an example of a **relative retention** time for two chromatogram peaks. It's the ratio of their adjusted retention times.

The formal name for the relative retention between two specified peaks is their **separation factor**. Using n-paraffins again as an example, the *separation factor* between n-hexane and n-pentane is about 2.0 on any liquid column.

The behavior of the n-paraffins is a special case of a general property shared by all peaks on all stationary phases. Any two peaks on a chosen liquid phase have their own characteristic separation factor that for most peaks is constant; it will not change when you change the column operating conditions.

This is useful to know. It means that changing the column length, temperature, pressure, liquid loading, or anything else will not change the separation factor between the two peaks. Unless you change the liquid phase, they will stubbornly stay in the same relative pattern whatever you do.

The reason that separation factors are fixed goes back to solubility. We already know that the solubility of a certain component in a certain liquid phase depends only on temperature and pressure. If either of those parameters changes, it has the same effect on all peaks in the column, so the ratio of their adjusted retention times doesn't change at all.

There are a few rare exceptions to this rule; for example, the separation factor between some olefin peaks changes with temperature.

We shall see that better separations can often be achieved by optimizing column operating conditions, but those improvements are mainly gained by reducing holdup time and peak widths, not by improving the separation factor.

For most peaks, the only way to change the separation factor is to select a different liquid phase. But even a different liquid phase won't change the separation factor unless the molecules of the two peaks are chemically different. For instance, all paraffins have a similar chemical structure, so their separation factors are about the same on all columns. That's why we could confidently state that adjacent paraffins have a separation factor of about 2.0 on any liquid column under any operating conditions.

The constant separation factor is another predictable pattern of peaks on the chromatogram. It's a property of the stationary phase, so when diagnosing peak positions, measure the adjusted retention time of each peak starting from the air peak time (t_M), as indicated in Figure 4.5; not from the sample injection mark.

Then, the separation factor (α) is:

$$\alpha = \frac{t'_R(B)}{t'_R(A)} \tag{4.1}$$

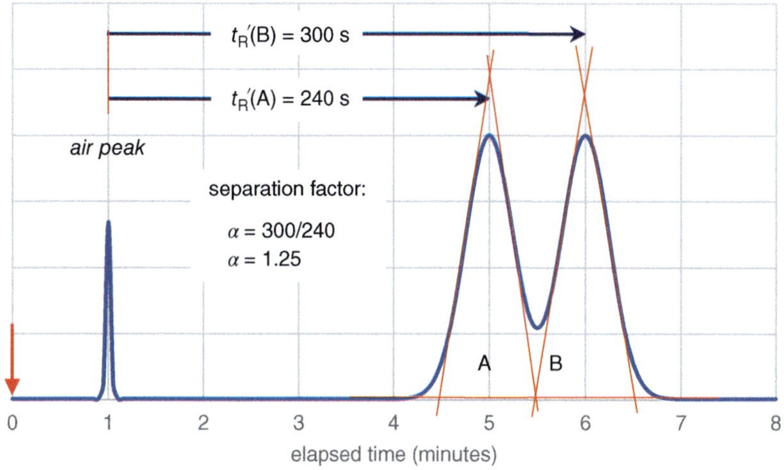

The separation factor is a property of the stationary phase, so the retention measurement needed is the adjusted retention time of each peak, starting from the air peak.

Figure 4.5 Measurements for Separation Factor.

You can tell from this equation that the separation factor is a measure of the column's ability to separate the two peaks, so it's sometimes called the selectivity of that stationary phase.

It's useful to observe the separation factor when selecting or troubleshooting a column to separate two components. Generally, the separation is easy when the separation factor is greater than 1.1 and becomes more difficult as it approaches unity. A separation factor of 1.0 indicates that the two components cannot be separated on that liquid phase whatever you do.

Separation and resolution

This introduction has used the word "separation" in its everyday sense of moving things apart from one another. That usage tends to imply a complete sorting of the things, so that all those of one kind are *over here* and all those of another kind are *over there*. Unfortunately, though, PGCs are always subject to an analysis time constraint, and, in the time available, it may not be possible to separate the molecules to that degree; there's usually some overlap between the peaks.

Given the commonality of peak overlap, separation becomes a relative term. When are two peaks truly *separated*? In theory, never. The number of analyte molecules present falls rapidly with increasing distance from the peak center, but it never reaches zero. So, if all separations suffer a minor degree of peak overlap, how do we know when the separation is enough? We need another variable to capture that.

Resolution

The extent of overlap between peaks is clearly a function of peak width. It is self-evident that wide peaks are more likely to overlap their neighbors than narrow peaks are. Therefore, an effective measure of column

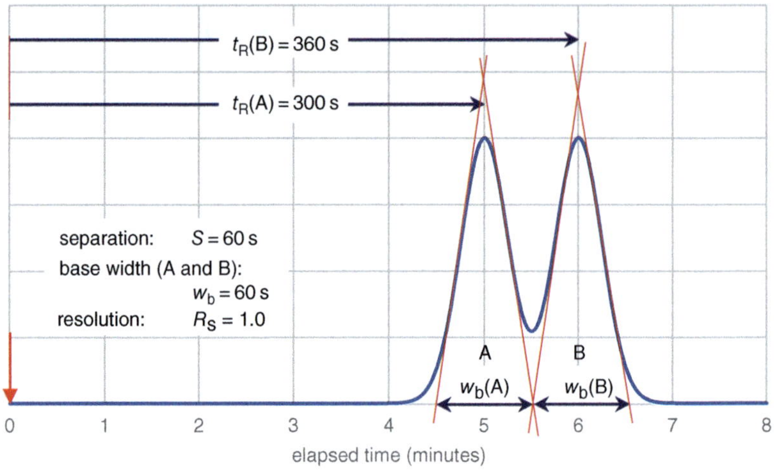

Figure 4.6 Measurements for Resolution.

The resolution between adjacent peaks is a property of the entire column system, so the retention measurement needed is the retention time of each peak, starting from the sample injection marker.

performance must include both peak separation and peak width. The concept of **resolution** satisfies this need.

To calculate the resolution between two adjacent peaks, A and B, measure the retention time (t_R) and triangulated base width (w_b) of each peak. Figure 4.6 shows the chromatogram measurements you will need. Again, use any measurement units you like; they will cancel out in the calculation.

Start by computing the separation between peaks (Δt_R) and their average base width (w_a):

$$\Delta t_R = t_R(B) - t_R(A) \tag{4.2}$$

$$w_a = \frac{w_b(B) + w_b(A)}{2} \tag{4.3}$$

Then calculate the resolution (R_s):

$$R_s = \frac{\Delta t_R}{w_a} \tag{4.4}$$

Some PGCs can calculate the resolution of selected peak pairs and issue a warning if the resolution starts to diminish.

Notice that the resolution shown in Figure 4.5 is exactly 1.0, which occurs when two equal and symmetrical *triangulated peaks* touch at the baseline.

From now on, it will be necessary to carefully distinguish between separation and resolution, and to use the correct terminology. We'll use *separation* in its generic sense of moving peaks apart from each other without implying totality.

Figure 4.7 confirms the effect of peak width on resolution when the two peaks don't move. The narrow peaks are well resolved and measurable, but the wider peaks exhibit a large overlap that will prevent an accurate measurement.

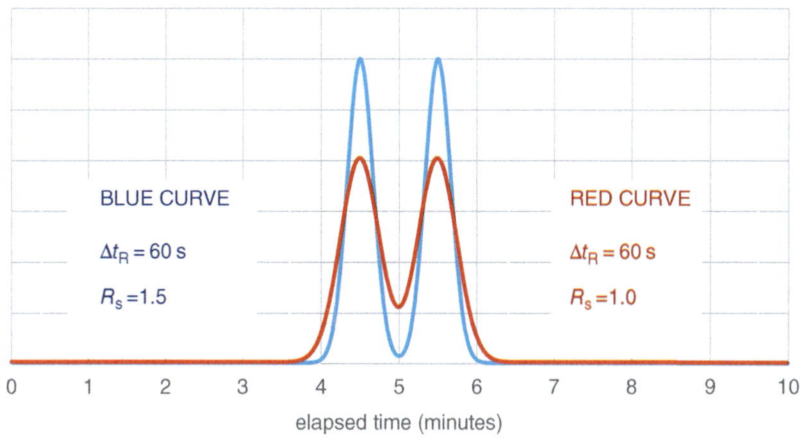

Two chromatograms are superimposed to show the effect of peak width on resolution. The blue peaks are perfectly resolved and easy to measure accurately. The red peaks have exactly the same separation but are wider than the blue peaks. The reduced resolution causes a high valley point between them, making accurate measurement more difficult.

Figure 4.7 Same Separation, Different Width.

Well-resolved peaks are essential for accurate measurement. The blue peaks in Figure 4.7 have a resolution of 1.5, and the valley between them almost touches the baseline. By definition, a **perfect resolution** equals 1.5 and is adequate for the accurate measurement of two peaks that are equal in size. For peaks that are unequal in size, more resolution is required. A resolution of 2.0 would create some flat baseline between equal peaks or make it possible to measure somewhat unequal peaks. From this, you can see that the resolution required for good measurement always depends on the relative size of the peaks.

In practice, chromatographers rarely calculate the obtained resolution because their experience allows them to judge whether the resolution is sufficient simply by looking at the valley height between the peaks. The wider peaks in Figure 4.7 have a resolution of 1.0, and the valley is too high for a good measurement of the peak areas.

You might be tempted to challenge that statement with an argument invoking computer baseline correction techniques. Fair enough. But it's always a mistake to expect software to make up for poor chromatography. Get the best resolution you can. Then use the computer to minimize the interference between adjacent peaks. Chapter 14 reviews the computer data-processing techniques used by a modern PGC.

Predictable patterns in resolution

For a given resolution, the degree of peak overlap is fixed and predictable.

The classical pattern of resolution seen when two equal-sized symmetrical peaks approach each other is illustrated in Figure 4.8. It's important that you see what's happening here. You'll often need to diagnose these patterns on chromatograms. The top trace ($R_s = 2.0$) shows two well-resolved peaks that would be easy to measure. The second one shows the so-called *perfect resolution* ($R_s = 1.5$) that is tight, but still okay for measurement.

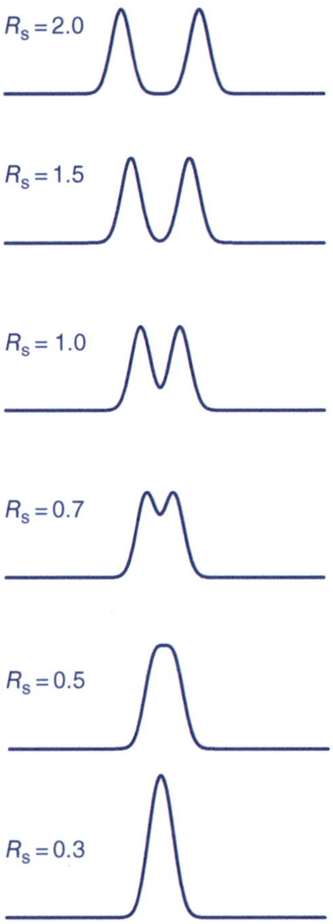

Showing the chromatograms obtained for various resolutions of two equal symmetrical peaks.

Figure 4.8 Patterns in Resolution.

When the resolution drops to 1.0, the peaks are difficult to measure. There are data-processing techniques that can provide a measurement, but it might not be accurate or reliable.

As the resolution falls below 1.0, the peaks start to fuse together. Eventually, measurement becomes impossible, but you should study these last three shapes carefully because you may need to recognize them during troubleshooting. At a resolution of 0.7 the two peaks are still visible, so it's clear that two separate components are present. But it's not so obvious when the resolution goes even lower.

The last two chromatograms show a single **fused peak**. Right now, we know there are two peaks hidden on these chromatograms, but when you are working alone in the field *you will not know*.

Recognizing fused peaks is a most important diagnostic skill. Possession of this skill separates the expert from the novice. Many diagnostic procedures start by recognizing a distorted peak, and end by deducing the cause of that distortion. If you aspire to be a PGC expert, start now to learn those tell-tale shapes.

Look carefully at the peak shape when the resolution is 0.5. It looks like a single peak, but there are some clues to tell you that it's not:

- The top of the peak is too flat – it has a domed top, not like the classic peak shape. If you follow the slope of the sides, it looks like the peak should have gone higher – but didn't.
- The peak is too wide. Look at a nearby peak to check it out; the fused peak is wider than other peaks on the chromatogram.

There's no excuse for misidentifying a pair of fused peaks as a single peak when the resolution is 0.5, but as the resolution falls lower it eventually becomes impossible to tell the difference. The last chromatogram is very difficult to diagnose, and only a very skilled chromatographer would notice that the displayed peak is really a combination of two different components. The best clue is its width; it's too wide to be a single peak at that location. Again, compare its width with other nearby peaks on the chromatogram.

For all the resolutions displayed in Figure 4.8, the two peaks are ideal; being equal in height and width, and perfectly symmetrical. While a little unrealistic, these simplifications are appropriate in a short introduction to peak topography. Thus, the peak shapes shown here are approximations of those that will appear on real chromatograms. Yet they still convey the essential message that peaks are not just bumps on a baseline. You should always be looking for the anomaly that portends trouble in the column system.

Real peaks tend to lean forward, vary in height, and get progressively wider. When such peaks merge together, the fused peak may exhibit additional nuances of shape that pose new challenges of diagnosis and identification. All this is ahead of us; there is no shortcut to becoming a process chromatograph guru.

Knowledge Gained

- *In this chapter, we consider the performance characteristics of a single gas-liquid column.*
- *When peaks are in the column, they are not spaced the same way as they are on the chromatogram.*
- *Inside the column, peaks are separated in space, but all peaks have been traveling for the same time.*
- *On the chromatogram, peaks are separated in time, but all peaks have traveled the same distance.*
- *A spatial separation occurs in the column, but a temporal separation is seen on the chromatogram.*
- *While inside the column, the width of each peak is a function of distance traveled.*
- *Therefore, peaks that migrate rapidly through the column are wider than those that move slowly.*
- *The temporal separation begins when the air peak exits the column.*
- *After the air peak, the additional time to each peak is the time it has spent in the liquid phase.*
- *More soluble peaks have not made much progress, so they have farther to go and are traveling slower.*
- *Thus, peaks that are equally spaced in the column become unequally spaced on the chromatogram.*
- *Early chromatogram peaks tend to cluster just after the air peak while later peaks move farther apart.*
- *On the chromatogram, the width of each peak is a function of its time in the column.*
- *Therefore, later peaks on the chromatogram are always wider than earlier peaks.*
- *Recognizable patterns of retention time and peak widths can help with peak identification.*
- *It's possible to say what an unknown peak certainly is not, but not to predict what it certainly is.*
- *Chromatographers use a shorthand notation for identifying peaks on the chromatogram.*
- *For a homologous series, the retention time doubles for each additional carbon atom in the molecule.*
- *The doubling rule predicts the retention time of any n-paraffin if the times of two of them are known.*
- *The retention time of any n-paraffin is predictable from the times of one n-paraffin and the air peak.*
- *Temperature ramping nulls the doubling rule and makes n-paraffin peaks equal in width and spacing.*
- *Temperature ramping allows complex separations, like the simulated distillation of gasoline.*
- *Temperature programming is uncommon in PGCs as column switching achieves most analyses faster.*
- *The separation factor between two peaks is the ratio of their adjusted retention times.*
- *The separation factor between two components tends to be constant on a chosen liquid phase.*
- *The separation factor for two adjacent members of a homologous series is about 2.0 on any column.*

- *Resolution is the time or distance between adjacent peaks divided by their average width (same units).*
- *A resolution of 1.5 almost touches the baseline and is perfect for the measurement of equal peaks.*
- *A resolution less than 1.5 may not be adequate for the effective measurement of both peaks.*
- *For equal peaks, the individual peak tops are visible in the fused peak down to a resolution of 0.7.*
- *A resolution of less than 0.7 yields a single fused peak having a distorted shape.*
- *More resolution is necessary when measuring unequal peaks.*

Did you get it?

Self-assessment quiz: SAQ 04

Q1. What is a spatial separation, and where would it occur?
Q2. What is a temporal separation, and where would it occur?
Q3. Under what circumstances would faster peaks be wider than slower peaks?
Q4. In the doubling rule, what is it that is doubling?
Q5. In Figure 4.1, estimate to the nearest minute the predicted retention time of the n-pentane peak.
Q6. What is temperature programming, and what is it used for?
Q7. What is the minimum resolution that allows accurate measurement of equal-size chromatogram peaks?

Check your SAQ answers with those given at the end of the book.

Student evaluation test: SET 04

Your instructor will provide answers to these questions.
This question set relates to separations on a single gas-liquid chromatographic column.

S1. When does the spatial separation of components begin?
Select the one correct answer:
- **A.** When the sample enters the column.
- **B.** When the air peak (if present) would enter the detector.
- **C.** When the first analyte peak enters the detector.
- **D.** When all the peaks are on the chromatogram.

S2. When does the temporal separation of components begin?
Select the one correct answer:
- **A.** When the sample enters the column.
- **B.** When the air peak (if present) would enter the detector.
- **C.** When the first analyte peak enters the detector.
- **D.** When all the peaks are on the chromatogram.

S3. After the temporal separation begins, what is the direct cause of the additional time for each peak to reach the detector?

Select the one correct answer:

A. The average time that the component molecules have spent in the gas phase.
B. The average time that the component molecules have spent in the liquid phase.
C. The average time that the component molecules have spent in the liquid phase minus the time they have spent in the gas phase.
D. The average time that the component molecules have spent in the liquid phase plus the time they have spent in the gas phase.

S4. Given these peak retention times:

$t_M = 50$ s
$t_R(A) = 100$ s
$t_R(B) = 110$ s

What is the separation factor between peaks A and B on this liquid phase?

S5. What is the resolution between two peaks of similar size when their separation is equal to the sum of their base widths?

S6. In a chromatographic separation do the peaks get progressively wider or narrower?

Select the two true statements:

A. While in the column, a peak that has migrated farther is typically wider than a peak that hasn't gone so far.
B. While in the column, a peak that has migrated farther is typically narrower than a peak that hasn't gone so far.
C. When on the chromatogram, an earlier peak is typically wider than a later peak.
D. When on the chromatogram, an earlier peak is typically narrower than a later peak.

S7. How do the peaks appear on the chromatogram when the column temperature is gradually increased during the analysis? Select all of the correct answers and no incorrect answer:

A. The later peaks are wider than the earlier peaks
B. The peaks are all about the same width
C. The n-butane (nC_4), n-pentane (nC_5), and n-hexane (nC_6) peaks are equally spaced
D. The time between the nC_5 and nC_6 peaks is double the time between the nC_4 and nC_5 peaks.

S8. For an isothermal separation, if the retention time of the air peak is 30 s and the retention time of propane peak (C_3) is 45 s, predict the retention time of the n-pentane (nC_5) peak.

Express you answer as an integer number of seconds.

S9. For an isothermal separation, if the retention time of the propane (C_3) and n-pentane (nC_5) peaks are 60 s and 120 s respectively, predict the retention time of the n-hexane (nC_6) peak.

Express your answer as an integer number of seconds.

References

Figures

Equations

4.1	$\alpha = \frac{t'_R(B)}{t'_R(A)}$	Separation factor for components A and B
4.2	$\Delta t_R = t_R(B) - t_R(A)$	Separation between adjacent peaks
4.3	$w_a = \frac{w_b(B) + w_b(A)}{2}$	Average peak width of adjacent peaks
4.4	$R_s = \frac{\Delta t_R}{w_a}$	Resolution between adjacent peaks

Symbols

α	separation factor
R_s	resolution between adjacent peaks
t_R	retention time
t'_R	adjusted retention time
Δt_R	separation between adjacent peaks
w_a	average peak width
w_b	triangulated base width

New technical terms

When first introduced, these words and phrases were shown in bold type. You should now know the meaning of these technical terms. If in doubt, consult the Glossary at the end of the book:

acetylenes
column system
diolefin
doubling rule
fused peaks
homologous series
hydrocarbon
isomers

isoparaffins
methylene
olefin
paraffin
perfect resolution
relative retention
resolution
separation factor
spatial separation
temperature programming
temporal separation

For individual chemical names, consult the Glossary.

Part Two

PGC analytics

1962 Bendix-Greenbrier Model 620

Figure 5.1 A Classic Process Gas Chromatograph. *Source: AAB Process Analytics. Reproduced with permission.*

"We are stuck with technology when all we really want is just stuff that works. How do you recognize something that is still technology? A good clue is if it comes with a manual."

Douglas Adams (2002)
1952–2001

Process Gas Chromatographs: Fundamentals, Design and Implementation, First Edition. Tony Waters.

Why study this?

Part Two of the book introduces the analytic unit of a process gas chromatograph (PGC) and will prepare you for further maintenance or applications training. The information presented here is also good background information for engineers and scientists who are new to process gas chromatographs and wish go on to more advanced studies.

These chapters are intentionally wide in scope but narrow in depth. They focus on the chromatographic devices found in the oven section of a PGC. To provide context, the text includes references to some manufacturers and model numbers of prior versions of process gas chromatograph.

If you're used to working with laboratory gas chromatographs, welcome to a new world! You'll find that a modern process gas chromatograph is a very different instrument, every part of it designed to work without attention in an industrial processing plant. Over many design iterations, the PGC analyzer has evolved into a dependable analytical machine that operates continuously by day and by night with near-perfect reliability in unpleasant and potentially hazardous environments.

5

Industrial gas chromatographs

"Process gas chromatographs work continuously and unattended in a harsh plant environment that may also be explosive or corrosive. They employ the same internal principles as their laboratory cousins, but with different objectives. Whereas laboratory gas chromatographs value the flexibility to measure dozens of components in many different samples, a PGC is dedicated to measuring a few analytes as quickly and reliably as possible in a sample that rarely changes".

Process analyzers

Introduction to process analysis

Most people who come to work with process gas chromatographs will have prior experience in process-plant instrumentation or laboratory analysis. Those from an industrial background will easily understand the safety measures employed by the PGC as well as its detector technology and signal processing techniques, but they may be challenged by the arcane chromatographic techniques used to separate analytes. Those from a laboratory background will more easily understand the intricate column systems used in the process instrument but may find its site-hardened automation difficult to grasp.

Part Two of the book attempts to bridge that gap and bring readers with differing backgrounds to a common level of knowledge, ready for more detailed study. Thus, you may find that some of the content is easy going, while other parts are challenging.

We start by introducing the wider field of process analytical measurement and identifying the niche occupied by the PGC within that discipline.

Process Gas Chromatographs: Fundamentals, Design and Implementation, First Edition. Tony Waters.

The measurement of quality

The fluid processing industries rely on measurements to keep track of what their processes are doing. Without measurement, process control is impossible. The many measurements made in an industrial process plant help to control plant safety, productivity, energy efficiency, and regulatory compliance.

There are two main varieties of process measurement:

- Traditional process instruments measure a specified *condition* of a process fluid, most often its temperature, pressure, level, or flow rate. These four measurements are invaluable for controlling *process stability*.
- Quality measuring instruments (QMI) measure a specified *quality* of a process fluid, such as its viscosity, flash point, or composition. These many measurements are vital for controlling *process efficiency*.

Industrial processes rely on many different **quality measuring instruments** to evaluate the quality of their raw materials, intermediates, valuable products, utilities, and wastes. Each type of **QMI** employs a unique principle of physics or chemistry to measure a desired quality of the process stream. The measured quality might be:

- A *physical property* of the process material – like the viscosity of a liquid stream.
- The *concentration* of a certain substance in the process material or in the environment – like the percentage of ethylene in a gas stream.

Although only the second type is truly analytical, we tend to refer to them both as **process analyzers**.

Process analyzers operate continuously without attention and report their data to the process control system. Fast and accurate quality information is often vital to the efficiency and safety of a plant, and may also be used to assure compliance with the terms of its operating permit. Most industrial processes can't run without it.

With all the principles of physics and chemistry to draw from, process analyzers employ a large variety of techniques to identify and measure the target analytes. The myriad of techniques can be confusing, and no one can truly profess to be expert at them all.

Some clarity accrues from grouping the QMI into families of similar analytical methods. This grouping may help you to appreciate the wide range of technologies used in process analysis, and to understand the key difference between the process gas chromatograph and other methods of analysis.

Here are the six generic variables measured by process analyzers. Each one of these variables is a surrogate for many different analytes, and the common trait of a large family of process analyzers. Any one of these variables requires unique technology and substantial technical interpretation before it truly represents a specific analyte:

- A unique **holistic property** of the process fluid, such as its flash point, density, or viscosity.
- A unique **physical property** of an analyte molecule, such as the paramagnetism of oxygen or the dielectric constant of water.
- An **electrochemical** phenomenon, perhaps an electrode potential or a galvanic current.
- The **photon absorption or emission**, as in infrared photometry, ultraviolet **spectroscopy**, or x-ray fluorescence, among others.
- The **physical separation** and subsequent measurement of molecules or ions, as achieved in gas chromatography or mass spectrometry.
- The outcome of a quantitative **chemical reaction**, as employed in wet-chemistry analyzers like those using titrimetry or colorimetry.

The six categories encompass a truly enormous knowledge base, and each one is a specialty in itself. It follows that the technologies employed by QMI are so varied and numerous that a detailed understanding of them all is beyond even an expert practitioner. Luckily, this book focuses on just one analytical technique: the science and art of process gas chromatography. Even so, it will be challenging to gain proficiency in this one technology. There's a lot to learn.

For those who intend to maintain process analyzers, a basic knowledge of physics and chemistry is helpful, but not essential. For more detailed work, such as PGC applications engineering or troubleshooting, it may become indispensable.

This field of study is all about measurement, and it's essential to grasp the basic principles of measurement science, including some elementary statistics. For an introduction, see *Measures, Units, and Calculations* (Waters 2013, 609–650).

Process gas chromatographs

Versatile and reliable

The process gas chromatograph (PGC) is without doubt the most versatile of the many process analyzers available to industry. Most process analyzers can measure only a specified property of the process material, but a PGC can measure the concentration of almost any substance in almost any fluid process stream – or even several substances in several process streams.

This versatility of application is the great strength of gas chromatography but can also be its downfall. It may seem easy to accomplish additional measurements, but the increased complexity of adding another analyte may reduce the overall reliability of a PGC enough to undermine its original purpose. Thus, it's vital to limit the specified analytes to those that are truly necessary for process control, eliminating all that are merely *nice to know*.

Your PGC will be more reliable if you limit the measurements it has to make. A PGC measuring one analyte in one stream will be almost perfectly reliable. Additional analytes or process streams are certain to reduce its ***availability****. This principle is anchored in statistics and is not open to debate.*

The focus on reliability may come as a surprise to those from outside the process control community. The pursuit of reliability is intense. In fact, a common title for an analyzer engineer at a processing plant is *Reliability Engineer*. This title emphasizes that each process analyzer is purchased for a specific purpose and financially justified by the consequences of not having that measurement. Sometimes those consequences are severe. The loss of an analytical measurement may cost the plant many thousands of dollars per hour, or even force a complete process shutdown.

The economics of fluid processing rely on continuous and efficient operation. Any deviation from that ideal is intolerable. So process measurement is quite different than laboratory measurement. When a process measurement fails, the process control system can't wait; it has to revert to a safe but inefficient mode of operation. That costs money. Therefore, all process instruments must be highly reliable. The process gas chromatograph has become a very reliable instrument due to its long experience of operating under the conditions found in process plants around the world. The knowledge gained from that accumulated experience and captured during the many PGC development cycles is now built into each new instrument.

PGC development

Over the years, manufacturers of PGCs have enjoyed a significant advantage not shared by the manufacturers of most other analytical instruments: they can use the same set of hardware devices to make different measurements. Application engineers configure standard equipment to measure the various analytes that process plants need, thereby creating a higher rate of production of each device than would otherwise be possible.

The PGC's flexibility of application is a strong market advantage that is not shared by process analyzers dedicated to measuring a single analyte. Discounting a few electrochemical sensors, the PGC has consistently outsold other kinds of process analyzer. Even after 60 years of production, the worldwide PGC vendors were still selling about 300 units per month (Clemons 2016).

The competitive market for PGCs has always been a strong incentive for further development. Over the years, more capital was invested in PGC development than in any other QMI technology. Due to this huge influx of cash, the technology rapidly improved, and new models frequently eclipsed old ones. Those working with PGCs during the latter part of the twentieth century witnessed rapid improvement in the performance of ovens, valves, columns, and detectors, but most of all in data processing capability.

If some process gas chromatographers from the 1960s were to suddenly confront a modern PGC, they would immediately recognize the basic hardware; they would find the valves, columns, and detectors awesome, yet still readily identifiable by function. But what would our time travelers make of the data-processing devices? These have changed so radically, they would be unrecognizable, their function incomprehensible.

Process chromatographs were early adopters of digital technology, and the rate of change was astonishing.

> *In 1966, just ten years after its debut, the PGC was the first process field instrument to communicate its measurements to the control room by serial data link (Clemons* 2017*). And, about ten years later, the Amscor Mark IV PGC was the first process field instrument to boast an integral microprocessor (Clevett* 1986*, 50–52).*

The author discovers a 1970s era Foxboro/Yokogawa Model 8110, still operating onsite in 2016. The author learned the PGC trade by engineering applications for this early model of process gas chromatograph.

Figure 5.2 A Very Old PGC with an Even Older PGC Engineer.
Source: Author's collection.

The history of the process gas chromatograph is relevant to the goals of this book. As a process chromatograph specialist, you are also a time traveler, but you will be going in the other direction. Your employer or client may ask you to calibrate, maintain, or troubleshoot PGC hardware that is quite different from the latest models; perhaps using cam timers and component-level electronics, for instance.

Thus, it's important in these introductory chapters on PGC hardware to review the various devices and techniques that were used in older models of process chromatograph. Reviewing older equipment is not just a history lesson; it will help you to understand why a modern PGC is built the way it is. And it's entirely possible that you will confront predigital equipment on site. For example, Figure 5.2 shows a circa 1970 PGC that we recently encountered on a jobsite in Asia, still operating in 2016.

The value of analysis

Analytical measurements are valuable to a process plant. The information they provide allows a plant to operate more efficiently, increasing the throughput of in-spec product or reducing operating costs, or both.

A PGC provides rapid indication of a change in the chemical composition of the fluid entering or leaving a process unit. This information is valuable. It allows a process operator or automatic controller to quickly correct for any production deviation by making appropriate adjustments to process operating conditions. The improved ability to control the process confers several potential advantages on the process owner:

- increased operational safety
- compliance with environmental regulations
- sustained product quality
- increased product throughput
- reduced energy and material costs

Any of these benefits would improve the overall profitability of a fluid processing plant. Typically, a process on analytical control can operate closer to a constraint, increasing production or saving cost. Common economic constraints are the need to meet product purity specifications and to comply with environmental regulations. To satisfy these important objectives, the concentration of certain impurities must not exceed specified amounts.

For example, consider the analytical control of an **ethylene splitter** tower. A PGC would measure the concentration of ethane in the ethylene overhead product and the concentration of ethylene in the ethane bottoms product. Then, knowing the concentration of ethane in the overhead stream, the plant would operate to maximize ethylene production by closely approaching but not exceeding the specification limit for ethane in the ethylene product (often 1,000 ppm). At the same time, the controller would minimize the loss of valuable ethylene in the bottoms stream.

Competing technologies

Gas chromatograph or spectrophotometer?

Process gas chromatography is a mature technology, and newer measurement techniques are sometimes more effective. Simple **photometers** have always been a cheaper alternative for the dedicated measurement of a single component. For example, an **infrared** or **ultraviolet absorption** analyzer is often a better choice than a PGC to measure simple molecules like carbon monoxide, hydrogen sulfide, or sulfur dioxide.

More recently, a significant market intrusion of new photometric technology has gradually chipped away at the original PGC market. These intruders include an alphabet soup of spectroscopies; near-infrared (NIR), Fourier-transform infrared (FTIR), nuclear magnetic resonance (NMR), tunable diode laser (TDL), quantum cascade laser (QCL), cavity ring-down (CRD), and Raman scattering spectroscopy (RSS).

This trend will continue. It's inevitable that advanced technologies will replace PGCs for some applications. But they will each have their own rocky road.

Although this book is about the process gas chromatograph, we recognize the process photometer as its main competitor. To better understand this healthy competition, Table 5.1 and the following paragraphs briefly compare the PGC with its archrival, the process **spectrophotometer**.

Of necessity, an **absorption photometer** measures the whole sample, and most molecules present contribute to the pattern of energy absorbance observed across the wavelength spectrum. The selectivity of measurement depends on finding a wavelength absorption band that is unique to each desired analyte. This approach is very successful for simple applications, though the measurement of analytes in a more complex mixture may suffer interference from the overlapping spectra of other components in the sample.

Table 5.1 Comparing the Process Photometer and the PGC.

Parameter	Process Photometer	Process Chromatograph
Technique:	Measures the absorption or emission of light by a whole sample using many different photometric techniques and wavelengths.	Separates the molecules of each analyte component and measures them separately.
Measurement:	Spectral characteristics of the sample fluid.	A property of the individual component molecules.
Selectivity:	By optical filter selected for key analyte wavelengths, or by chemometric modeling.	By physical separation of the molecules prior to measurement.
Interference:	Non-analyte molecules may interfere with measurement.	Non-analyte molecules should not interfere.
Accuracy:	Good.	Excellent.
Speed of Response:	Very fast.	Slower, but possibly adequate.
Typical application:	A single analyte or an inferred physical quality of a process gas or liquid stream.	Several analytes or a computed physical quality of a process gas or liquid stream.
Traditional multistream:	Uses separate analyzers.	Uses sequential stream selection.
Newer multistream:	Simultaneous processing of signals transmitted by fiberoptic cables from separate sensors.	Concurrent separation and analysis of injected samples from different streams.
Maintenance:	Low demand.	Moderate demand.
Typical installation:	At or near process tap.	Remote from process tap.
Installed cost:	Simple device: low cost. Complex device: high cost.	Simple job: moderate cost. Complex job: high cost.

In contrast, a PGC separates and directly measures individual analyte molecules. So, if applied correctly, the PGC is immune to variations in the concentration of other substances in the sample; unlike photometry, no interaction occurs between sample components.[1] Therefore, process gas chromatography may be the superior technology when the plant needs an accurate measurement of a few individual substances.

A persuasive argument for photometry is its speed of response. The response of a simple photometer is virtually instantaneous, although a more complex device may need a little time for data processing.

Data processing is now a key feature of any advanced spectrophotometer. Many photometric analyzers infer their measurements by a sophisticated

[1]This is a simplification. In more advanced work, we shall see that some minor interactions between components are possible. Such interactions are rare, but it's occasionally found that a change in the concentration of one component affects the retention time of another component or even the sensitivity of the detector to that other component.

pattern recognition procedure known as **chemometrics**. From a thorough calibration procedure, the integral microprocessor discovers key spectral wavelengths that correspond to the target variable and creates a mathematical model that correlates the intensity of those key wavelengths to the value of each desired analyte. To achieve a robust model, the calibration sample set must fully represent the expected variability of the process. In some applications, this may demand up to 100 calibration samples.

When the process variation is predictable and adequately modeled, the inferential method will faithfully follow trends in the monitored parameter and will do so with a faster response and a lower maintenance workload than a PGC.

When the process composition falls outside the purview of the calibration set, a correlation method of analysis may suffer significant measurement error. Often, the processor will detect the error and generate an **outlier** notification. Even so, an inferential method of analysis may not be the best choice when the process operators need to know the exact concentration of a desired analyte, especially when the stream composition is subject to random change.

Many correlation photometers use direct insertion probes and claim to eliminate the troublesome sampling system. While one can feel the allure of a simple probe that provides an instant and continuous measurement of stream composition, this technology is still in its infancy. As yet, photometric probes are not immune to fouling at the interface between the probe and process fluid. A traditional sampling system may still be required.

Speed of response

As noted above, an advantage of a spectrophotometer is its speed of response. Usually, the spectral processing is faster than a conventional chromatographic analysis. And a single microprocessor can concurrently analyze the signal from several optic probes, thereby providing rapid updates of many measurements.

PGCs have been multiplexed almost from their inception, but always on the messy fluid side. Not any longer. Instead of the long cycle time needed to analyze several process streams sequentially, a PGC can dedicate a sample injector valve to each process stream. By analyzing streams concurrently and then multiplexing its detector signals, a PGC mimics the architecture of its photometric competitor. Combined with the speed advantage of parallel chromatography (discussed in Chapter 9), a modern PGC can often rival the speed of the spectrophotometer.

In evaluating the response time of an analytical instrument, you should recognize that speed of analysis is not the only concern. For process control, instantaneous response is unnecessary; instead, the measurement must respond faster than the process lag time, which might be several minutes. Provided the PGC satisfies this criterion, it's suitable for the application. On the other hand, the time delay due to sample transport is critical, and many

PGCs are retarded by inadequate sampling systems, making them unsuitable for the application (Waters 2013).

The outlook

The ongoing developments in wavelength discrimination and signal processing make it virtually certain that spectroscopic methods will become more accurate and less susceptible to interference. As their capability improves, spectrometric instruments will continue to replace chromatographs for certain applications. This diversity should be encouraged, not feared. In the history of metrology, there has always been an *optimal technology* extant for each measurement. That history should remind us that future innovations will inexorably supersede our current *best practice.*

The PGC percentage of the process analyzer market is starting to decline, which is partly due to the cost-efficiency of modern PGCs handing several applications concurrently and partly due to additional competition from spectrophotometers. Yet PGC shipments are holding steady, reflecting a lower percentage of a larger market.

The 30,000 dirty chromatographs out there are not in imminent danger of being replaced by shiny new spectrophotometers.

The PGC analytics unit

Introduction

Before proceeding to study the PGC Analytics Unit, let's review what we already know. Part One has explained how gas chromatography works, so we can now summarize the necessary hardware. What should you expect to find in the analytical section of a process gas chromatograph?

Figure 5.3 provides a figurative and grossly simplified answer. The six colored zones in Figure 5.2 represent the different technologies that together make up the PGC analytics unit and determine its overall performance. This book devotes a separate chapter to each of these zones.

> *Lest you get the impression that process chromatographs are easy, Figure* 5.3 *shows what it's really like.*

The PGC analytics unit is an analytical transducer: it accepts process fluid as input and in the shortest possible time outputs a chromatogram signal optimized for the measurement of the desired analytes.

The measurement itself is performed by a microcomputer in the PGC control unit by techniques studied in Part Three of this book.

The following six chapters discuss the six technologies you will encounter inside the analytical section of a PGC analyzer. Each chapter introduces the principles and practical techniques used in one of the colored zones of Figure 5.2. Following is a short preview of the knowledge they convey.

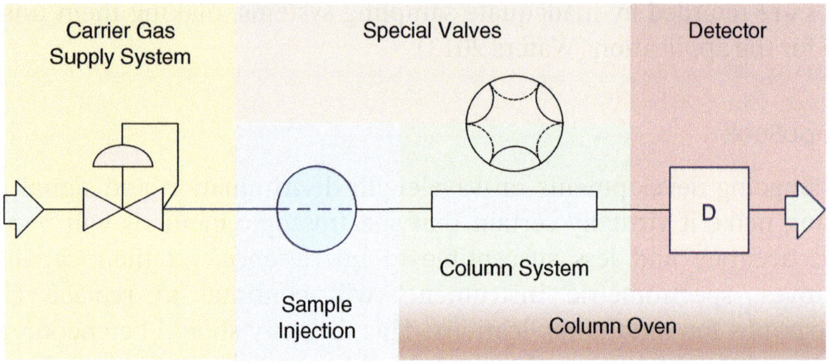

Figure 5.3 Process Gas Chromatograph Functions.

This simplified block diagram shows the main functions of the analytical section of a process gas chromatograph. The six colored zones identify the technologies covered in the next six chapters of the book.

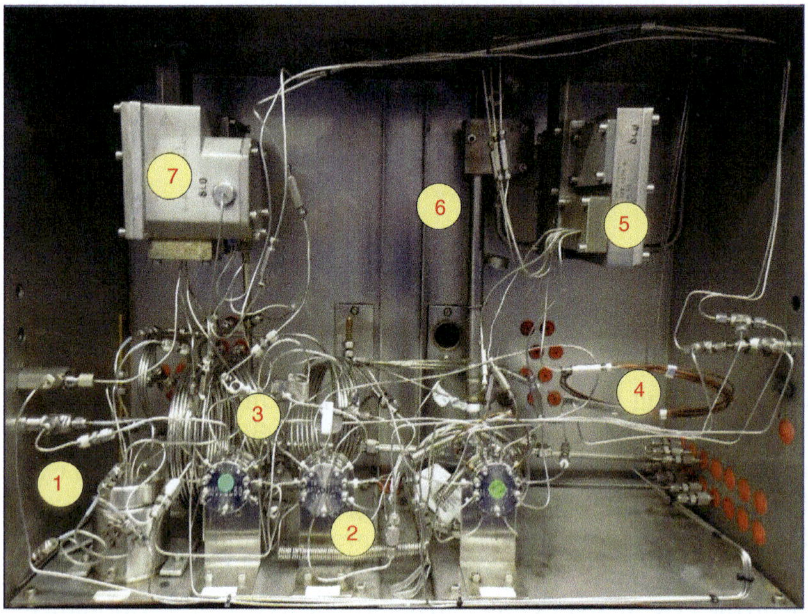

Figure 5.4 A Typical PGC Column Oven.
Source: Author's collection.

Note the extensive use of 1/16-inch o.d. tubing to minimize connection volumes. Chromatographic devices: 1. sample injector valve; 2. three column valves, one also acting as a gas sample injector; 3. vertically coiled ⅛-inch o.d. columns; 4. horizontal coiled 1/16-inch o.d. micropacked column; 5. block containing eight thermal conductivity detectors with space above for a duplicate; 6. hot-air oven heater; 7. flame ionization detector in an explosion-proof housing.

Carrier gas supply system

The carrier gas must be selected to optimize the overall performance of the gas chromatograph. In particular, the choice of carrier gas can have a large effect on column and detector efficiency. The pressure of the carrier gas is controlled by a pressure regulator, sometimes mounted inside the oven for improved stability.

Chapter 6 examines the candidate gases and the logic for selecting one as the carrier of choice. It also looks at the requirements for carrier gas purity and some methods of pressure and flow control.

Sample injection system

The injected sample may be a gas or a liquid, but the procedures differ. We know that the injected volume must be highly repeatable because the measurements depend on it. But how can a PGC achieve that desired precision?

Chapter 7 outlines several ways to precisely inject a gas or liquid sample. When analyzing a gas sample, temperature and pressure are important because they both affect the number of molecules injected. When analyzing a liquid sample, we're more concerned about the possibility of gas bubbles in the sample volume and ensuring very rapid vaporization of the injected liquid.

Chromatographic valves

A peak is a local concentration of analyte molecules in the carrier gas. When in the gas phase, separated from other peaks, the peak has pure carrier gas on either side of it. This untenable situation drives the analyte molecules to gradually spread by **diffusion** into the carrier gas, thereby increasing the peak width over time and reducing the resolution between the peaks.

To minimize this source of peak broadening, all gas pathways external to the columns must be extremely small. The connections to, from, and between columns must be smooth, clean, and free from unpurged niches, cavities, and void spaces where further diffusion might occur. This is particularly important in the design of the sample injector and column valves.

Chapter 8 reviews the special chromatographic valves that PGCs use for sample injection and for switching peaks into different columns.

Column system

In Part One, to gain understanding, we focused on the function of a single column. In process analysis, though, a single column is rarely sufficient. Most PGCs have several columns switched in or out of circuit by column valves. These special valves are usually similar to those used for sample injection.

Chapter 9 introduces the common functions performed by column systems and gives examples of some basic column arrangements found in many PGCs.

Detectors

Older PGCs had only one detector, mainly because a second detector would have meant a complete duplication of the electronics. Microprocessors changed all that. A PGC can now have as many detectors as it needs, all digitally multiplexed to one computer. Each detector outputs an analog chromatogram signal, ideally with a stable baseline. Some PGCs immediately digitize the detector signal and transmit the digital signal to a microcomputer in the control unit for further processing.

Chapter 10 introduces the available PGC detectors and details the operation of the most common types.

Temperature-controlled ovens

To obtain reproducible chromatograms, the column temperature must be very closely controlled, either isothermally or to a preset program. The column oven of an isothermal PGC may now achieve a temperature stability of ±0.02 K. Thermal conductivity detectors are usually installed in the column oven. Other detectors may reside in a separate temperature-controlled compartment.

Chapter 11 reviews the evolution of PGC ovens that are suitable for operating in potentially hazardous process environments. It describes the three main types of heating: the airless oven, the air-bath oven, and direct column heating.

Knowledge Gained

- *A process instrument (PI) measures a specified condition of the process stream.*
- *Typical PI measurements include temperature, pressure, level, and flow rate.*
- *A quality measuring instrument (QMI) measures a specified quality of a process or environmental fluid.*
- *One type of QMI measures a physical property of the fluid, such as its density or viscosity.*
- *Another type of QMI measures the concentration of selected substances in the fluid.*
- *Both types of QMI are called process analyzers, but this is strictly true only of the second type.*
- *Based upon the type of technology used, there are six categories of process analyzer.*
- *The PGC is a concentration analyzer that measures the percentage of selected substances in a fluid.*
- *For the best reliability, one should minimize the number of measurements made per PGC.*
- *PGCs use custom configurations of standard devices to separate and measure the desired analytes.*
- *Due to advances in technology, new PGCs are very different from older models still working in plants.*
- *Analytical measurements are very valuable as they allow a process to operate closer to a constraint.*
- *Analytical control increases the production of on-spec product and may save operating costs.*
- *Other measuring technologies compete with PGC, but none has its versatility of measurement.*
- *A spectroscopic analyzer provides a faster response than a PGC, and often requires less maintenance.*
- *Some spectrometric analyzers have no moving parts and are thereby more robust than a PGC.*
- *A spectroscopic analyzer may experience complex interferences due to spectral overlap.*
- *Spectroscopic methods of analysis may infer the measured quality from a mathematical model.*

- *The accuracy of an inferential method rests on how well the calibration set represents the process.*
- *A PGC may be more accurate and suffer less interference than a spectroscopic analyzer.*
- *A well-applied PGC can accurately measure more substances in a sample than a spectrometer can.*
- *PGCs are experiencing a declining market share, but unit shipments are holding steady.*
- *Overall, the analyzer unit accepts process sample and outputs a repeatable chromatogram signal.*
- *The analyzer unit employs six distinct technologies to achieve precise and accurate measurements.*
- *The carrier gas system must supply a controlled pressure of an adequately pure carrier gas.*
- *The sample injection system must inject a very repeatable volume of sample.*
- *The injector and column valves must be highly reliable and not contribute to peak broadening.*
- *The column system must adequately separate the analytes within the desired cycle time.*
- *The detector must generate a stable and repeatable chromatogram signal to enable precise results.*
- *The column oven must maintain close temperature stability in a hazardous process environment.*

Did you get it?

Self-assessment quiz: SAQ 05

Q1. To improve the reliability of a PGC, should we reduce the number of analytes it measures or the number of streams it analyzes, or both?

Q2. What technology did PGCs introduce to process field instruments that had never done before?

Q3. Since a PGC can measure almost any analyte in almost any process fluid, why is it not used for simple measurements, such as the concentration of carbon dioxide in a gas stream?

Q4. Comparing PGCs with photometric process analyzers, which of the following features is a potential advantage of using a process gas chromatograph?

- **A.** A PGC may use a direct-insertion measurement probe.
- **B.** PGC measurements are less likely to suffer interference from the varying concentration of other substances in the sample.
- **C.** A PGC is likely to provide more accurate measurements when measuring several substances in the process sample.
- **D.** A PGC usually has a faster response to process change.

Q5. Why can a PGC easily measure several components in a process stream that might be difficult for a wavelength absorption analyzer to measure?

Q6. In what way is PGC technology becoming similar to process spectrophotometry?

Q7. How does a wavelength absorption analyzer use chemometrics to measure a physical property of a stream when there is no known relation between the physical property and the absorption of light?

Check your SAQ answers with those given at the end of the book.

Student evaluation test: SET 05

Your instructor will provide answers to these questions.

S1. What kind of process analyzer is a titrator?
Select the one correct answer:

A. It's an electrochemical analyzer.
B. It's a physical property analyzer.
C. It's a photometric analyzer
D. It's a wet-chemistry analyzer.

S2. What kind of process analyzer is an instrument that measures the absorption by the process fluid of selected wavelengths of light?
Select the one correct answer:

A. It's an electrochemical analyzer.
B. It's a physical property analyzer.
C. It's a photometric analyzer
D. It's a wet-chemistry analyzer.

S3. Which of the advantages listed below could accrue from installing a PGC at a fluid processing plant?
Select all the correct answers:

A. increased operational safety
B. compliance with environmental regulations
C. sustained product quality
D. increased product throughput
E. reduced energy and material costs

S4. How might it be possible to improve the reliability of a PGC?
Choose all that apply:

A. Eliminate all the measurements that are merely "nice to know."
B. Measure only the analytes necessary for process control.
C. Reduce the number of PGCs by using more multistream analyzers to perform the same analyses.
D. Eliminate multistream sample selection by measuring only one stream per sample injector valve.

S5. PGCs were the first process field instruments to embrace the advantages of digital data transmission. According to the text, in about which year was the first PGC offered for sale with an integral microprocessor?
An integer answer is required.

S6. Which of the following devices is the one most unlikely to be inside a PGC oven?
Choose one device:

A. A carrier gas pressure regulator.
B. A carrier gas flowmeter.
C. A sample injector valve.
D. A column valve.
E. A chromatographic column.
F. A detector.
G. A heating device.

S7. According to the text, why is the PGC percentage of analyzer sales declining?
Choose all that apply:

A. Too many PGC applications have failed to work in the field.

B. PGCs can now handle several applications concurrently.
C. Some prior PGC applications are now performed by spectrophotometers.
D. The market for process analyzers is expanding.

S8. Photometers using chemometric data processing are sometimes a better choice of process analyzer than a PGC. What conditions would tend to favor the chemometric method of analysis?
Choose all that apply:

A. The desired measurement is difficult to accomplish by measuring the individual components – like octane number, for instance.
B. The process variation is predictable and adequately modeled.
C. The process needs an exact measurement of a single component in an unpredictable background mixture.
D. The process fluid is a dirty liquid.

S9. Assuming that both techniques would give a good response to the desired measurements, when is a process gas chromatograph probably a better choice than a process photometric analyzer?
Choose all that apply:

A. When a very fast response time is necessary.
B. When the PGC must separately measure several components of the sample.
C. When the PGC must measure a single component such as carbon dioxide in a gas stream.
D. When the PGC must individually measure all the components of the sample.

References

Cited

Adams, D. (2002). *The Salmon of Doubt: Hitchhiking the Galaxy One Last Time*, 115. London, UK: Pan Macmillan.

Clemons, J.M. (2016). Chromatography in process analysis. In: *Encyclopedia of Analytical Chemistry* (ed. R.A. Meyers). Chichester, UK: Wiley https://doi.org/10.1002/9780470027318.a2107.pub3.

Clemons, J.M. (2017). Personal communication received December 14, 2017 with photograph of computer-controlled analyzer data system by Bendix & Real Time Systems Inc. on Greenbrier Instruments booth at 1966 Chicago ISA Exhibition.

Clevett, K.J. (1986). *Process Analyzer Technology*. New York, NY: Wiley.

Waters, T. (2013). *Industrial Sampling Systems*. Solon, OH: The Swagelok Company.

Table

Figures

5.3 Process Gas Chromatograph Functions
5.4 A Typical PGC Column Oven

Symbol

K A kelvin, the SI unit of absolute temperature, equal to one Celsius degree.

New technical terms

When first introduced, these technical terms were shown in bold type and you should now understand what they are:

absorption photometer
availability
chemical reaction
chemometrics
diffusion
electrochemical
ethylene splitter
holistic property
infrared absorption
outlier
photometer
photon absorption
photon emission
physical property
physical separation
process analyzer
quality-measuring instrument
QMI
spectrophotometer
spectroscopy
ultraviolet absorption
wet chemistry

This chapter compares process gas chromatography with many other methods of process analysis, particularly photometric techniques. Unless you wish to explore those other process analyzers, you won't need to know any details of the analytical techniques listed below. If the need arises, you'll find more details in the Glossary at the end of the book.

cavity ring-down
colorimetry
dielectric constant
electrode potential
flash point
fluorescence
Fourier transform
galvanic current
infrared
mass spectrometry
near infrared
nuclear magnetic resonance
paramagnetism
photon absorption
photon emission
quantum cascade laser
Raman spectroscopy
titrimetry
tunable diode laser
ultraviolet
viscosity
X-ray fluorescence

6

Carrier gas system

Choice of carrier gas

Carrier gas purpose

The important purpose of the carrier gas system is to provide a stable transport and detection background for the components of the sample. A change in flow rate may adversely affect the detector signal and component retention times. Most process chromatographs rely upon constant flow rate to maintain detector zero and span settings and to properly locate the components of interest on a fixed time base.

The carrier gas literally does *carry* sample molecules through the columns. Since the gas is only a transport medium, a gas chromatograph can use any stable, non-reactive gas. Later, we shall see that the gas used as a carrier affects the resolution of peaks, due mainly to viscosity and density differences. But this effect is small and is rarely important in column design.

Choice of carrier gas

While column performance is sometimes a consideration, a column designer will usually choose a carrier gas to optimize the sensitivity or stability of the detector. The detector has an unenviable task: to differentiate a small number of solute molecules from an enormous number of carrier gas molecules. The carrier gas can assist or hinder that task.

Typical carrier choices include hydrogen, helium, nitrogen, or argon, although PGCs have occasionally used carbon dioxide or compressed air in the past. Table 6.1 lists the relevant properties of the most common carrier gases.

Hydrogen carrier gas

When using a detector that responds to a change in a physical property of the gas passing through it, such as the thermal conductivity detector (TCD),

Process Gas Chromatographs: Fundamentals, Design and Implementation, First Edition. Tony Waters.

Table 6.1 Properties of Common Carrier Gases.

Property:	Viscosity µPa·s	Density kg/m^3	Thermal Conductivity mW/m·K
Carrier Gas:		*illustrative values at 1 bara and 300 K*	
Hydrogen	8.9	0.081	186.6
Helium	19.9	0.160	155.7
Nitrogen	17.9	1.123	26.0
Argon	22.7	1.601	17.7

Source: Viscosity and Thermal Conductivity values from Huber and Harvey (2018ab). Density values calculated.

we choose a carrier gas that differs greatly in that property from the analytes. The thermal conductivities of hydrogen and helium are quite high, about 7.0 and 5.8 times higher than nitrogen, so hydrogen or helium are often the best options for the carrier gas in process applications that use the TCD.

When using hydrogen or helium carrier gas, the analyte molecules have a lower thermal conductivity than the carrier gas, so the detector electronics is set up to give a positive response to a reduction in thermal conductivity.

Hydrogen is a low-viscosity carrier gas that moves quickly through the columns, reducing analysis time. It's also plentiful and less expensive than helium.

Another factor favoring hydrogen for the TCD is its use in other nearby detectors. The other two detectors popular in PGCs are the flame ionization detector (FID) and the flame photometric detector (FPD), which both need hydrogen as fuel. In the past, PGCs with flame detectors used nitrogen or helium carrier gas, and added hydrogen fuel at the detector, but many PGCs now use hydrogen carrier to avoid the need for an additional nitrogen or helium supply. It then makes sense to use the same hydrogen supply as carrier for any adjacent application using a thermal conductivity detector.

Hydrogen carrier gas is a good carrier gas for open tubular columns with thin liquid films, as it provides faster separations than helium or nitrogen, but with slightly less efficiency – a somewhat reduced plate number. When using a thick-film or packed column, the improvement in speed is not so great.

Helium carrier gas

Helium is an efficient carrier gas – yielding higher plate numbers – but it doesn't have the analysis speed of hydrogen (it's more viscous than nitrogen). Helium is a non-reactive gas, so it's often preferred for its safety. But helium is becoming scarce, so it's more expensive than hydrogen and sometimes in short supply. Many plants have converted their PGCs to use hydrogen carrier because of unreliable helium supplies.

Nitrogen carrier gas

Nitrogen is the preferred carrier gas for measuring hydrogen, often by a separate thermal conductivity detector, and is sometimes effective for measuring other components at percentage levels. It was the traditional choice of carrier gas for a flame ionization detector, although hydrogen is now common. Nitrogen gives good column efficiency, but slower analysis times than hydrogen.

Argon carrier gas

Argon is another suitable carrier gas for the measurement of hydrogen. It's also a good choice for low-level oxygen measurement; the separation of argon from oxygen is difficult, and the argon carrier gas renders it unnecessary.

Chapter 10 discusses the common PGC detectors in more detail.

Mixed carrier gases

For a while, it was fashionable to use a mixture of gases as carrier gas. The most common application was the measurement of hydrogen and other components with a thermal conductivity detector.

As noted above, hydrogen has a higher thermal conductivity than helium. All other substances have lower thermal conductivity. From this, we deduce that a hydrogen peak should go negative when a TCD runs on helium carrier and is set up to produce positive peaks for other substances. Unfortunately, it doesn't. The hydrogen peak is positive for low concentrations and then turns negative as the hydrogen concentration increases (Snavely and Subramaniam 1998).

Using a mixture of hydrogen and helium as the carrier gas ensures a negative hydrogen peak at all concentrations. Inversion of the electronic chromatogram signal then gives a positive peak, allowing the chromatograph to measure hydrogen and other components on a single detector.

But mixed carrier gases failed to work well in process applications, solving one problem while causing another. The problem came down to the use of multiple columns – almost universal in process chromatographs. When a column valve switches carrier gas from one column into another, the hydrogen rushes into the new column ahead of the helium. The detector then responds to the separation of carrier gases, upsetting the chromatogram baseline and interfering with the measurement of peaks. For these reasons, we soon abandoned the technique.

Carrier gas purity

The carrier gas must be pure and dry. It should not contain reactive impurities like oxygen or water, as these will often damage the stationary phase within the columns. Also to be avoided are impurities that will invoke a response from the detector and raise the chromatogram baseline, like hydrocarbons do when using a flame ionization detector. Impurities like these may cause troublesome noise on the detector signal that makes it difficult to measure the analyte peaks. They can also cause baseline upsets and measurement errors when switching columns or replacing gas cylinders.

Carrier gas impurities that are non-reactive and not detected usually don't pose a problem. When using an FID, for example, the presence of argon in the nitrogen carrier gas can be tolerated, as argon is neither reactive nor detected, but the presence of hydrocarbons could be devastating.

In summary, the need for purity is selective and might depend upon:

- the analytes and their range of measurement.
- the type of stationary phase used within the columns.
- the detectors used.

Analytical effect of impurities

The analytical requirement is to exclude from the carrier gas any substance you intend to measure. If the carrier gas contains a significant amount of an analyte, the measurement of that analyte will suffer an error equal to its concentration in the carrier gas. Figures 6.1a and 6.1b illustrate this effect; if a PGC measures a sample containing 1 ppm of methane using a carrier gas contaminated with 1 ppm of methane, the methane peak will disappear, and the analyzer will report a value of zero for that component.

> *To understand this strange result, imagine what would happen if you injected a sample of the contaminated carrier gas. Clearly, that sample would contain 1 ppm of methane, but it can have no effect on the detector because it's identical to the carrier gas. The injection doesn't change anything.*

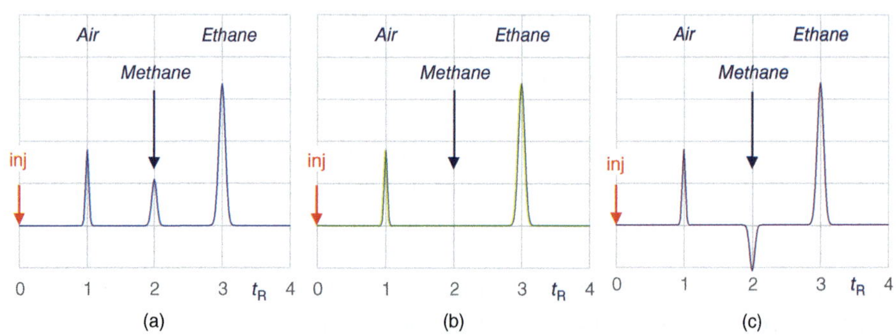

(a) Sample has 2 ppm methane. (b) Sample has 1 ppm methane. (c) Sample has no methane. Illustrative chromatograms for a carrier gas contaminated with 1 ppm methane: (a) When the sample gas contains 2 ppm of methane, a 1 ppm methane peak appears. (b) When the sample gas contains 1 ppm of methane, there is no detector response for methane. (c) When the sample gas contains no methane, the methane peak is negative. The ethane peak is unaffected by the amount of methane in the carrier gas.

Figure 6.1 Effect of Impure Carrier Gas.

Figure 6.1c shows what happens when the sample contains less methane than the carrier gas does; in this example, none at all. We see a negative peak eluting at the position of methane, caused by an absence of methane. In this example, the injected sample volume contains no methane, but the carrier gas on both sides of it contains 1 ppm of methane, which travels through the column at the normal speed of methane. Therefore, the *absence of methane* moves through the column at the same speed as methane and produces a negative peak on the chromatogram at the normal retention time of methane.

A negative peak – often called a **vacancy peak** – appears on the chromatogram whenever the sample contains less of an impurity than the carrier gas does. This might be an important thing to know when troubleshooting an unexpected negative peak on a chromatogram. You can see a real vacancy peak in Figure 6.2.

The presence of methane in the carrier gas has no effect on the measurement of ethane. An impurity in the carrier gas suppresses the zero only for that impurity, not for other measured peaks.

Of course, a carrier gas that contains 1 ppm methane will be perfectly adequate for analyzing methane on a range of 0–1 vol% since it will cause only a minute zero error in the measurement.

In summary, the required purity of the carrier gas depends on the components measured and the desired measurement ranges. As a general guide,

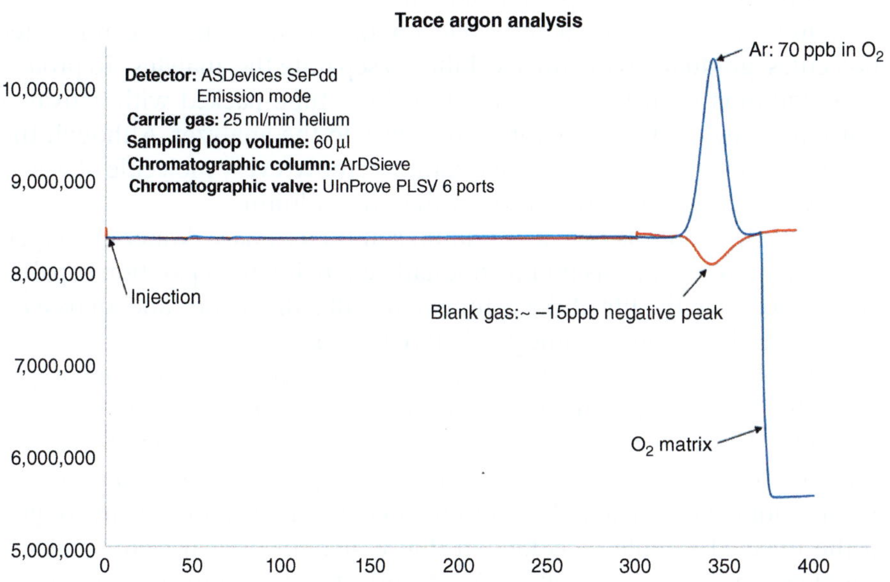

The red chromatogram is an analysis of a blank gas containing less argon than the carrier gas does. The consequent reduction of argon caused by sample injection migrates down the column at the same speed as argon giving a 15 ppb negative peak at the retention time of argon, as confirmed by the real 70 ppb argon peak on the blue chromatogram.

Figure 6.2 A Vacancy Peak in a Real Chromatogram.
Source: Analytical Sensing Devices, Ltd. Reproduced with permission.

a purity of 99.996 % is desirable in most applications, but an FID at high sensitivity may need a purity of 99.999 %.

Damaging effect of impurities

Constant exposure to traces of oxygen or water at the elevated temperature inside the column may damage the stationary phase, resulting in a gradual loss of separation between peaks. To evaluate the risk, you will need to know a little more about the different types of column.

Recall that the stationary phase within the column might be a solid or a liquid. Columns that use a solid stationary phase are always susceptible to deactivation by absorbing moisture from the carrier gas. You must protect these columns by drying the incoming carrier gas before use.

There are two types of solid stationary phase: molecular sieves and active solids. A PGC will most likely use a **molecular sieve** column when it needs to separate any combination of hydrogen, oxygen, nitrogen, and methane peaks.

> *Molecular sieve columns are uniquely capable of separating oxygen and nitrogen by molecular size, but they can't separate oxygen from argon except at subambient temperatures.*

Early PGCs used a few other solid stationary phases, mainly the active adsorbent solids: charcoal, silica gel, and alumina. But **adsorbent solids** are troublesome and now rarely used in columns, superseded in most applications by more stable proprietary column packings.

A molecular sieve or active solid column will strongly retain water molecules, gradually reducing its ability to separate the analytes. To protect these columns from moisture, install a dryer tube packed with activated molecular sieve in the carrier gas line, close to the analyzer. Although the external molecular sieve dryer tube cannot remove every molecule of water, it's a very effective dryer when colder than the column.

Most PGC columns now use a liquid stationary phase. Water or oxygen can slowly react with this liquid, gradually changing its separation capability. To maximize the life of the columns, install a dryer tube and an oxygen absorber in the carrier gas line just before the analyzer.

Ironically, it's possible to overdo the purification of the carrier gas when the PGC is using **porous polymer** columns. These column packings naturally absorb and retain a certain amount of oxygen from the carrier gas. An oxygen absorber trap on the carrier gas can starve the porous polymer for oxygen. Then, the column will permanently absorb any oxygen in the injected sample. It is known that a 40 ppm oxygen peak vanished completely because of this effect. Apparently, this loss of measurement can also happen for low ppm amounts of other light gases (Eidt 2018, personal communication).

When analyzing for less than 100 ppm of any component, it's always advisable to protect the analyzer from detrimental impurities present in the carrier gas. As noted above, an impure carrier gas can directly affect the accuracy of analyte measurement. It might also cause a noisy baseline, which would then interfere with the integration of peak areas. In addition, spurious baseline disturbances may appear after column switching. Hydrocarbons are often the worst offenders, so use an activated charcoal adsorption tube to remove hydrocarbons from the carrier gas.

> *Note the need to also remove hydrocarbons from the combustion air consumed by a flame ionization detector. The best way to clean the combustion air is by installing a catalytic oxidizer in the air supply line.*

Not all impurity traps are equally efficient. A typical absorber has a transparent polycarbonate tube and can remove moisture, oxygen, or hydrocarbons down to about one part per million (ppm). To remove multiple impurities, simply connect the desired absorbers in series – oxygen, then moisture, then hydrocarbons – or install a combination absorber for all three. Most absorbers, except those using activated charcoal, incorporate an indicator band that changes color to indicate impending saturation.

For ultra-trace analyses that require extremely pure carrier and detector gases, you might need a high-efficiency absorber that uses a metal or glass tube and can remove contaminants down to part-per-billion (ppb) levels. For the ultimate sensitivity, pass the gas through an electrically heated device that permanently bonds the impurity molecules to the surface of a getter alloy.

Maintenance of gas cleaners

A neglected trap will eventually reach saturation and become a source of carrier gas contamination and poor analyzer reliability. Set up a maintenance schedule to ensure regular dryer replacement, at least annually. A dual-tube system with shutoff valves will allow you to change one dryer while the other is in use. Make sure you can purge the new dryer with carrier gas before it goes online.

You can reactivate a molecular sieve dryer by removing it from service and then backpurging with pure nitrogen for several hours, while heating to about 300°C. Allow the tube to cool before disconnecting the purge, then immediately cap the ends to prevent the ingress of moisture.

Carrier gas supply system

A PGC may draw its carrier gas from a single gas cylinder or from several cylinders manifolded together at a central point. When large numbers of gas chromatographs are in use, a central cylinder area is convenient and will reduce costs. For new installations, consider an elevated cylinder handling facility that provides a convenient dock for truck deliveries.

To ensure continuity of supply, it is best to have two supply manifolds for each gas. Each supply manifold may comprise a single gas cylinder or several

This device comprises three pressure regulators. The lower dual regulator accepts two carrier gas supplies and automatically switches to the reserve supply when the initial supply pressure drops to a preset threshold value. The upper regulator controls the carrier gas pressure to the PGC.

Figure 6.3 Automatic Changeover Regulator. *Source: Image courtesy of Swagelok Company. Reproduced with permission.*

cylinders ganged together at full pressure. Install a two-stage pressure regulator on each manifold, with both regulators feeding the common supply line. Set one regulator at a slightly higher pressure than the other one, so that cylinder will empty first. The other regulator will automatically switch in when the line pressure starts to fall.

Figure 6.3 is an example of a proprietary switchover device that also allows the operator to vent any residual pressure when changing gas cylinders.

Don't specify single-stage regulators because they have limited ability to hold a constant output pressure when their inlet pressure changes, and it will. Two-stage regulators will deliver superior regulation as the tank pressure declines.

Carrier gas regulators must not be of the venting kind. Surprisingly, atmospheric gases will diffuse back up the vent port and contaminate the carrier gas.

Contamination of the carrier gas supply by moisture or oxygen may cause slow column damage, reducing the life of a column. In addition, serious baseline upsets may occur, particularly during column switching.

To guard against contamination, fit pressure regulators with metal diaphragms. An elastomer diaphragm will permit oxygen and moisture to permeate into the carrier gas from the outside air. You don't want that. In addition, the elastomer may contaminate the carrier gas by adsorbing and then releasing impurities into the gas stream.

For the same reason, never use polymer tubes or flexible hoses to connect the regulators to the carrier gas supply line. Moisture and oxygen from the outside air will permeate into the carrier gas even though it's under pressure. The driving force for **permeation** is the **partial pressure** of an impurity, not the total pressure of the gas. Since the partial pressure of moisture and oxygen molecules in the carrier gas is close to zero, their leakage path is always inward.

For a different reason, even convoluted metal hoses are undesirable. Over long periods of operation, their irregular internal surface may accumulate dust particles that alternately adsorb and desorb impurities in the carrier gas. An observed symptom of this malady would be a disruption of the PGC baseline occurring each morning as sunlight heats the flexible hose, returning to normal in the cool of the night.

Carrier gas supply line

Never use polymer tubing or welded steel pipe for any part of a carrier gas supply line. Instead, use good-quality stainless steel tubing in a single continuous run and minimize the number of tube fittings. Before installation, clean the tubing internally by flushing with a polar and nonpolar solvent (acetone and hexane are suitable) and then remove all traces of the solvents with a clean nitrogen purge.

Do not use a **halogenated solvent**.

Do not allow the ends of the cleaned tubing to touch the ground.

During installation, clean all the fittings in a similar way. Minimize the use of screwed fittings and seal them with PTFE tape, not pipe dope. When applying the tape, leave the first thread bare, or small slivers of PTFE may form and blow down the line, possibly blocking the carrier gas regulator inside the analyzer. It's a wise precaution to install a small 0.5 μm filter at the analyzer connection to guard against this possibility.

After installation, test the carrier gas supply line for leaks. First, purge and fill the line with helium or hydrogen, depending on the type of leak detector you are using, and set the purge gas to the normal working pressure. Then, search for leaks. Start by sealing the system and waiting for several hours to ensure the trapped pressure doesn't leak out. This procedure will detect severe leaks that you might otherwise miss. Then, scan all the tube fittings with an electronic leak detector. It's best to avoid the use of a proprietary soap solution as it may allow water ingress into the line. After sealing any found leaks, purge the line with carrier gas before connecting it to the analyzer.

The carrier gas line pressure is usually about 200 kPa above the column head pressure to provide optimum conditions for the internal pressure regulator. With less pressure differential, the regulator may not recover quickly from a sudden flow demand during column switching. When that happens, the baseline may waver due to pressure variations as the regulator tries to regain control and return to stable operation.

Pressure regulation

It's far easier to control pressure than flow rate, so PGCs typically control their carrier gas head pressure. When a column is at constant temperature, its flow resistance changes very slowly, if at all. Keeping a constant pressure differential across the column is a very effective way to ensure a constant rate of flow.

A constant pressure differential also allows some automatic compensation of peak retention time. A slight increase in column temperature, for instance, tends to reduce the retention time of a peak, but it also increases carrier gas viscosity and thereby causes a slight reduction in the flow rate, which tends to increase the peak retention time. Thus, the two effects of temperature change oppose each other, providing partial compensation (Annino and Villalobos 1992, 86).

Mechanical pressure regulators

Traditionally, PGCs always used a mechanical regulator to control the carrier gas inlet pressure and, due to their high reliability, some still do. Mounting the regulator inside the column oven or in a separately heated and insulated compartment will improve its performance.

Of course, there would be no advantage gained by installing a pressure regulator inside a temperature-programmed oven. Also, the newer

PGC ovens, even when isothermal, can operate at an oven temperature that is higher than the maximum working temperature of the regulator used. Those PGCs may enclose their pressure regulators in a separate compartment held at constant temperature.

Electronic pressure controllers

After PGCs adopted microprocessors, it was evident that the digital control of pressure would offer several advantages so, for certain applications, PGCs now use **electronic pressure controllers** instead of mechanical pressure regulators. The traditional way was to have a single pressure regulator to set the column head pressure, and then use needle valves or capillary-tube restrictors to drop that pressure to the levels required by the different columns. Today, a PGC may employ multiple electronic pressure regulators to supply the carrier gas to each column at exactly the right pressures.

A PGC may also use electronic pressure controllers to control the fuel gas and air supplies needed by flame detectors.

It's easy to set the electronic pressure regulators from a maintenance display, or to program them to follow an automatic sequence. The processor then stores the set points in memory and restores them on demand.

The PGC controller can measure a column pressure in one mode of operation, and exactly replicate that pressure when the columns switch to another mode of operation.

It's even possible to have a flow-programmed separation in lieu of (or as well as) a temperature programmed one. To gradually increase the carrier gas flow rate, the controller ramps up the column inlet pressure on a preset schedule. Hence, the technique is known as pressure programming. It's another way to reduce analysis time but not often invoked: the majority of PGC applications still run with constant flow rate.

Flow regulation

Flow regulators, which rely on maintaining constant differential pressure across a restriction, are not suitable for use in a PGC. Maintaining a constant pressure across the columns provides a more constant carrier flow. Pressure regulators can quickly provide a burst of additional gas, which a column system may sometimes need. Flow regulators can't do that. With a flow regulator, the detector could take over a minute to stabilize after column switching.

Measuring the carrier gas flow rates

Carrier flow is more important than pressure but is more difficult to measure. Typically, the flow rate is set between about 10–50 mL/min for conventional packed columns and 1–5 mL/min for capillary columns, in accordance with the application data.

It's not good practice to install a flowmeter at the detector vent, where it might affect the detector baseline. Installing a flowmeter on the carrier gas supply is ineffective because it would then operate at full column head pressure, which would reduce the reading to near zero. Consequently, gas chromatographs don't have carrier gas flowmeters.

The best way to measure detector vent flow rate is with a soap-film flowmeter, a device that measures the time taken by soap bubbles to travel up a calibrated tube. The soap-film flowmeter is more accurate when you subtract the effect of saturation by water in the tube (about 2 %). It is less accurate with hydrogen as that gas readily diffuses through the bubbles.

Several electronic flowmeters are available to measure a carrier gas flow. Some of these use the bubble technique with electronic sensing and calculation, while others are fully electronic. They all provide a direct readout of the flow rate (Hinshaw 2004).

An advantage of a soap film flowmeter (with or without digital readout) is that it will never lose its calibration. A soapless digital flowmeter is more convenient to use but will gradually lose its calibration and may need annual servicing.

Setting the flow rates

Typically, the desired flow rate is set by adjusting the carrier gas pressure going via the sample injector to the first column. The pressure applied to the inlet of a column is known as the *column head pressure*. When the PGC uses electronic pressure regulators, the column head pressure is under software control.

We shall see that most PGCs use several columns with special column valves to divert selected peaks from one column into other. A multiple-column system may need one or more additional carrier gas flows that must be set and accurately maintained. Over the years, PGCs have used different devices to control these important flow rates. For adjusting column vent flows, most PGCs use miniature needle valves or short lengths of capillary tube. For adjusting column inlet flows, older PGCs again employ a valve or capillary, or even a dummy column, but a newer PGC will sometimes use an electronic pressure controller.

Each column system requires a different procedure for setting the carrier flow rates, so always follow the given instructions.

Generally, the procedure starts with the column valve in a position that puts two or more columns in series. The technician adjusts the column head pressure to achieve the desired analysis time and measures the resulting detector vent flow. Most PGC column valves have only two positions. The technician switches the valve to its other position and adjusts the appropriate flow restrictor or electronic pressure regulator to get the same detector vent flow rate.

Never reset the carrier flow of a chromatograph that is working well:

> *If you are satisfied with the chromatogram and the cycle time, there is no need for further adjustment. The pressure and flow readings shown in the manual are only for reference and are not mandatory settings.*

We discourage the minor adjustment of pressures and flows. The flow resistance of the columns may have changed slightly, but this is not an immediate problem if the separation is still adequate. But a significant change in column performance might indicate a more serious problem such as a leak in the column system, which you must fix.

Optimum flow rate

As you might expect, a column works best at an intermediate carrier flow rate, not too fast and not too slow. The theory of gas chromatography illuminates the effect of carrier flow rate – or more correctly, its **velocity** – on the resolution of peaks. It turns out that the optimum flow rate is rather low, so most PGC columns run faster than their optimum speed to achieve a shorter analysis time.

Knowledge Gained

- *The carrier gas provides a stable peak transport and detection medium.*
- *Most PGCs rely on constant carrier gas flow rate to identify peaks by their retention times.*
- *Most PGCs rely on constant carrier gas flow rate to get a stable baseline for measurement.*
- *Since the carrier gas is merely a transport medium, a PGC could use any stable, non-reactive gas.*
- *The carrier gas has only a small effect on column efficiency, so it's chosen mainly to suit the detector.*
- *Hydrogen or helium carrier is common with a thermal conductivity detector (TCD).*
- *Nitrogen or argon carriers give the best sensitivity when measuring hydrogen with a TCD.*
- *Flame detectors previously used nitrogen or helium carrier gas, but hydrogen is now a common choice.*
- *The combustion air supply to a flame ionization detector must not contain any hydrocarbons.*
- *As a carrier, hydrogen is plentiful, low in cost, fast in analysis, and likely used as fuel by another detector.*
- *As a carrier, helium is effective and safe, but is becoming expensive, with shortages of supply.*
- *To measure hydrogen, PGCs once used a mixture of helium and hydrogen as carrier, but not anymore.*
- *The carrier gas must be pure and dry; it should not contain oxygen, moisture, or hydrocarbons.*

- *The carrier gas should not contain impurities that would raise the detector baseline.*
- *The carrier gas should not contain a significant amount of any intended analyte.*
- *The presence of an analyte in the carrier gas will suppress the detector response for that peak only.*
- *When the concentration of an impurity is the same in sample and carrier, there's no detector response.*
- *When the sample contains less of an impurity than the carrier gas does, the impurity peak is negative.*
- *A negative peak that is solely due to an impurity in the carrier gas is known as a vacancy peak.*
- *A carrier gas impurity has no effect on measuring other substances.*
- *A carrier gas impurity may cause detector noise or baseline upsets when switching columns.*
- *Two gas cylinders with regulators will automatically use all of one cylinder before switching in the spare.*
- *On the carrier supply manifold, use two-stage non-venting pressure regulators with metal diaphragms.*
- *All carrier gas lines should be stainless steel tubing precleaned with polar and nonpolar solvents.*
- *Traditionally, PGCs have used mechanical pressure regulators, but electronic ones are now common.*
- *Flow regulators can't supply the flow demands of column switching, so are not suitable in a PGC.*
- *A soap-film flowmeter or its electronic equivalent is necessary for measuring carrier gas flows.*
- *Carrier flow velocity is a very important parameter in the theoretical performance of a column.*

Did you get it?

Self-assessment quiz: SAQ 06

Q1. Calculate the thermal conductivity of hydrogen and helium relative to nitrogen.
Q2. List the advantages and disadvantages of using hydrogen as a carrier gas.
Q3. List the advantages and disadvantages of using helium as a carrier gas.
Q4. List the advantages and disadvantages of using nitrogen as a carrier gas.
Q5. Why are flow controllers not used in a PGC to control carrier gas flow?
Q6. Imagine that you are measuring 0–5 ppm methane on a flame ionization detector and the methane peak on the chromatogram goes negative. Which one of the following is the most likely explanation?
Select one option:

A. There is a negative quantity of methane in the sample gas.
B. There is 1 ppm of methane in the sample gas.
C. The carrier gas contains more methane than the sample gas does.
D. The carrier gas contains less methane than the sample gas does.

Q7. As a general rule, when purchasing a cylinder of a carrier gas, what purity is typically needed for a PGC measuring ppm level analytes?
Select one option:

A. 99.9 % purity
B. 99.99 % purity
C. 99.999 % purity
D. 99.9999 % purity

Check your SAQ answers with those given at the end of the book.

Student evaluation test: SET 06

Your instructor will provide answers to these questions.

S1. Which of the pure gases listed below is the one *least* likely to be suitable as carrier gas for a PGC?
Select one option:

A. Oxygen
B. Nitrogen
C. Hydrogen
D. Helium

S2. Which of the impurities listed below must be absent from the combustion air supply for a flame ionization detector?
Select all that apply:

A. Nitrogen
B. Oxygen
C. Hydrocarbons
D. Moisture

S3. Which one of the following solvents should *definitely not* be used to clean the tubing before it's installed in the PGC carrier gas supply line?
Select one option:

A. Methanol
B. Acetone
C. Hexane
D. Trichloroethylene

S4. On occasion, a PGC applications engineer might choose a carrier gas for any of the good reasons given below. But, of the four reasons given, which one is mandatory?
Select one option:

A. To suit the needs of the detector.
B. To get the shortest analysis time.
C. To optimize the column performance.
D. To eliminate the detector response for the major component.

S5. Which of the features listed below is a potential advantage of using hydrogen as a carrier gas for a PGC using a thermal conductivity detector?
Select all that apply:

A. It provides more sensitivity than nitrogen carrier gas for measuring hydrocarbons.
B. It will likely provide a faster analysis time.
C. It is plentiful and inexpensive.

D. It might already be in use as fuel for a flame detector (FID or FPD) in the same analyzer.
E. It is chemically inert and inherently safe.

S6. Which of the features listed below is a potential advantage of using helium as a carrier gas for a PGC using a thermal conductivity detector?
Select all that apply:
A. It provides more sensitivity than nitrogen carrier gas for measuring hydrocarbons.
B. It will likely provide a faster analysis time than hydrogen or nitrogen carrier.
C. It is plentiful and inexpensive.
D. It might already be in use as fuel for a flame detector (FID or FPD) in the same analyzer.
E. It is chemically inert and inherently safe.

S7. Which of the gases listed below would give a good linear response and reliable performance when used as the carrier gas for a PGC measuring hydrogen with a thermal conductivity detector?
Select all that apply:
A. Argon
B. Nitrogen
C. Helium
D. A mixture of hydrogen and helium

S8. Imagine that you are measuring 0–5 ppm methane on a flame ionization detector and you're asked whether a particular gas supply is pure enough to be used as the carrier gas. All the specifications listed below would be important, but which **one** is the most important for getting an accurate measurement?
Select one option:
A. The concentration of oxygen in the proposed carrier gas.
B. The concentration of methane in the proposed carrier gas.
C. The concentration of ethane in the proposed carrier gas.
D. The concentration of moisture in the proposed carrier gas.

S9. What is a vacancy peak?
Select one option:
A. A peak that is missing from the chromatogram at the retention time of an expected component peak.
B. A valley between two peaks on the chromatogram.
C. A negative peak at the expected retention time of a component, due to the injected sample having less of that component than is present in the carrier gas.
D. A negative peak at the expected retention time of a component, due the negative detector response for that component.

References

Cited

Annino, R. and Villalobos, R. (1992). *Process Gas Chromatography: Fundamentals and Applications*. Research Triangle Park, NC: Instrument Society of America.
Hinshaw, J. (2013). Measuring Gas Flow for Gas Chromatography. In *LCGC North America* 31 (3, March 1): 210–217. Accessed December 30, 2018 at http://www.chromatographyonline.com/measuring-gas-flow-gas-chromatography

Huber, M.L. and Harvey, A.H. (2018a). *Thermal Conductivity of Gases*. Accessed December 31, 2018 at https://ws680.nist.gov/publication/get_pdf.cfm?pub_id=907540

Huber, M.L. and Harvey, A.H. (2018b). *Viscosity of Gases*. Accessed December 31, 2018 at https://ws680.nist.gov/publication/get_pdf.cfm?pub_id=907539

Snavely, K. and Subramaniam, B. (1998). Thermal conductivity detector analysis of hydrogen using Helium carrier gas and HayeSep® D columns. *Journal of Chromatographic Science* 36 (April): 191–196.

Table

Figures

Symbols

bara	A bar absolute, a metric unit of pressure defined as 100 kPa and adopted as the "standard atmospheric pressure," equal to about 14.5 psia.
kg/m^3	A kilogram-per-cubic-meter, the standard international unit of density.
Pa	A pascal, the standard international unit of pressure.
Pa·s	A pascal-second, the standard international unit of dynamic viscosity: 1 Pa · s = 1000 cP (centipoise).
ppb	A part-per-billion of concentration.
ppm	A part-per-million of concentration.
psia	A pound-per-square-inch absolute, an American unit of pressure.
W/m·K	A watt-per-meter-kelvin, the standard international unit of thermal conductivity.

Note: The unit symbols defined herein may carry a prefix to adjust the size of the unit. Common prefixes are: k (kilo) meaning one thousand, m (milli) meaning one thousandth, or μ (micro) meaning one millionth.

New technical terms

When first introduced, these words and phrases were in bold type. You should now know the meaning of these technical terms. If in doubt, consult the Glossary at the end of the book:

adsorbent solid	**permeation**
electronic pressure controller	**porous polymer**
halogenated solvent	**vacancy peak**
molecular sieve	**velocity (of carrier gas)**
partial pressure	

For information on the names of chemical compounds, refer to the Glossary.

7

Sample injection

"The start of a chromatographic analysis is the injection of the sample. Everything else depends on it. The goal is to inject the same volume of sample every time. More specifically, we should always inject the same number of molecules. Any variation in the amount of sample injected worsens the repeatability of measurement: and yet, we might have a way to mitigate even that effect ..."

Introduction

Since we can't count or weigh the molecules going into the column; the best we can do is to inject a constant volume of them. To achieve that constancy, the physical sample volume must be repeatable – which is more difficult for very small samples – and each injection must be at the same temperature and pressure.

If the sample temperature or pressure varies, the number of molecules present in the sample volume will also vary and the measurements won't be repeatable.

Classical PGCs use mechanical valves for sample injection. Figure 7.1 pictures a typical example. This one is a slide valve, one of many types of valve reviewed in Chapter 8. Although mechanical devices tend to be unreliable, a valve is still the easiest way to inject a constant physical volume of sample fluid.

A PGC can be set up to accept gas or liquid sample injection. After many decades of development, injector valves for gas samples have greatly improved, with some now rated for several million operations. Injector valves for liquid samples can't yet match that reliability in service because process liquids tend to be more aggressive than process gases are.

PGCs using capillary columns may use pressure-balance techniques or solid-state devices to perform valveless sample injection. The next chapter describes these special techniques.

A liquid sample must be fully volatile. Most samples will instantly vaporize upon injection, but some may take a little longer. Of course, all molecules

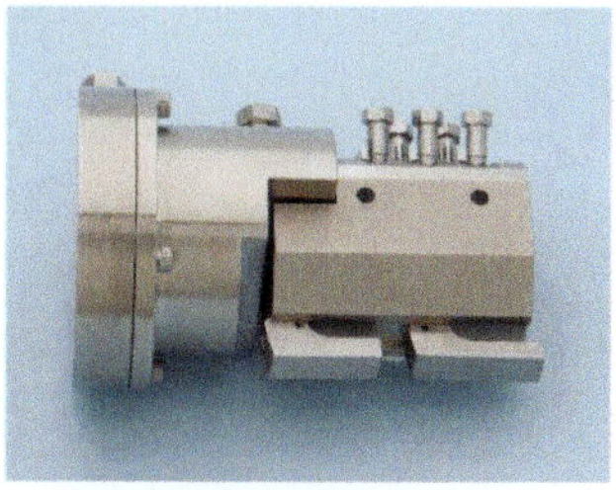

A typical two-position air-actuated sliding-plate injector valve with all ports on one side for easy access. Two levers on the bottom of the valve allow quick slider removal and replacement without disconnecting tubes. The valve may have eight or ten ports with many optional slider materials.

Figure 7.1 Typical PGC Sample Injector Valve.
Source: ABB Process Analytics Reproduced with permission.

Process Gas Chromatographs: Fundamentals, Design and Implementation, First Edition. Tony Waters.

in the sample must enter the vapor phase before they can progress along the column. Truly non-volatile components will accumulate in the injector valve, connecting tubes, or columns, and will eventually degrade the analyte separations.

Sample injector valves should inject the sample when energized. Then, a loss of electric power or actuating gas will not cause a spurious injection that might damage or contaminate one of the columns.

Injecting gas samples

Gas sample volume

For gas samples, typical sample volumes for packed columns are in the range of 0.1–0.4 mL, although larger volumes may be necessary for parts-per-million (ppm) analysis. Smaller samples down to about 30 μL may be necessary for capillary columns. For milliliter sizes, the physical sample volume is simply a short length of tubing attached to the injector valve and known as the **sample loop**.

Figure 7.2 is a schematic representation of a typical six-port gas injector valve. In the "sample fill" condition, sample gas flows continuously through the sample loop, thus ensuring a fully up-to-date sample composition. A sample flow of 30–50 mL/min is usually adequate to purge the sample loop.

A much larger flow is necessary to minimize time delay in the sample transport line, typically more than 1 L/min. This high flow rate cannot pass through the sample loop, so it's diverted to disposal in the **sample conditioning system**, prior to reaching the PGC.

The proper design of systems to extract, transport, and condition the sample before it reaches the injector valve are of vital importance to every PGC. Poor sample system design is often the root cause of PGC downtime. For complete design details, consult Waters (2013).

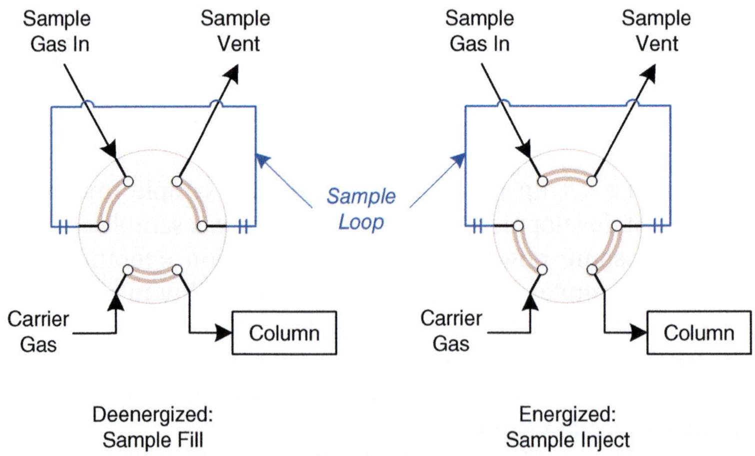

Showing the carrier and sample gas flow paths in a six-port rotary or diaphragm valve. In a rotary valve, the valve seat (pink) rotates 60° when the valve is energized. Other valves may have differing geometries but are functionally identical. Valves like this are also used for switching columns.

Figure 7.2 Gas Sample Injection.

When the sample valve moves to the "sample inject" position, carrier gas enters the sample loop, first compressing the sample in proportion to the pressure change, and then pushing it onto the column as a narrow plug. This **plug injection** is important. The sample molecules must stay close together as they enter the column, otherwise the peaks will spread, becoming wider and more difficult to separate.

In practice, the carrier gas compresses the sample gas in both directions, since the carrier gas in the column is at the same pressure as the carrier gas from the regulator. So, at the instant of injection, the carrier gas momentarily reverses out of the column causing a small pressure pulse that might spook a sensitive detector. Since the carrier gas compresses the sample volume from both directions, it starts its journey from the approximate center of the sample loop. Realizing this, the valve should stay in the inject position long enough to completely transfer the sample gas into the column.

The above discussion applies to typical injector valves that have two positions, like those shown in Figure 7.2. However, there are some diaphragm valves that allow the ports to open or close independently (AFP 2008). With this kind of valve, it's possible to pressurize the gas in the sample volume *before* injecting it into the column. If using this technique, the sample gas compresses from only one direction, and the sample instantly enters the column when the next port opens.

To minimize mixing, it's better to use a long thin sample loop than a short fat one. Table 7.1 lists some popular tubing sizes. Most PGCs now use a sample loop made from $^{1}/_{16}$-inch or $^{1}/_{8}$-inch tubing but an older PGC might use a larger size. If it's necessary to avoid adsorption of ppm-level polar molecules on the inside wall of the sample loop, use silicon-treated tubing.

Gas sample temperature

Isothermal PGCs usually have the sample injector valve in the column oven to hold it at constant column temperature, but some recent models

Table 7.1 Sample Loop Tubing Sizes.

Nominal Size	Outside Diameter	Wall Thickness	Internal Diameter	Unit Volume
			mm	mL/m
1/16″	0.063″	0. 020″	0.572	0.26
	0.063″	0.014″	0.876	0.60
1/8″	0.125″	0.035″	1.40	1.5
	0.125″	0.028″	1.75	2.4
	0.125″	0.020″	2.16	3.7
3 mm	3 mm	0.7 mm	1.6	2.0
	3 mm	0.5 mm	2.0	3.1
4 mm	4 mm	1.0 mm	2.0	3.1
	4 mm	0.7 mm	2.6	5.3

feature an optional temperature-controlled oven for the sample valve. When selected, this option allows the injector valve to operate at optimum temperature, which may not be the same as the column temperature. For example, mechanical valves tend to be more reliable when not too hot. Also, some samples suffer a chemical reaction if overheated; butadiene, for example, will polymerize at elevated temperature, so flowing that sample gas into a hot oven is not a good practice.

If the sample is a gas or vapor, it's easy to see the effect of a temperature change on measurement **accuracy**. When the injector operates at 60 °C (333 K), a change of one third of a degree (± 0.33 K) causes ± 0.1 % change to the sample amount, and this contributes ± 0.1 % to the overall uncertainty of the measurement.

Even with precise temperature control, it's important for the sample gas to reach injector temperature. To heat the incoming gas to the temperature of the sample injector valve, some PGCs have an inlet preheating coil while others have the incoming sample tubing in contact with a hot metal mass. It's worth checking. You may find your PGC has its sample inlet and outlet connections reversed, so the heating coil is on the outlet side. It doesn't do much good there. Running without the preheater may make an analyzer sensitive to sample flow rate, and that will result in a loss of measurement **precision**.

Do not exceed the specified rate of sample flow. If the flow rate is too high, it can exceed the heat transfer capability of the heat exchange device. The gas will then be cooler, and the sample volume will contain more molecules. To avoid making the measurements dependent on flow rate, do not exceed the sample flow rate recommended by the manufacturer.

Gas sample pressure

The pressure of a gas sample is critical. If the gas pressure in the sample loop increases, the injected sample volume will contain more molecules and all the measured values will increase; and *vice versa*. The analyzer response is directly proportional to the absolute pressure of the gas in the sample loop.

Most PGCs inject a gas sample at atmospheric pressure. To achieve this, older instruments allow the small sample flow to vent directly to the atmosphere. But even with this direct exit, the gas pressure in the sample loop may change slightly with flow rate, depending on the flow resistance of the exit path.

Most jurisdictions no longer allow the direct venting of gas samples. One location even required us to pump a stack sample back into the stack! Most PGCs now vent sample gas into a flare header or collection tank, and invariably suffer vent pressure variations.

The usual way to mitigate the vent pressure effect is by **atmospheric referencing**, sometimes called atmospheric balancing. A few seconds before each injection, the PGC actuates a valve to block the incoming sample flow and connect the sample loop directly to the outside air. Within a few seconds, the gas pressure in the sample volume equalizes with the atmosphere and is ready for injection.

It's a wise precaution to install a coil of narrow-bore tubing in the atmospheric vent line to provide a long path for any atmospheric gases diffusing back into the sample. Gas diffusion is surprisingly fast and can contaminate the sample. Set the referencing time to just a few seconds to avoid this happening.

PGCs use various valve systems to achieve atmospheric referencing; Figure 7.3 shows two examples. The common arrangement shown in Figure 7.3a is not the best way to do it because the atmospheric referencing valve (ARV) blocks the sample flow when referencing, and this may cause the gas pressure to increase in the supply line upstream of the valve. Figure 7.3b shows a better valve system that achieves pressure balance without disrupting the sample flow.

Atmospheric referencing works for gas samples that are slightly above or slightly below atmospheric pressure. If the sample is under partial vacuum, a reverse flow of outside air will flow into the system during atmospheric referencing. To prevent this backflow of air from reaching the sample volume, insert an additional coil of 1/8-inch tubing in the sample vent line.

Referencing doesn't compensate for natural variations in atmospheric pressure, which can be significant. Modern PGCs may have an absolute pressure sensor to monitor the true pressure of the sample, and digital routines to compensate for any variation found. As with any sensor or alarm, the sensing device must be very reliable. To prevent an unwarranted adjustment to the measured analyte concentration, the sensor and its software system should be at least 100 times more reliable than the PGC itself.

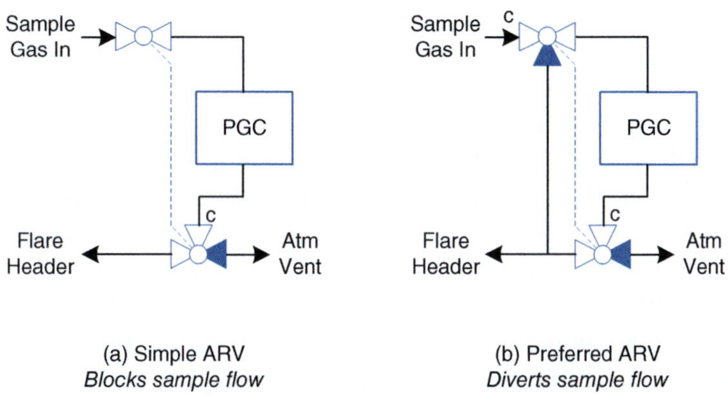

In both designs, the PGC actuates the ganged valves a few seconds before sample injection. This stops the sample flow and references the captured gas sample to the atmosphere. The design in (b) is preferred because it doesn't interrupt the sample flow. When the "flare header" line is under partial vacuum, install a coil of tubing between the analyzer outlet and the vent valve to prevent air from reaching the sample loop.

Figure 7.3 Atmospheric Referencing Systems.

Injecting liquid samples

Less preferred

It's inherently more difficult for a process gas chromatograph to analyze a liquid sample than to measure a gas sample. The difficulties relate to the **repeatability** of the extremely small sample size, the need to avoid gas bubbles, and the need for fast and complete vaporization of the injected sample. In addition, a liquid sample may require a higher column temperature or even temperature programming. Therefore, when both options seem equally valid, we prefer to sample a process vapor stream rather than a process liquid.

Vaporizing a liquid sample

When it's necessary to analyze a liquid sample with a gas chromatograph, there are two options; to vaporize a continuous flow of the liquid and inject a vapor sample, or to inject a liquid sample and allow it to vaporize after injection.

Volatile liquids containing only minor impurities are easy to vaporize, and vapor injection is best. More complex mixtures often have a wider boiling range, so the more volatile substances tend to evaporate first, followed by those less volatile. This process is known as **fractionation**, and typically produces variable measurements that cycle out of phase with each other (Waters 2013, 419–433).

When external vaporization is not a good option, consider direct liquid injection. For most applications using liquid injection, the sample must be fully volatile and leave no residue after the liquid sample vaporizes. In addition, the liquid must not react or decompose when exposed to the temperature of injector or column.

Ideally, the liquid sample should vaporize instantly after injection, so its molecules enter the column in the vapor phase. In some uncommon applications, though, the sample injector deposits a less-volatile liquid sample directly on the column and allows it to vaporize there. This usually works out okay, because the heavy molecules don't move down the column quickly, so not much peak spreading occurs. When running on a temperature program, the heavy molecules hardly move at all until the column gets hot.

The need for the sample to vaporize limits gas chromatography to analyzing only stable liquids that boil below about 350 °C. Yet it's possible to measure even higher boilers in a sample when they are present at low concentrations. Although this is a very large range of chemicals indeed, there are many liquids too involatile for analysis by gas chromatography.

In practice, most of the PGCs using liquid sample injection are analyzing samples that boil below 150 °C, approximately equivalent to nonane (C_9). Such samples rapidly vaporize, allowing isothermal separation of the analytes. For a sample of higher boiling point, temperature programming may be the better option.

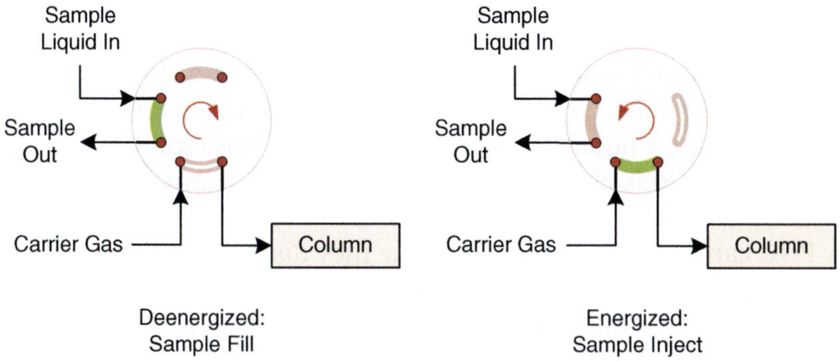

Figure 7.4 Liquid Sample Injection.

Liquid sample volume

Liquid injector valves are often of similar construction to gas injector valves but designed to inject a much smaller sample. The goal is to inject about the same number of molecules as gas sampling valves do, and that's why liquid sample volumes are so small. For $C_3 - C_4$ hydrocarbons, for example, a volume of liquid contains about 300 times as many molecules as the same volume of vapor.

Chromatographers often talk about injection volumes in microliters, which to most people is not a familiar unit. If you keep in mind that one microliter (μL) equals one cubic millimeter (mm^3), a microliter is easier to visualize.

When using packed columns, a liquid sample valve may need to inject less than 1 μL. Mostly, this is easy to do, since mechanical valves can reliably inject as little as 0.035 μL. Yet even this small amount is too much for a narrow-bore capillary column, and for those we must resort to the delicate sample-splitting technique discussed in the next chapter.

Because of the small volume needed, injector valves for liquid samples cannot have an external sample loop. Instead, the sample volume is a slot machined into the valve rotor, or a small hole in the slider. Figure 7.4 shows the typical flow paths through a rotary or diaphragm valve.

Liquids having a high boiling point, and those that decompose when heated, are difficult to inject successfully. Equally difficult are liquids of wide-boiling range. For these applications, the PGC industry has developed special plunger valves like the one shown in Figure 7.5. These valves take a small sample from a cold liquid stream and quickly transfer it into a hot zone, thereby ensuring rapid vaporization. The next chapter reviews these and other chromatographic valves.

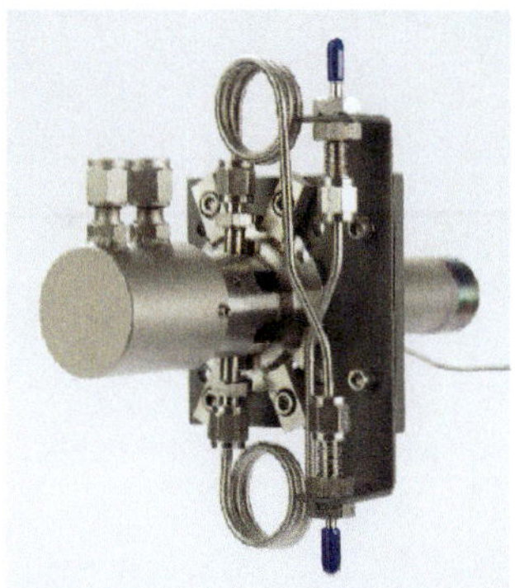

This liquid injector valve mounts through the oven wall, with the sampling section outside and the vaporizing section inside the column oven. Air actuates a piston that injects the sample into the vaporizer section. The sample size is typically between 0.13 μL and 0.50 μL. The tubing loops on the inlet and outlet ports are just for stress relief.

Figure 7.5 Plunger Valve for Liquid Sample Injection. *Source: ABB Process Analytics Reproduced with permission.*

Liquid sample temperature

To ensure rapid vaporization of the injected sample, it may be necessary to heat the injector valve. Yet the process fluid in the metered sample volume must be completely liquid. Heating a liquid may release some dissolved gas, and when a liquid is close to its bubble point temperature, it may start to bubble.

A PGC can't tolerate bubbles because they displace liquid from the tiny sample volume, so the injected volume contains fewer molecules. A visible symptom of bubbles in the injected sample is the random occurrence of low peak areas for every peak on the chromatogram.

To avoid bubbles, you will need to increase the sample pressure or reduce the sample temperature. Having a higher sample pressure is less of a problem now than it used to be, but you may need to install a new sample injector valve; a modern injector can tolerate ten times the pressure than its antecedents could.

Sample temperature depends on the location of the valve. In early PGCs, the sample injector valve was in the column oven. In many applications this worked well enough, but the column oven is too hot for light hydrocarbon liquids like propane or butane, and it may not be hot enough to rapidly vaporize a heavy oil. Injector valves now have their own separate temperature-controlled enclosure, controlled at the optimum temperature for the application.

The ability of a modern PGC to accept a higher sample pressure and optimize the injector valve temperature extends the range of process liquids it can analyze. For example, we previously had to prevaporize a light hydrocarbon stream and analyze it as a vapor. Now, we can inject many such samples in their native liquid state, thus avoiding the oft-troublesome procedure of vaporizing a continuous flow of the process liquid.

It's often thought that liquid sample temperature has negligible effect on the performance of a PGC, but this is not entirely true:

> *Temperature change has two effects on a liquid sample injection; it affects the number of molecules injected and their subsequent rate of sample vaporization.*

Most liquids expand slightly with an increase of temperature, and the resulting drop in density means there are fewer molecules in a constant volume of hot liquid. For example, Table 7.2 shows the effect of temperature and pressure on the density of liquid hexane. From this, a temperature change of 7 °C is enough to cause a 1 % change of liquid density, resulting in 1 % error in the measurement of every analyte.

Note, however, that the physical volume of the sampler may also expand with temperature, nullifying the effect of liquid expansion.

When heating the liquid sample valve, you should ensure that the sample liquid comes to full injector temperature before injecting each sample. It takes more energy to heat a liquid stream than a gas stream, so the flowing liquid may not warm to injector temperature before it arrives at the valve.

Table 7.2 Temperature, Pressure, and Liquid Density.

Temperature (K)	Pressure (bara)	Density (kg/m^3)	Compression Rate
293.15	1142.47	671.6	0.00082 % per bar
293.15	3236.19	797.2	
303.00	1	650.9	−0.15 % per kelvin
323.00	1	631.9	

* *Compression rate is average density increase per bar or kelvin.*
Illustrative data for n-hexane from Borzunov et al. (1970, 146–152) – pressure and Mekhtiev et al. (1975, 64–100) – temperature.

Then, any change in the flow rate will also change the liquid temperature in the injector, causing small variations in the number of molecules injected. To avoid this problem, reduce the sample flow rate until the measurements become stable.

Be particularly careful with the calibration sample; at one jobsite, we saw errors in calibration because the calibration liquid was colder than the process liquid.

A cool injector valve may also lead to slow vaporization, causing wide peaks on the chromatogram. If your peaks are wider than expected, reduce the liquid flow rate to about 10 mL/min and try again.

It might also be important to have a low rate of sample flow when the injector valve is in the column oven. The additional thermal load from a high liquid flow could upset the oven temperature control.

Liquid sample pressure

As you can see from Table 7.2, a big change in liquid sample pressure doesn't much affect its density, nor the number of molecules contained in a constant volume. Nevertheless, sample pressure is of vital importance in a liquid sampling system; it must be high enough to prevent bubbling, but low enough to prevent valve leakage.

For volatile samples prone to bubbling, it might be effective to block the liquid exit just before sample injection. Blocking the sample flow can cause the liquid pressure in the injector valve to rise slightly, thereby inhibiting bubble formation.

In early PGCs, we had to keep liquid sample pressure below carrier gas pressure for fear of liquid leaking into the columns. Thankfully, those days are past, and we now have liquid injector valves that can operate at high sample pressures, some as high as 70 barg. This is a tremendous advantage as it allows the PGC to inject a volatile liquid at high pressure, negating the need for prior vaporization.

Of course, the high pressure must come from somewhere, and we would never recommend a sample pump, as they tend to be unreliable. Ideally, the process tap should be on a high-pressure process line, and the sample transport line should not drop too much pressure.

When injecting a high-pressure liquid, the physical volume of the sample slot may change slightly with pressure. To maintain a constant sample volume, ensure that the sample liquid remains at constant pressure. If the process pressure is variable, regulate the sample pressure entering the injector valve, but don't drop the pressure more than necessary.

Ideally, the calibration liquid should be at the same pressure as the sample liquid.

Other techniques

Sample splitting

Some capillary columns demand a very small sample volume. For a gas sample, one can use a conventional sample injection valve having an internal slot or hole as the sample volume, instead of a sample loop. But a liquid sample injection would require an impossibly small sample volume, so a special plunger valve is available for liquid applications. As detailed in the next chapter, this special liquid injector has a heated vaporizing chamber and an optional flow splitter that vents most of the injected sample, allowing only a small portion to enter the column.

Remote sample injection

When the sample fluid is highly toxic or is difficult to transport by normal means, it may be possible to inject the sample remotely. In this unusual technique, the sample injector valve is located close to the process sampling tap. After injection, the carrier gas transports the slug of injected sample to the remote analyzer through a long length of $^1/_{16}$-inch tubing. Upon arrival at the analyzer, the sample enters the column, which separates the analytes in the normal way (Converse and Conrad 2004).

The primary risks in this approach are that the injection band will broaden excessively before it reaches the column or that the peaks will begin to tail. Therefore, the ideal sample transport line would be short and narrow, and fabricated from surface-deactivated tubing.

Normalization

Normalization is a special calibration procedure further discussed in Chapter 14. Although normalization is not a sample injection technique, it's relevant to the present discussion as it provides a way to mitigate the effect of variation of the injected sample volume.

In effect, the normalization technique estimates the total number of molecules injected and adjusts the measurements accordingly. The idea is to measure the concentration of everything in the sample and then factor the results, forcing them to sum to 100 %.

For example, if the concentrations of all components sum to 97.3 %, The PGC will divide each concentration by 0.973 to ensure the total is exactly 100.0 %.

Before normalization, the PGC should check that the total of all measurements is within a preset limit of 100 %. If not, the processor should abort the normalization procedure and issue an alarm. It would be a mistake to normalize results unlikely to be representative of the whole sample analysis.

Normalization can be a powerful technique to improve PGC method precision, as it effectively eliminates all variability associated with changes in the amount of sample injected. However, normalization has its own peculiar errors. It assumes that the PGC has measured all substances present in the sample, and that the calibration is perfect for every one of them. In practice, neither of those assumptions are likely to be true, so always apply normalization with care.

Paradoxically, the inappropriate use of normalization can reduce measurement accuracy by trying to improve it.

SCI-FILE: *On Analytic Units*

A fable

Suppose we send a sample to the laboratory and ask how much of a certain analyte it contains. The analyst carefully measures the amount of analyte in the sample and sends a report. Let's say the result is 12.543 mg.

Even if it's very accurate, that measurement is useless because the sample size is unknown. The report tells us exactly how much analyte the sample contains, but it tells us nothing about the process that the sample came from.

To make the measurement useful, the analyst should show the amount of analyte as a fraction or percentage of the sample size. Only then, can it tell us something about the process fluid.

To obtain a percentage, the size of the sample and the amount of analyte it contains must be in the same units. In the above example, suppose the sample size was 568.44 mg. Once we know that, we can divide the weight of analyte by the weight of original sample, and then multiply the ratio by 100 to get a percentage by weight:

$$conc = 100 \times \frac{12.543\ \text{mg}}{568.44\ \text{mg}} = 2.204\ \%\ \text{wt}.$$

When expressed as a ratio to the original sample size, the analysis result becomes useful. Assuming good sampling technique, it now accurately represents the concentration of the analyte in the process fluid.

Constant sample size

Sample injection into a PGC is analogous to the fable recounted above. It's the same as sending a sample to the laboratory and getting a numerical response. The PGC doesn't know the sample size, so it can't calculate the analyte percentage from the detector response.

To overcome this drawback, the PGC adopts a strategy that would be difficult to follow in the laboratory. A PGC sample injector always sends precisely the same amount of process sample to its "analyst" (*i.e.* into its columns and detector) and monitors the response. Then, during the process of calibration, it sends instead a sample of known composition, and stores the value of each measured component.

Provided that each sample injection is exactly the same amount, the PGC can use the stored value for each component in the calibration sample to calculate the percentage of that component in the process samples.

In other words, the PGC compares a process sample to its last calibration sample and reports the difference. It has no absolute calibration at all.

Therefore, the performance of the PGC sample injector affects both precision and calibration. Without precise sample injections, a PGC can be neither precise nor accurate.

Different ratio units

Since concentration is always a ratio, the basic way to express it is simply a fraction. The fundamental unit of concentration is thus the mole fraction, typically shown as a decimal. So, if half the molecules in the sample are methane, the mole fraction (χ) of methane is 0.50 whatever the other molecules are:

$$\chi = 0.50$$

Chemical engineers prefer mole fraction and use it to state process composition. Clearly, the mole fractions of all stream components must accumulate to unity. To convert mole fraction to mole percent, multiply by 100:

$$100\chi = 100 \times 0.50 = 50\,\%$$

From this, we see that the percent sign % is simply a factor with the constant value 0.01 (no units):

$$50\,\% = 50 \times 0.01 = 0.50$$

For measuring smaller concentrations, we define other factors such as "ppm" or "ppb," where ppm is a part per million (10^{-6}) and ppb is a part per billion (10^{-9}). Thus:

$$50\text{ ppm} = 50 \times 10^{-6} = 0.000\,050$$
$$50\text{ ppb} = 50 \times 10^{-9} = 0.000\,000\,050$$

All the above ratios are based on moles – the actual number of molecules present. To know that, we would have to count them, but we don't know how to do that yet. Instead, we can measure their volume, or their weight.[1] Thus, three kinds of ratio unit are possible:

- Mole fraction, percent, ppm, or ppb.
- Volume fraction, percent, ppm, or ppb.
- Weight fraction, percent, ppm, or ppb.

In all these ratios, the analyte measurement must be in the same units as the sample size measurement. Then, the units cancel out, and the final measurement is a unitless quantity.

Unfortunately, a ratio measurement based on mole, volume, or weight has quite different values, even when it's the same sample. Therefore, you should never report an analysis as simply percent, ppm, or ppb. The report must clearly specify that the ratios are by mole, by volume, or by weight.

Because of the potential confusion about units, some analysts prefer to retain the original units, rather than canceling them out.

On this basis, for instance:

- 2.3 % (vol) = 23 mL/L
- 2.3 ppm(mol) = 2.3 μmol/mol
- 2.3 ppb(wt) = 2.3 μg/kg

This approach goes even further for the analysis of aqueous solutions. Such analyzers often measure their concentrations in mixed units such as gram per liter (g/L). Informally, "1 ppm" in an aqueous solution often means 1 mg/L. Since the density of water is about 1 kg/L, this approximates to 1 ppm(wt).

Injected quantity

From the above discussion, it should be clear that the quantity of sample taken for analysis must be constant *in the units of the measurement:*

- For a **mole percent** analysis, we would need to always inject the same number of molecules.
- For a **volume percent** analysis, we would need to always inject the same volume of sample.
- For a **weight percent** analysis, we would need to always inject the same weight of sample.

[1] Laboratory analysis is often done by weight because it's easy to weigh stuff.

In practice, all PGCs inject a constant volume because it's the easiest parameter to keep constant. So, all PGCs really measure volume percent. Yes, really, they do.

To calibrate a PGC in mole percent or weight percent requires some careful thought.

Luckily, the number of molecules in a gas sample at constant temperature and pressure is almost exactly proportional to its volume. If analyzing a gas sample, therefore, it's okay to calibrate in mole percent.

Weight percent is not so easy. To accurately measure weight percent, the PGC would have to inject samples of predictable *weight*. For a constant volume sample injector, this means that the sample density must be constant.

A gas or liquid sample that contains a high percentage of one component will have a nearly constant density, regardless of composition. In this case, a volumetric injection would have an adequately constant weight, making it possible to calibrate in weight percent.

This is also true for process samples that contain a high percentage of similar substances, like the C_4 hydrocarbons, for example. But the density of samples with a wider range of components may vary with composition, leading to measurement error.

Lastly, be aware that some liquids change their volume and hence their density when mixed together, making calibration difficult in any units. For such samples, a volumetric injection won't guarantee the same number of molecules in each injection, so a custom non-linear calibration may be necessary.

Conversion of units

It's sometimes necessary to convert an analysis given in weight percent to the equivalent values in volume or mole percent. This conversion is possible only if the full analysis is known and adds up to approximately 100%.

For a gas sample, the conversion is easy. First, retrieve the molar mass of each component from the internet. Then follow Table 7.3, which illustrates the math using simple numbers. It's easy to implement this table on a spreadsheet.

If starting with weight percent, divide the percentage of each component by its molar mass (as a memory aid, think of it as eliminating the weights). If starting with mole or volume percent, multiply by the molar mass instead. Then, recalculate the percentage of each component as shown in Table 7.3.

The results might surprise you. Volume percent and weight percent can be very different from each other!

The calculation of mole percent in Column E of Table 7.3 is exact. For volume percent, the table relies on the ideal gas laws. Low-pressure gas samples certainly behave as ideal gases, and their volume percent is equal to their calculated mole percent. But if you are looking at a high-pressure process or calibration gas, that may not be true, and you would need to consider the compressibility factor of each component gas in the mixture.

For liquids, you can use Table 7.3 to convert between weight percent and mole percent. Divide weight percent by molar mass to get mole percent and multiply mole percent by molar mass to get weight percent.

To get liquid volume percent from Table 7.3, it should be possible to use the liquid density of each component in place of its molar mass, and this may work for "ideal liquids" like two similar hydrocarbons. For other liquids, though, volume percent is an unreliable measurement because the total volume of a liquid mixture may not be the same as the sum of its component volumes.

Yet a PGC still injects a constant volume. In the rare case that the liquid mixture is non-ideal, conversion to weight or mole percent may be impossible, and the response to certain components may depend on the concentration of other components present in the sample.

In some applications, it might be feasible to measure the other components and adjust the analyte calibration to allow for their presence – perhaps by using a chemometric technique. Good luck with that.

Table 7.3 Conversion of Percentage Units for Gas Samples.

	A	B	C	D	E	F	G
1	**CONVERT UNITS:**		**Weight % to Volume %**		**Volume % to Weight %**		
2	**Sample Analysis**	*Analyte Molar Mass*	*By Weight or by Mass*	*Analyte Quantity*	*By Mole or by Volume*	*Analyte Mass*	*By Weight or by Mass*
3	*Units:*	*g/mol*	*%*	*Mol/100g*	*%*	*g/100mol*	*%*
4	Hydrogen	2.0	30.0	15.0	60.0	120.0	30.0
5	Helium	4.0	36.0	9.0	36.0	144.0	36.0
6	Hydrogen Sulfide	34.0	13.6	0.4	1.6	54.4	13.6
7	Carbon Monoxide	28.0	14.0	0.5	2.0	56.0	14.0
8	Sulfur Dioxide	64.0	6.4	0.1	0.4	25.6	6.4
9	TOTALS:		100.0	25.0	100.0	400.0	100.0
	Spreadsheet Calculation:	*Row n:*	*Enter values*	*Dn = Cn/Bn*	*En = 100Dn/D9*	*Fn = En*Bn*	*Gn = 100Fn/F9*
		Row 9:	*sum(C4:C8)*	*sum(D4:D8)*	*sum(E4:E8)*	*sum(F4:F8)*	*sum(G4:G8)*

Enter the weight % analysis in column C or the volume percent analysis in column E. The total in C9 or E9 must be close to 100 % or you cannot proceed. Enter the molar mass of each component in column B to the accuracy desired (approximate values are used here for ease of understanding). The extreme difference between weight and volume percent in this example is due to the wide range of components chosen and is somewhat atypical of most samples.

Knowledge Gained

Practice

- *Most PGCs use mechanical valves for gas sample injection, and these are now highly reliable.*
- *A PGC can inject a gas sample or a fully-volatile liquid sample.*
- *Liquid injection valves are less reliable, as process liquids tend to be more aggressive than gases.*
- *For a gas injector, the sample volume is an adjustable length of small-diameter tubing.*
- *The sample volume for a liquid sample must be about 300 times less than for a gas sample.*
- *Liquid sample volume depends on the hole or slot size in the valve slider, rotor, or plunger.*
- *To ensure repeatable analysis, the temperature and pressure of the injected sample must be constant.*
- *Some PGCs preheat the sample to constant temperature before it enters the injector valve.*
- *Briefly connecting a gas sample to the outside air ensures constant pressure before injection.*
- *A PGC may employ a barometric sensor to compensate for changes in atmospheric pressure.*
- *A PGC may inject a liquid sample as a vapor (after vaporization) or in its natural liquid state.*
- *Always inject a sample with a wide boiling range as a liquid, using a special injector valve.*

- *A liquid injector may include a* ***sample splitter*** *to vent most of the injected sample to waste.*
- *PGCs can analyze samples boiling up to 350 °C, but most PGC column ovens operate at 150 °C or less.*
- *Many PGCs have the injector in the column oven, but some models control its temperature separately.*
- *Liquid sample volume changes only slightly with temperature or pressure changes.*
- *Rarely, the injector is located at the process tap and the carrier gas transports the sample to the PGC.*
- *Normalization of results can sometimes compensate for variations in injected sample volume.*

Theory

- *A percentage is a ratio of two values and has no unit of measurement.*
- *The "%" sign is just a multiplier, it's a constant having the value 0.01.*
- *The value of a ratio measurement depends on the units being ratioed: volumes, moles, or weights.*
- *For the same sample, volume percent, mole percent, and weight percent may all have different values.*
- *To measure volume percent, an analysis must start with a known or constant sample volume.*
- *To measure mole percent, an analysis must start with a known or constant number of molecules.*
- *To measure weight percent, an analysis must start with a known or constant sample weight.*
- *A PGC measures a small volumetric sample of the process fluid.*
- *As a volumetric device, a PGC inherently measures the analytes in units of volume percent.*
- *In a gas sample, mole percent is essentially equal to volume percent, so mole percent calibration is okay.*
- *Convert weight percent of a gas to mole percent or vice versa using the molar mass of each component.*
- *Molar mass also works for converting between liquid weight percent and liquid mole percent.*
- *The total volume of a liquid mixture doesn't always equal the sum of the component volumes.*
- *Since all PGCs inject a constant volume, liquid samples may require a custom calibration.*

Did you get it?

Self-assessment quiz: SAQ 07

Q1. What kind of sample would use an external sample loop?
Q2. What does "%" mean?
Q3. What are two necessary conditions of the sample to allow an accurate calibration of a PGC in volume percent concentration units?

Q4. What is a necessary condition of the sample that would allow an accurate calibration of a PGC in weight percent concentration units?

Q5. How is it possible that the flow rate of the sample gas might affect the operation of a PGC?

Q6. If a liquid sample has a wide boiling range, would it be best to inject it as a liquid, or vaporize it and inject the vapor?

Q7. Consider an equal-volume gas mixture of the first four paraffins:

Methane	CH_4	25 %vol.
Ethane	C_2H_6	25 %vol.
Propane	C_3H_8	25 %vol.
n-Butane	C_4H_{10}	25 %vol.

Convert this analysis to weight percent.
Assume the atomic mass of hydrogen is 1.0 and the atomic mass of carbon is 12.

Check your SAQ answers with those given at the end of the book.

Student evaluation test: SET 07

Your instructor will provide answers to these questions.

S1. Would it be better for the sample injector valve to inject the sample when energized, or when deenergized?
Select one option:

- **A.** Inject at the instant the valve energizes.
- **B.** Inject at the instant the valve deenergizes.
- **C.** Neither of the above.

S2. What is a sample loop?
Select one option:

- **A.** It's a way to divert a large sample flow so that it doesn't all have to pass through the sample injector valve.
- **B.** It's a coil of tubing on the inlet side of a gas sample injector valve to allow the gas to reach oven temperature before injection.
- **C.** It's a bypass circuit that allows the flow of a sample fluid to continue while the sample injector valve is injecting the sample.
- **D.** It's a short length of tubing between injector valve ports that determines the volume of sample gas the valve injects.

S3. If a process gas chromatograph has a sample volume of 0.1 mL, what kind of sample fluid would it most likely need?
Select one option:

- **A.** A gas or vapor sample.
- **B.** A liquid sample.

S4. Compared with a 0.15 mL gas injection of C_3–C_4 hydrocarbons, what is the equivalent sample volume if injected as a liquid?
Select the one best option:

- **A.** About 0.15 μL of liquid.
- **B.** About 0.5 μL of liquid.

C. About 1.0 μL of liquid.
D. About 4.5 μL of liquid.

S5. Which statement below best explains the function of an atmospheric referencing valve?
Select one option:

A. Ensures that an injected gas sample is always at exactly the same pressure.
B. Ensures that an injected gas sample is always at about the same pressure.
C. Ensures the injection of a constant volume of a liquid sample.
D. Prevents the formation of bubbles in a liquid sample volume.

S6. Which two of the statements below are valid reasons why a PGC is now less likely to vaporize a liquid sample flow and then inject the vapor than was common in the past?
Select two options:

A. Process plants now have fewer applications for process liquid analysis than they used to.
B. In a modern process gas chromatograph, the injector valve can have an integral heater to vaporize an injected liquid sample.
C. It's no longer common for process gas chromatographs to measure light hydrocarbon streams.
D. Sample injector valves now have a higher pressure rating than they used to have.

S7. If a well-calibrated PGC measures all components in a process sample and the sum of those measurement values is 96 %vol., what is the normalized concentration of an analyte whose original measurement is 24 %vol.?

S8. If using a calibration gas cylinder containing 12.0 % by weight of hydrogen (H_2) and a balance of propane (C_3H_8), what percentage of hydrogen would you expect a PGC to measure in this gas when it's correctly calibrated to measure hydrogen as volume percent?

In this question, assume the atomic mass of hydrogen is 1.0 and the atomic mass of carbon is 12.

S9. PGCs inherently measure volume percent. When the PGC injects liquid samples, which of the circumstances listed below would *always* allow accurate calibration of the analytes in any chosen concentration units; volume, mole, or weight percent?

This is a challenging question. You may have to apply knowledge gained in other sections of the course. Treat each option independently; you don't need to consider potential interactions between them.
Select all that apply:

A. When the volume of the sample mixture differs from the sum of the volumes of the individual liquid components.
B. When the PGC groups several components together and measures them as a single analyte.
C. When the liquid sample density is essentially constant and the PGC doesn't measure one or more of the main components of the sample.
D. When the PGC is measuring only a single impurity in a 99.99 % pure process liquid on a range of 0–100 ppm.

References

Cited

AFP (2008). *DV3-series positive port shut off diaphragm valve for analytical instrumentation*, Product Brochure. Thetford Mines, QC, Canada: Analytical Flow Products, a division of Mécanique Analytique Inc.

Borzunov, V.A., Razumikhin, V.N., and Stekolnikov, V.A. (1970). Bestimmung der Dichte von n-Hexan und Wasser bei Drücken bis 10000 kg/cm^2. *Teplofiz. Svoistva Vesh. Mater.*

Converse, J.G. and Konrad, K.K. (2004). Relayed remote discrete sampling. In: *Proceedings: ISA Analysis Division Symposium – Louisville 2004*. Research Triangle Park, NC: International Society of Automation.

Mekhtiev, S.I., Mamedov, A.A., Khalilov, S.K., and Aleskerov, M.A. (1975). Experimentelle Untersuchung des Einflusses von Octylmethacrylat auf Viskosität und Dicht der Kohlenwasserstoffe, *Izv. Vyssh. Uchebn. Zaved. Neft Gaz.*

Waters, T. (2013). *Industrial Sampling Systems*. Solon, OH: Swagelok Company.

Tables

Figures

Symbols

$[A]$	The concentration of analyte A.
bara	A bar absolute, a metric unit of pressure defined as 100 kPa and adopted as the "reference atmospheric pressure," equal to about 14.5 psia.
g/mol	A gram per mole, the unit of molar mass.
kg	A kilogram, the standard SI international unit of mass.
kg/m^3	A kilogram per cubic meter, the standard SI international unit of density.
K	A kelvin, the standard SI international unit of absolute temperature, equal to one Celsius degree.

L	A liter of volume, being one thousandth of a cubic meter.
m	A meter, the standard SI international unit of length.
min	A minute of time.
mol	A mole, the standard SI international unit of quantity ($\sim 6.022*10^{23}$).
ppb	A part per billion of concentration.
ppm	A part per million of concentration.

NOTE: Unit symbols can have a modifying prefix, commonly: k (kilo) meaning one thousand, m (milli) meaning one thousandth, or μ (micro) meaning one millionth.

New technical terms

When first introduced, these new words and phrases were in bold type. For further information, refer to the Glossary at the end of the book.

accuracy
atmospheric referencing
fractionation
mole percent
normalization
plug injection
precision
repeatability
sample conditioning system
sample loop
sample splitter
volume percent
weight percent

For information about chemical names, consult the Glossary.

A liter of volume, being one thousandth of a cubic meter

A meter, the standard international unit of length

... time

8

Chromatographic valves

"The incorporation of several special-purpose valves sets a process gas chromatograph apart from most laboratory gas chromatographs. For autonomous process analysis, a special valve is necessary to reliably inject a sample on demand. In addition, as we'll soon see, a PGC needs similar valves to switch selected peaks into different columns".

Valve technology

Evolution

The late 1950s was a time of vibrant PGC technology advancement. The original developers of PGC hardware were the research engineers at end-user companies like Union Carbide, Phillips Petroleum, Imperial Chemical Industries, and Shell Oil. These pioneers soon realized that sample injection by ordinary instrument valves was inadequate for all but the simplest percent-level separations. To improve the performance of their prototype PGCs, they evolved three major valve mechanisms for injecting samples: linear valves, rotary valves, and diaphragm valves. The linear category includes spool valves, slide valves, and plunger valves. The descendants of those early valves reside today in every online process gas chromatograph.

The sample injector is in constant contact with the process fluid, so the selection of contact materials is of vital importance. In many applications, stainless steel, and Teflon™ are adequate – the latter sometimes impregnated with glass fiber or carbon. For more corrosive samples, an injector valve may incorporate parts made of **Hastelloy™** or **Monel™** (Clemons 2016).

Nearly all chromatographic valves are binary devices having just two positions. Most operate from instrument air, but a few use carrier gas pressure. Special-purpose PGCs destined for remote areas, such as pipeline monitoring stations, may use valves with electric solenoid actuators. Most valves have a spring return, but a few use twin actuators to drive the valve

Process Gas Chromatographs: Fundamentals, Design and Implementation, First Edition. Tony Waters.

in both directions. A PGC may use several of these valves to inject samples and switch gas flows between columns.

The sample injector valve is a critical component in a process gas chromatograph. Consider our expectations:

- It must inject precisely the same volume each time it operates, without noticeably interrupting the carrier gas flow.
- It must have low internal volume and a clean flow path devoid of cavities that might occlude sample molecules.
- It must never leak, even in continuous contact with a potentially corrosive or abrasive process fluid.

It would be hard enough for any mechanical device to comply with such stringent demands, yet we also expect the valve to operate day and night without failure for at least one year. For a one-minute analysis, that's 500,000 cycles.

Any valve technology that satisfies these requisites for effective sample injection will also qualify as a means for switching columns. So, for production economy, PGCs tend to use the same type of valve for column switching as they use for sample injection.

The strange effect of competition

Clearly, the reliable and precise injection of samples is a prime requirement for any automated gas chromatograph. Realizing this, the early PGC vendors tended to oversell the merits of their preferred style of injector valve to such an extent that each brand name became synonymous with a standard valve technology, leaving no possibility of retreat and constraining their development options for decades. Yet all their chromatographs worked, and for a long time there was never any defining performance distinction between them; they all flourished in direct competition during the last four decades of the 20th century. Given this evidence, it would be reasonable to conclude that the three major valve technologies are essentially equivalent.

Of course, each valve technology has it limits and experiences occasional failure. The sample injector valve is a critical component; it's the single point of contact between the raw process fluid and the complex internal instrumentation of the PGC. Some failures were inevitable. In retrospect, though, it's probable that each vendor's fixation on using a single valve technology for all applications worsened the failure rate.

It's easy to replace a valve, but the labor and parts are expensive, not to mention the downtime. The cost of valves and their ease of replacement opened the door to third-party competition, notably from Valco Instruments and Analytical Flow Products. These companies developed superior valves, often using diaphragm technology, and some end users retrofitted them into their existing process chromatographs.

While third-party valve suppliers were gaining experience in the retrofit market, PGC vendors were busily incorporating the latest column technology into their new PGC models. We shall see that this new column technology demanded performance beyond the capabilities of the traditional injection and column valves. By the turn of the millennium, the slide valve was heading for extinction, and the rotary or diaphragm valve (often from those same third-party sources) became the valve of choice.

Valve types

Solenoid instrument valves

Early PGC developers devised custom assemblies of readily-available **solenoid valves** to inject gas samples. But they soon realized that the unpurged passages (called **dead legs**) that are present in any assembly of off-the-shelf instrument valves cause gross peak widening, which spoils the separation of peaks. Thus, it soon became evident that PGCs need special valves for efficient sample injection and it was safer and more convenient to use air pressure to power these new valves. Nevertheless, you may find that a PGC dedicated to the analysis of natural gas in pipelines uses proprietary valves powered by electric solenoids.

Spool or piston valves

The **spool valve**, sometimes called a **piston valve**, is a linear-motion device used in many early models of PGC for gas sample injection and column switching. The Bendix C5 valve is a good example (Bendix 1972). If you look back to Figure 5.1, it's easy to spot four of these spool valves installed in a process gas chromatograph by their distinctive black actuators.

Figure 8.1 shows how the routing of carrier and sample gas through the valve depends on the position of the spool, which is a grooved rod fitted with several O-rings to seal off annular spaces between the valve ports. To operate the valve, a pneumatic or electric actuator moves the spool by a small fixed distance to change the flow routing through the device, thus injecting a sample or switching columns. A spring (or another actuator) returns the spool to its rest condition.

Their inherently large internal volume precluded the use of spool valves to inject liquid samples; they were suitable for gas samples only.

While initially successful for injecting large gas samples into wide-bore columns, the spool valve failed to provide the plug injection needed for smaller samples. The gas path through the valve was an irregular shape that was difficult to purge, so the valve never achieved a true plug injection. Instead, it tended to disperse the pack of molecules on their way to the column, mixing them with a little carrier gas. Sadly, a dispersion like that always makes the chromatogram peaks wider and more difficult to separate.

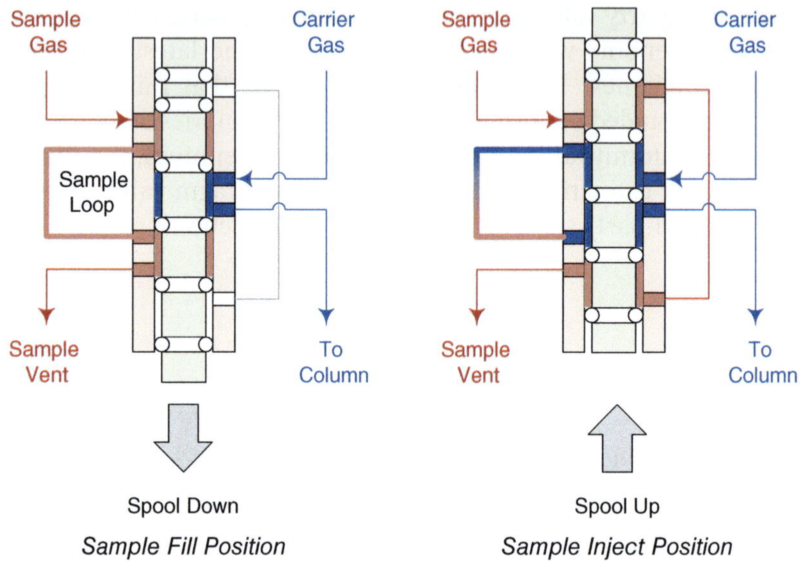

In the sample fill position, the sample gas flows through the sample loop and out to vent. Meanwhile, the carrier gas is flowing into the column. When the spool moves to the inject position the carrier gas flows through the sample loop, carrying the sample gas into the column.

Figure 8.1 Function of a Spool Valve.

The same issue occurred in the use of spool valves for column switching. As PGCs started to use packed columns with smaller internal diameters, the large internal volume of the valve caused unacceptable peak broadening.

Due to these shortcomings, the spool valve is now obsolete. Yet in its heyday, the valve was quite popular, so you might find one at an older jobsite. The Greenbrier subsidiary of Bendix continued to use their six-port and eight-port C5 valves until they changed to sliding-plate valves (Bendix 1975). Their choice of slide valves set a long-lasting precedent for their successors, ABB Process Analytics.

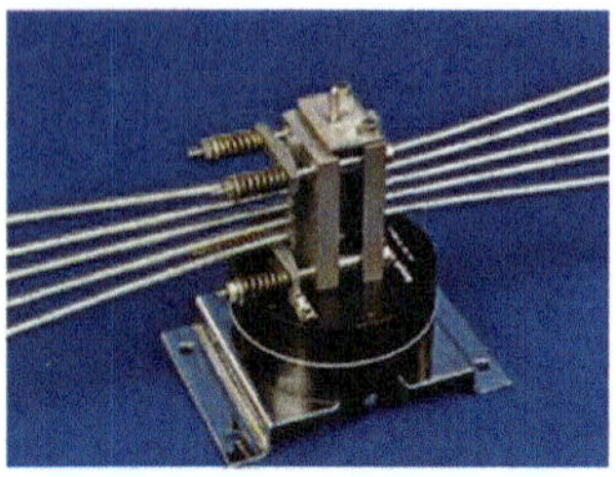

A Servomex Model 400 slide valve for sample injection or column switching (Servomex 1985).

Figure 8.2 Early Six-Port Slide Valve.
Source: Servomex Group Ltd. Reproduced with permission.

Slide valves

The **slide valve** (also called a sliding-plate valve) is also a linear-motion valve, but with smaller and smoother flow paths than the spool valve. An early success was the slide valve developed by Monsanto and thoroughly tested onsite (Wall 1961). Figure 8.2 shows a typical slide valve.

The history of the slide valve goes back to the first process gas chromatographs. When Beckman acquired Watts in 1957, they inherited the spool valve originally designed by Union Carbide. Beckman soon replaced the spool valve with a slide valve of their own design, and they continued to use slide valves throughout their long production of PGCs (Crum 1961; Turner and Villalobos 1961).

The slide valve was also the long-time favorite of Foxboro and ABB, and quickly rose in popularity to become one of the three major valve technologies adopted by the process gas chromatograph manufacturers.

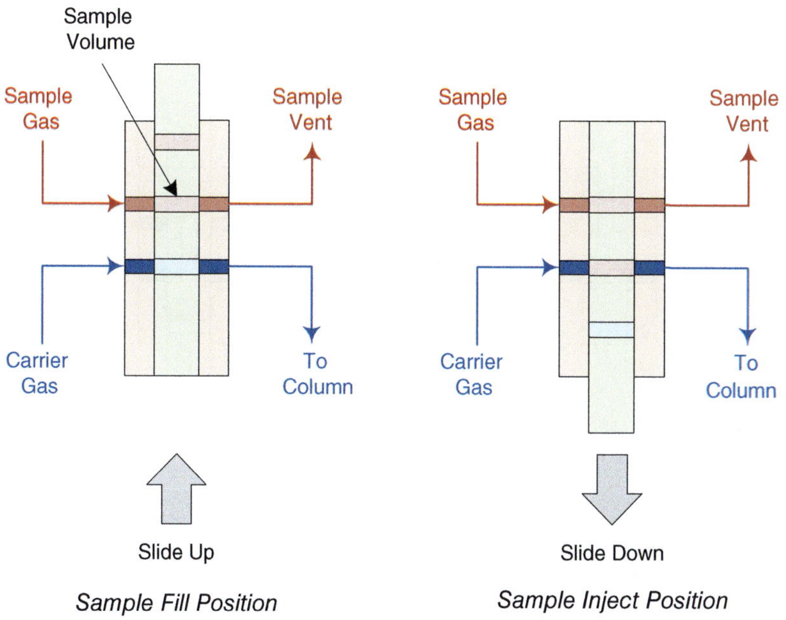

A typical slide valve used as a four-port liquid injector. The sample volume is simply a hole in the slider that fills with sample. This style of valve has no slots in the slide, only holes, thereby providing excellent flow alignment and a very clean injection pathway.

Figure 8.3 Slide Valve for Liquid-Sample Injection.

Figure 8.3 illustrates a simple four-port slider valve, set up for liquid injection. The size of the hole drilled through the slider determines the sample volume. Note the perfect alignment of ports and the complete absence of dead legs. These valves are good, but expensive to produce.

As you can see in Figure 8.3, a typical slider valve for liquid injection traps a small volume of liquid when in the sample fill position, which can be problematic as the trapped liquid might corrode the valve or polymerize in the hole, blocking the flow passage. It would be possible to add two extra connections to purge the liquid from the stationary hole, but we don't know of any slide valve that does that.

Figure 8.4 shows the gas-sample version of the injector. This is a six-port valve with a pattern of holes and slots machined into the slider to connect adjacent ports. The flow paths are similar to those found in classic rotary and diaphragm valves, so all three valves are functionally equivalent.

Looking at Figures 8.3 and 8.4, you will again notice that in both positions of the valve one slider hole contains a small trapped volume. In the inject position, the bottom hole in the slider houses a trapped volume of carrier gas (shown in blue). It's essential that this gas doesn't escape. If the carrier gas leaks out of the trapped volume and air replaces it, turning the valve off will inject a small sample of air into the column. The injected air may then form a spurious peak on the chromatogram.

The slide valve can have many configurations.

Early Beckman valves used a polymer slider moving between two highly polished stainless-steel plates, with the 1/16-inch connector tubes welded

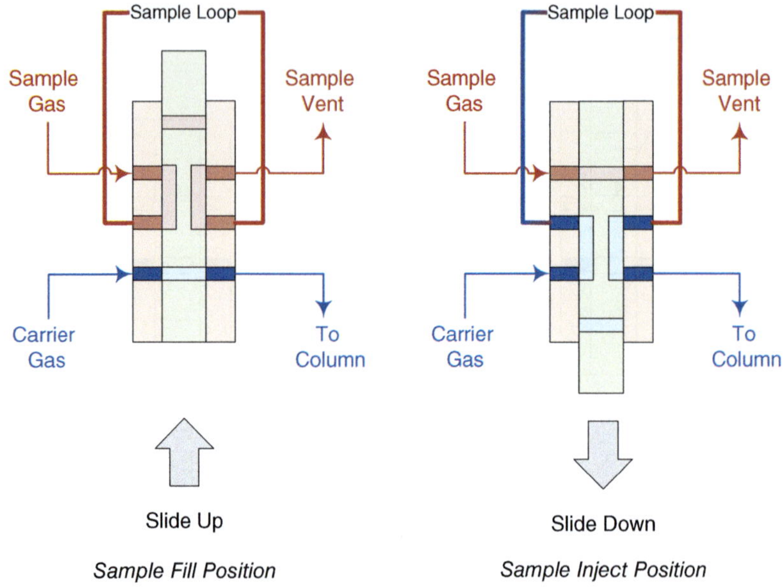

Figure 8.4 Slide Valve for Gas-Sample Injection.

Showing a six-port gas sample injector with an external sample volume. *In the sample fill position, gas sample purges through the sample loop. When actuated to the sample inject position, carrier gas enters the sample loop and pushes the process sample into the column. The ten-port slide valve has a similar geometry, but with five ports on each side of the slider instead of three. Without the sample loop, these valves are also good for column switching.*

to the plates. After having some problems with the dimensional stability of the polymer, they changed to a metal slider running between Teflon™ pads (Deming *et al.* 1978).

Most subsequent valves featured a slider of stainless steel or exotic alloy moving between stationary polymer seal glands. Foxboro even had a valve whose sliding surfaces were optically-flat ceramic blocks so hard they would grind away any particles that got between them (Annino et al. 1976).

In some slide valves, all the tube connectors are on one side and the fluids don't pass through the slider. Instead, the fluids flow through slots machined into the polished face of the slider. The slots connect adjacent ports and are quite similar to the slots seen in rotary valves. These valves allow fast replacement of the slider without disconnecting tubes.

Scratches on the slide or seat have always been the chief problem experienced with slide valves. Good filtration is essential. The smallest scratch can cause a leak of carrier gas into the sample volume or, even worse, a leak of liquid sample into a column.

Without their sample loops, six-port and ten-port slide valves were also good for column switching. Chapter 9 has some examples of those applications.

Thousands of slide valves remain in service around the world. Its long history of use attests to the reliability of sliding-plate technology, and yet the valve is now slipping into obsolescence because the latest rotary and diaphragm valves have smaller and smoother internal pathways more suited to the needs of ultra-fast separations using capillary columns. And they cost less to make.

Rotary valves

Union Carbide was one of the first processing companies to build and install a prototype PGC online. Between 1954 and 1956, they experimented with three different injector valve designs. Their original injector employed five discrete pneumatic valves, but this didn't work well, so they designed and fabricated a prototype **rotary valve**. This was probably the first rotary valve in an online PGC (Fellows 1957).

Back in 1956, Union Carbide engineers could not have known that their rotary valve design was destined to become one of the three major valve technologies used in process gas chromatographs. And yet their original home-made rotary valve was unsuccessful, so they resorted to a spool valve for their trailblazing PGC design (Spracklen 1957).

Monsanto had more luck with their rotary valve technique and in collaboration with Perkin-Elmer successfully used one in a 1957 prototype PGC installed at their Texas City plant (Helms and Claudy 1958, 272). It was not long thereafter that Shell Oil designed and built a successful rotary valve for injecting liquid samples as small as 1 μL (Penther and Hickling 1961).

Perkin-Elmer adopted rotary valve technology for their pioneering 1956 Model 178 prototype PGC, and for its descendant; the 1957 Model 184 production PGC. The Perkin-Elmer rotary valves were electrically actuated, and had options for six, eight, or sixteen ports (Helms and Noren 1957).

The 1957 CEC Model 26-202 PGC also used rotary valves (Karp 1961).

But Yokogawa is the true champion of rotary valves. Their first PGC was the 1966 Model 8110, which married the 1962 Foxboro PGC (via a technology-sharing agreement) to an updated programmer of their own design. Since the Foxboro analyzer used proprietary sliding-plate valves, the first PGCs sold by Yokogawa also employed those valves (Yokogawa 1966), but Yokogawa soon replaced them with rotary valves of their own design.

The Yokogawa valve pictured in Figure 8.5 is versatile, as it comes with single or dual rotors (both four-port or six-port) for gas sample injection and column switching. It's also an easy valve to maintain; you can replace the small rotor seats without undoing any tube connections.

Yokogawa has had so much success with their rotary valve that they have hardly changed the design at all. The same model of valve inhabits every Yokogawa PGC shipped since 1969.

You have already seen some typical applications of PGC rotary valves, as we used them in Chapter 7 to illustrate gas and liquid sample injection. In that chapter, Figures 7.2 and 7.4 are applications for the common six-port version, but almost any number of ports is possible; PGC rotary valves can have three, four, six, eight, ten, twelve, fourteen, or sixteen ports.

The valve stator is a flat metal disc drilled with small holes to form the ports. For fluid connections, each hole connects to a 1/16-inch pigtail tube or

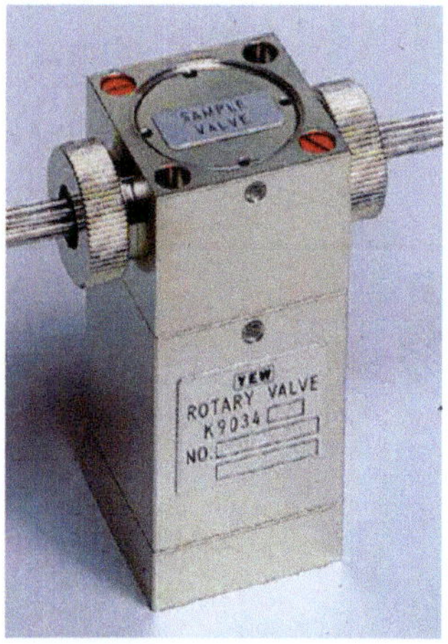

The dual rotors rotate 60° when actuated by air pressure. Yokogawa uses the same valve for gas-sample injection and column switching. A similar valve for liquid-sample injection has an internal sample volume and a 90° rotation.

Figure 8.5 Rotary Gas-Sample Injector Valve.
Source: Yokogawa Electric Corporation. Preproduced with permission.

Showing the slots in the rotor seat (black disc) and the holes in the matching stator.

Figure 8.6 Example of a Valve Rotor and Stator.
Source: Yokogawa Electric Corporation. Preproduced with permission.

to a miniature tube union. The valvemaker polishes the surface of the stator to optical flatness.

Figure 8.6 shows the stator and seat of the Yokogawa valve. This seat is available in carbon-filled or virgin polytetrafluoroethylene (PTFE). Other valvemakers have used a polished metal rotor with a PTFE gasket.

The port openings in the stator of a rotary valve usually form a circular pattern, which eliminates the trapped volume inherent in a linear valve. When the valve is at rest, small slots in the rotor seat connect alternate adjacent ports. When actuated, the rotor rotates by the fraction of 360° necessary to connect each port to its opposite adjacent port: a four-port valve rotates 90°, a six-port valve rotates 60°, an eight-port valve rotates 45°, and a ten-port valve rotates 36°.

The slots in the rotor of a rotary valve function like the slots in the slider of a slide valve. Neither of these align with the same perfection as the holes in the slider of the liquid injector valve shown in Figure 8.3, yet experience proves their good functionality.

A rotary valve for injecting a liquid sample does not have a full circle of ports. The valve rotates 90° and uses one of the rotor slots as the sample volume. When the valve has only four ports, as in Figure 7.4 of Chapter 7, one of the rotor slots contains trapped liquid, which is the same concern we noted for slide valves. The trapped liquid might cause downtime by corrosion or polymerization. To eliminate this risk, Yokogawa adds two ports to the valve just for purging the liquid from the unused slot. Figure 8.7 shows the result; it's a six-port valve with 90° rotation, an unusual combination.

For injecting liquid samples, we have found by experiment that the slot geometry of a rotary valve is less effective than the hole geometry of a slide valve shown in Figure 8.3. Presumably, this is partly due to mixing in the slot and partly due to slower vaporization; the polymer has lower thermal conductivity than the metal has, so less-volatile samples take more time to

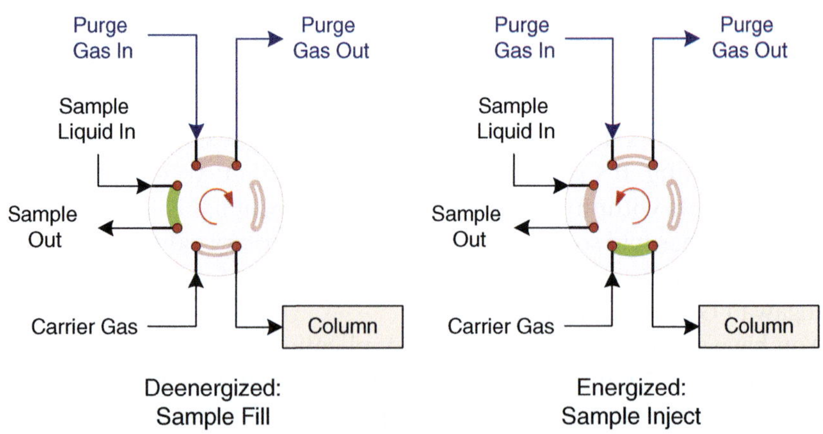

When using a four-port 90° rotary valve to inject liquid samples, the unused rotor slot is full of trapped liquid sample that might damage the valve. Adding the two purging connections shown in blue and an extra slot in the rotor allows a purge gas to flush the liquid from the unused rotor slot, avoiding the risk of valve damage.

Figure 8.7 Six-Port Rotary Liquid-Sample Injector.
Source: Adapted from Matsuura (1991).

vaporize. The slower injection is not a problem with packed columns but can be a disadvantage when using fast capillary columns.

In practice, the above observations may not be relevant, as local jurisdictions may not approve the use of rotary valves for handling hazardous liquid samples. The most common alternative is to use a plunger valve (described below) for the injection of liquid samples. Anyway, a plunger valve might be mandatory when capillary columns are in use.

A recent development in rotary valve technology is the valve pictured in Figure 8.8. The sliding flow channels in this miniature valve are replaceable inserts. To block inboard, outboard, or cross-port leakage, the valve face has leak-interception grooves purged by carrier gas (ASDevices 2018). A later section of this chapter discusses the various techniques of leak mitigation in more detail.

The 2018 μInProve™ ISV miniature rotary valve has options for four, six, eight, ten, twelve, or fourteen ports and may have a pneumatic or electric actuator. Without the actuator, the six-port version is only 28.5 mm wide and 33.5 mm tall, yet it features some very advanced features.

Figure 8.8 Miniature Multiport Rotary Valve. *Source: Analytical Sensing Devices, Ltd. Reproduced with permission.*

Diaphragm valves

The **diaphragm valve** is the third major valve technology used in PGCs for sample injection and column switching. This valve originated in the experimental PGCs installed at Shell Netherlands in 1956 (Hooimeijer et al. 1958). Phillips Petroleum further developed the valve, correctly perceiving the potential of the diaphragm technique to increase the speed and reliability of PGC measurements (Karasec and Ayers 1960; Broerman 1964). Now, after many design innovations by valve manufacturers, the technology has certainly delivered on its promise; for gas samples, one PGC valvemaker claims 20 ms overall switching time, and more than ten million operations between failures (Siemens 2007).

In addition to speed of operation and minimal maintenance, a goal of the design was to improve reliability by eliminating the sliding surfaces found in linear valves and rotary valves. The ports of a diaphragm valve connect in the same way as the ports of a rotary valve do, so the valves are functionally identical, and their flow diagrams are similar. Their difference lies in the method employed for routing the flow. Instead of a rotor turning on a stator, this valve has a polymer diaphragm lying flat across the ports in the valve cap.

There are two versions of the diaphragm valve, based on the method used to seal the unwanted flow paths; one uses plungers driven by pneumatic pistons to press on the diaphragm, and the other just uses air pressure without the pistons and plungers. Figure 8.9 pictures the latter valve. It has no moving parts, except for the flexion of the diaphragm itself.

To examine the function of a diaphragm valve, consider a typical six-port plunger version like the one pictured in Figure 8.10 (AFP 2008). The flow paths in this valve are the same as in the rotary valve first seen in Chapter 7, Figure 7.3.

Figure 8.11 shows a typical mechanism. Pressure from small plungers pressing on the diaphragm can seal the flow path between adjacent ports. Typically, alternate plungers work together as a set; thus, a six-port valve has two sets of three plungers. When the valve is in the rest condition, spring

The Siemens Model 50 valve uses direct air pressure on the diaphragm to seal the desired ports, with no other moving parts. When instrument air is not available, the valve can operate on carrier gas pressure. The valve is suitable only for gas sample injection or column switching.

Figure 8.9 Pressure-Seal Diaphragm Valve. *Source: Siemens Analytical Products and Solutions. Reproduced with permission.*

pressure on one set of plungers closes three flow paths, while spring pressure in the opposite direction opens the other three flow paths, thereby forming the connections shown in the deenergized state of Figure 7.3.

The integral actuator is a dual-piston pneumatic cylinder. Applying air pressure between the pistons moves the two pistons in opposite directions. A mechanical linkage applies the motion of each piston to separate sets of plungers. Different spring pressures ensure that the open flow paths close first, before the other flow paths open. The flow connections are then in the energized state of Figure 7.3.

Upon release of air pressure, the valve springs return the pistons to their rest condition, again momentarily closing all six ports to prevent crossflow.

An alternative design of diaphragm valve has independent sealing pistons centered over each port; each piston is separately powered by air pressure. In this design, the piston presses on the area of the diaphragm that is directly covering the port, rather than blocking the flow path between ports. This direct action provides a positive closure without creating the unpurged cavities at pinch points that may occur in other valves. Since the ports are independent, this valve may have any desired number of ports, and it may operate them in any desired sequence. To program a sequence of port operations, preset the actuation pressure for each piston and then apply a sequence of gas pressures to a common manifold powering all the pistons (see AFP 2008).

The APF diaphragm valve is available with six or ten ports for gas sample injection and column switching. It can also have an internal sample volume for liquid sample injection.

Figure 8.10 Piston/Plunger-Seal Diaphragm Valve. *Source: Analytical Flow Products, Ltd. Reproduced with permission.*

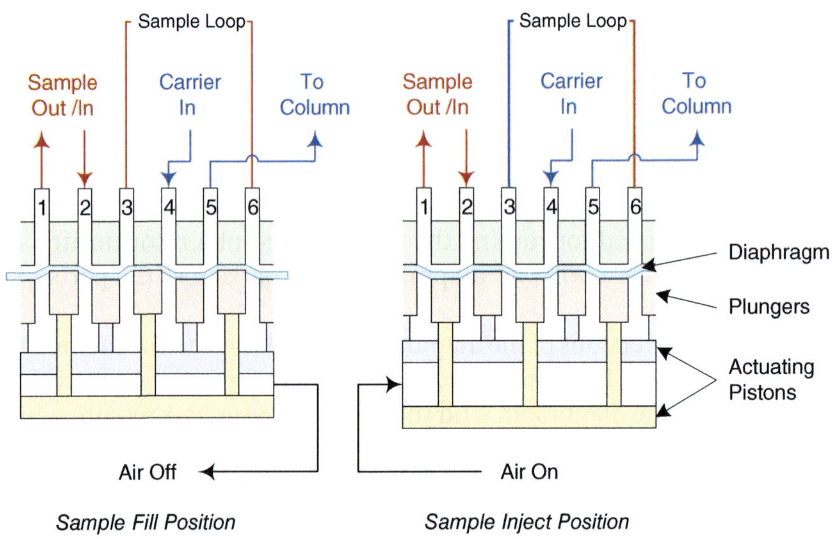

Schematic of a six-port sample injector valve with an external gas sample volume. The hole pattern in a diaphragm valve is circular, but for clarity it's shown here in linear fashion. In reality, the #6 port is adjacent to the #1 port. In the rest condition, springs push the two sets of pistons together, opening flow paths 2–3, 4–5, and 6–1 and closing paths 1–2, 3–4, and 5–6. Actuation air pressure forces the two sets of pistons apart, momentarily closing all flow paths, and then opening paths 1–2, 3–4, and 5–6. Most PGCs use the same valve without the external sample loop for column switching.

Figure 8.11 Diaphragm Valve-Switching Mechanism. *Source: Adapted from Valco (2016).*

Some diaphragm valves can also provide the atmospheric referencing function described in Chapter 7. They use an additional air inlet to block sample flow just before injection, thereby allowing the gas sample to equilibrate to atmospheric pressure. This feature eliminates the need for an atmospheric referencing valve and allows faster injection (see Siemens 2007).

There's not much motion in a diaphragm valve, which makes it fast; some valves switch in 10 ms and achieve 150 ms from the initial signal to completed operation. The mechanical parts move less than one millimeter, so there's not much wear. In addition, the valve has a low internal volume, making it an ideal choice for column switching as well as for sample injection.

The small movement of the actuator consumes little driving gas, so it's common to see diaphragm valves powered by carrier gas pressure when instrument air is not available.

The fast action and small internal volume of a diaphragm valve commend it for use with capillary columns, but not all valves are suitable for such delicate duty. Classical versions of the valve may have small unpurged dead legs at diaphragm pinch-points that cause excessive peak broadening when operating at the low flow rates typical of capillary columns.

Material selection for a diaphragm valve focuses on the valve head and diaphragm. The valve body and plunger mechanism are not in contact with the sample and can be stainless steel or anodized aluminum. The valve head is usually stainless steel, but more corrosion-resistant materials are available, such as Monel™, Hastelloy™, titanium, PEEK™, or ceramic. Use a silicon-treated valve head when measuring ppb-levels of polar molecules. The diaphragm may be Teflon™ or polyimide, or a layered combination of different polymers. When measuring parts-per-billion concentrations there are some concerns, discussed below, about the permeability of the membrane.

Late-model diaphragm valves are easy to maintain, with parts accessible without removing the valve from the chromatograph.

Plunger valves for liquid injection

Greenbrier was the first PGC company to develop a **plunger valve** (Karp 1961) for liquid sample injection, and similar valves have been available ever since (Bendix 1973; Villalobos 1975). This special valve is necessary when the sample has a high boiling point or a wide boiling range, and the standard sample injector valve can't reliably inject it. It's also appropriate when the liquid sample is unstable and must be kept at a lower temperature than the column oven.

A plunger valve like the one pictured in Figure 8.12 is ideal for injecting a liquid sample on a narrow-bore column. The valve takes a tiny sample from a cool liquid flow and transfers it to a heated zone where it quickly vaporizes into the hot carrier gas flow. To achieve this, the valve installs in the wall of the column oven, with the liquid sample flow outside the oven and the vaporizing section inside.

To accurately meter a small sample volume, the valve depicted in Figure 8.12 has a circular groove in the plunger rod. In other valves, the

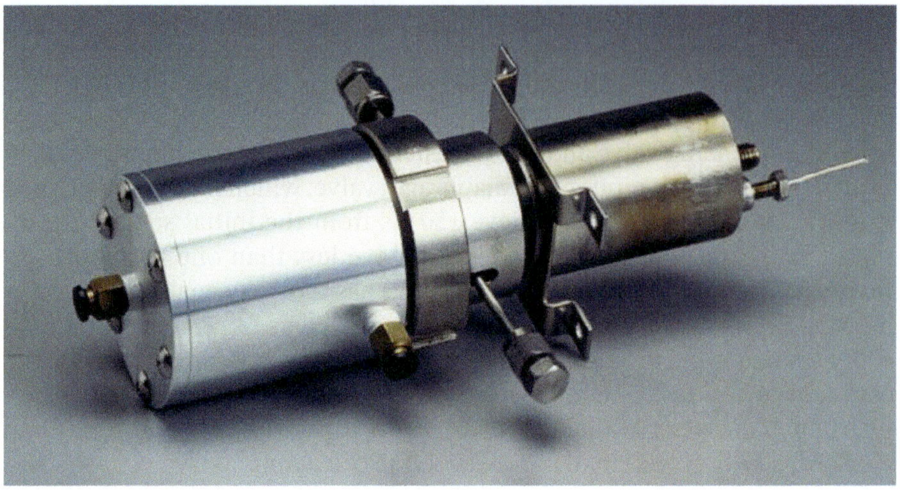

Figure 8.12 Plunger Valve for Liquid Sample Injection. *Source: Yokogawa Electric Corporation. Reproduced with permission.*

This valve mounts through the wall of the column oven, so the sample fill section is outside the oven and the hot vaporizer section is inside the oven.

sample volume is a small hole through the rod. The plunger, sometimes called the stem or quill, may be as small as $^1/_8$-inch diameter, and typical sample volumes range from 10 μL down to 0.035 μL.

Yet even this small sample is too large for a narrow-bore capillary column, so a plunger valve might include a **sample splitter** to consistently divert most of the vaporized sample to vent. Comparing the column and vent flows gives the **split ratio** and determines the amount of sample that enters the column.

Figure 8.13 illustrates the operation of a plunger valve. In the valve rest position, cool liquid sample flows around the groove and out to disposal. Inside the oven, carrier gas flows through the vaporization chamber and into

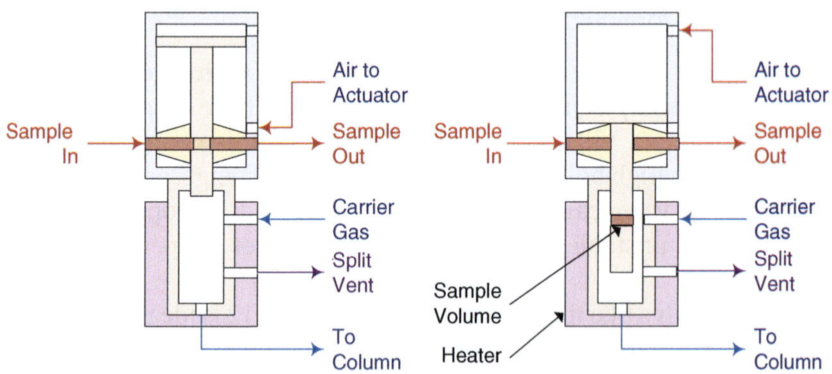

Figure 8.13 Typical Plunger Valve Operation. *Source: Adapted from Siemens (2011, 4).*

The sample volume is a small hole or circumferential groove in the plunger. This volume fills with liquid sample in the cold zone and then moves into the hot zone by downward movement of the plunger. For packed columns, the whole vapor sample might enter the column, but capillary columns need a smaller sample. To achieve this, the vaporized sample splits and most of it flows to waste via the split vent; only a small fraction of the sample enters the column via the bottom connection.

the column. Upon actuation, the plunger moves the sample groove into the hot zone for vaporization. After allowing about 5–10 s for the injected sample to fully vaporize, the valve is deactuated, and the plunger returns to its rest position.

To optimize valve reliability, the sample-wetted plunger seals consist of Teflon™ layers with different glass contents, or of an alternative polymer compatible with the sample liquid. In addition, the vaporization chamber may have a glass liner to minimize the adsorption of the vaporized sample molecules.

Some plunger valves have direct heating of the plunger itself. Plunger heating rates of up to 250 K/min are possible, enabling the analysis of very high-boiling samples.

Table 8.1 summarizes the properties and usage of chromatographic valves.

Table 8.1 Summary of PGC Chromatographic Valves.

Valve Function:		Sample Inject		Column Switch		Original Valve Manufacturer	
Valve	Type	Gas	Liquid	Packed	Capillary	PGC Vendor	Third Party
Linear	Solenoid	✓	✗	✓	✗	15, 19	
	Spool	✓	✗	✓	✗	9	
	Slide	✓	✓	✓	✓	1, 5, 8, 9, 11, 17	6
	Plunger	✗	✓	✗	✗	1, 18, 21	13
Rotary		✓	✓	✓	✓	3, 21	4, 7, 20
Diaphragm	Air-seal	✓	✗	✓	✓	18	
	Pistons	✓	✓	✓	✓	2, 5, 10, 16, 18	4, 20
Valveless Switching		✓	✓	✓	✓	18	
MEMS		✓	✗	✗	✓	12, 14, 18	
4-port	Any	✗	✓	✓	✓	*K*	✓ often used
6-port	Any	✓	✓	✓	✓	*E*	✓ rarely used
10-port	Any	✓	✗	✓	✓	*Y*	✗ never used

NOTE: The table portrays the most common applications. It is not exclusive.
Key to valve manufacturers: 1 = ABB Process Analytics, 2 = ABB Totalflow, 3 = AGC Instruments, 4 = Analytical Flow Products, 5 = Applied Automation*, 6 = Dinfa B.V., 7 = ASDevices, 8 = Beckman Instruments*, 9 = Bendix-Greenbrier*, 10 = Envent, 11 = Foxboro*, 12 = Honeywell-Elster, 13 = MAT Germany*, 14 = Qmicro B.V., 15 = RMG Messtechnik, 16 = Rosemount-Daniel, 17 = Servomex*, 18 = Siemens A.G., 19 = STF BACS Russia, 20 = Valco Instruments, 21 = Yokogawa.
[a] now obsolete

Other switching techniques

There are ways to inject samples or switch columns without using mechanical valves. These special techniques are a convenient way to meet the low-volume needs of capillary columns.

Miniature PGCs may employ **MEMS** solid-state valves for sample injection and column switching. These minute devices have had some success on ultra-clean process streams, such as those found in natural gas pipelines, but have not yet proved reliable in general process plant applications.

PGCs using capillary columns sometimes employ the **live-switching** technique, also called a Deans Switch (Deans 1965, 1968). This valveless way to switch columns has two advantages over mechanical valves: extremely low volume and no moving parts. To change the direction of carrier gas flow, an electronic pressure controller applies a closely-controlled pressure to a "live tee," thereby forcing the column flow to move in the desired direction. In practice, the live tee can have several inlets and outlets, and it will require some special skill to set up the system to work reliably.

These advanced switching techniques are beyond the scope of this PGC primer.

Valve leaks

About leaks

Mechanical injector valves always incorporate a sealing surface and therefore risk many potential leaks between carrier gas, sample fluid, and the outside air.

For liquid samples, it is not unusual for an injector valve to fail due to blockage or damage from solids present in the process fluid. Corrosion or solvent damage might also occur because of a chemical reaction between the sample and contact materials.

For gas samples, injector valves tend to be more reliable; if given a clean sample, they will cycle through many millions of operations before needing maintenance. However, there's always the potential of leakage by diffusion across the sealing surfaces.

In the past, crossport leakage was not a noticeable problem. The high flow rates passing through the valve diluted any leak to an undetectable level. For example, a typical leak specification for a diaphragm valve is 1 μL/min, and with a carrier flow rate of 50 mL/min, this leak rate would give a contamination of 20 ppm.

Over time, columns became smaller and carrier gas flow rates went down. At a carrier flow of 1 mL/min, the same leak would contaminate the carrier gas by 1000 ppm, enough to ruin any analysis.

A valve leak can occur in either direction. The force driving a diffusion leak comes from the population density of the leaking molecules on either side of the leak path. Each kind of molecule leaks from a place where there are many of those molecules to a place where there are fewer of them. For example, consider a sample of methane gas that has no oxygen in it. There are always a lot of oxygen molecules in the outside air, and these will leak into the methane sample at any opportunity they get, thus contaminating the sample. Meanwhile, there are many methane molecules in the sample, and very few of them in the outside air, so the methane molecules will leak in the other direction, going from the sample gas into the outside air. As strange as it may seem, the two gases will leak in opposite directions through the same leak path.

The minor leakage of air into the sample volume may have a deleterious effect on PGC reliability. If oxygen or moisture leaks into the carrier gas, it may gradually damage the columns or spoil a low-level measurement of these gases. It will not, however, affect the measurement of other gases in the sample, such as ethane.

A moment's thought will reveal a surprising truth: increasing the sample pressure will not reduce the rate of oxygen ingress! There are still no oxygen molecules in the sample, whatever the sample pressure. So, the inward rate of oxygen leakage remains the same.

> *Diffusion leaks do not depend on the partial pressure of other gases in the sample. Increasing the sample pressure will not prevent outside molecules from diffusing into the sample; it makes no difference at all!*

Technically, the leak rate of each substance is proportional to the difference in its **partial pressure** on either side of the leak path. To compute the partial pressure of each substance in a gas sample, multiply its concentration by the pressure of the sample at that location.

From the above discussion, we can expect that linear and rotary valves may allow inward contamination of the carrier gas or sample volume by atmospheric gases. It's also true that outward loss of sample molecules may occur, but the number lost is unlikely to be significant.

Diaphragm valves have a different leak profile than the other valves because they don't employ a sliding seal. Consequently, they are unlikely to suffer leaks to and from the atmosphere, but they may have another problem. Gas molecules can pass through a diaphragm, a process called **permeation**. The driving force for permeation is differential partial pressure, the same as it is for diffusion leaks, and the rate of permeation increases with temperature.

The rate of permeation depends on the material of the diaphragm and the type of molecules involved. Metal diaphragms are usually impervious, but all polymer diaphragms allow a high permeation rate; it's well known, for instance, that moisture readily permeates through a Teflon™ diaphragm. To avoid the diffusion leakage of atmospheric gases, diaphragm valves sometimes allow a purge gas to sweep air out of the operating mechanism behind the diaphragm.

Crossport leakage can occur in any kind of PGC valve and follows the same rules. Inside a sample injector valve, the gas sample flow path runs close to the carrier gas flow path, so crossport leakage is probable. As discussed above, the excess carrier pressure will not stop diffusion leakage across the seat. Should the leak rate of sample gas into carrier gas become significant, the chromatogram baseline may drift or become bumpy, though it's more likely that the inverse leak of carrier into sample will be dominant, particularly with hydrogen or helium carrier gas. If a leak dilutes the sample, it will reduce all peaks on the chromatogram.

When a valve has an integral actuator, as diaphragm valves do, it's also possible for the valve actuation gas to leak into the sample. When troubleshooting, keep that in mind.

Valve leak mitigation

Some rotary or diaphragm valves use leak barriers to prevent trouble from minor leaks. Two levels of protection are available (see AFP 2007).

The first barrier is a pair of concentric grooves in the rotor or stator that catch any leakage from or to the atmosphere. As evident in Figure 8.14, the barrier grooves encircle both inside and outside the valve ports intercepting any nascent leak. Typically, a carrier gas flow of about 5 mL/min flows round the groove to purge out the intercepted molecules. If desired, a detector can monitor the purge vent gas to provide a leak alarm.

Some valves have a second barrier to protect against crossport leakage. Valves so equipped have additional grooves to intercept leakage between ports. The extra grooves typically connect to the inner and outer concentric grooves to ensure good purging with carrier gas.

Diaphragm valves don't suffer much diffusion leakage at the sealed edges of the diaphragm, but atmospheric air may permeate through the diaphragm. This is a minor effect that is unlikely to be significant in a normal flowing system. But there are column systems that hold the carrier stationary for several minutes, perhaps allowing enough inbound

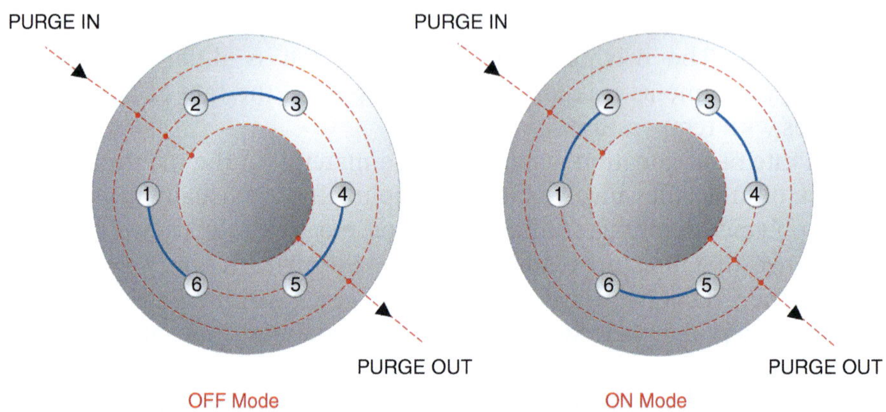

Figure 8.14 Purging Grooves to Intercept Leakage. *Source: Analytical Flow Products, Ltd. Reproduced with permission.*

Leak interceptor grooves are a feature of some rotary or diaphragm valves. In this example, two concentric grooves (shown in red) encircle the six small valve ports and intercept any leakage going to or from the outside air. A purge gas (usually carrier) flows through the grooves and may also purge the mechanism behind the diaphragm, preventing permeation leaks. If desired the purge gas may flow to a detector to provide a leak warning.

permeation to form additional peaks on the chromatogram. Inbound contamination might also cause significant error when injecting a gas sample for measuring parts-per-million levels of nitrogen, oxygen, or water. In such cases, the choice of diaphragm material is important.

PTFE is quite permeable to water vapor and would not be the best choice for the diaphragm when measuring moisture at low concentration. In addition, PTFE tends to emit minute amounts of halogen compounds that might interfere with a measurement that's using an **electron capture detector**. Optional materials for diaphragms may include PEEK™, polyimide, multiple polymer films, metalized polymer, or even metal alloy. However, a composite material having a thin metal film sandwiched between two polymer films is better for reducing permeation leakage (AFP 2008, Siemens 2003).

Some valves have a connection to purge the actuator side of the diaphragm with a small flow of carrier gas, eliminating the atmospheric source of contamination. Another use for this connection is for applying a partial vacuum to the back of the diaphragm to balance the pressure of a sample at subambient pressure. A regular diaphragm valve will not work at subambient pressure if atmospheric pressure on the other side of the diaphragm prevents it from opening.

Knowledge Gained

- *PGCs use special chromatographic valves to inject samples and to switch columns.*
- *Most chromatographic valves are binary devices operated by electric or pneumatic actuators.*
- *A sample injector must reliably inject the same volume of sample each time it operates.*
- *PGCs may use the same type of valve for sample injection and column switching.*
- *Valve types include linear valves (spool, slide, and plunger), rotary valves, and diaphragm valves.*
- *Spool valves, now obsolete, had a sliding rod and O-ring seals to inject gas samples or switch columns.*
- *Most slide valves use a polished metal or polymer slider moving on a polymer or polished metal plate.*
- *Sliders may be single or double-sided with holes or slots to route the sample fluid and carrier gas flow.*
- *Most rotary valves rotate an appropriate fraction of 360°, then return to their rest condition.*
- *Most rotary valves use a polymer rotor seat that rotates and seals against a polished metal stator.*
- *The stator has a circular pattern of ports connecting to tube pigtails or miniature tube unions.*
- *The rotor seat has machined slots to route the sample fluid or carrier gas flow between ports.*

- *Rotary and slide injector valves may trap fluid in an unused flow path that can damage the valve.*
- *Adding purging ports to a sample injector enables the removal of fluid trapped in an unused flow path.*
- *Diaphragm valves have flow paths that are similar to rotary valves but employ no sliding motion.*
- *The valve cap has a polished surface with a circular pattern of holes to provide inlet and outlet ports.*
- *Most diaphragm valves have six or ten ports that connect to pigtail tubes or miniature tube unions.*
- *A polymer diaphragm lies flat against the valve cap covering the valve ports.*
- *Applying pressure on the diaphragm can seal the flow paths between adjacent ports.*
- *The sealing force of diaphragm valves come from air-powered plungers or direct air pressure.*
- *In the rest condition the force seals alternate flow paths leaving the other flow paths open to flow.*
- *Actuation air moves two sets of pistons, closing the open paths and opening the closed paths.*
- *During actuation, the open flow paths always close before the closed flow paths open.*
- *More versatile valves have independently actuated plungers aligned to directly close each port.*
- *Diaphragm valves consume little actuation gas and sometimes use carrier gas for pneumatic actuation.*
- *For analyzing wide-boiling liquid samples, PGCs use a special injector valve that installs in the oven wall.*
- *The plunger valve takes a cold sample, then quickly moves it into a hot zone for fast vaporization.*
- *Sample injection or column switching is also possible using a solid-state valve or live-tee.*
- *All valves have the potential to leak between ports or between a port and the atmosphere.*
- *Diaphragm valves have a good seal to atmosphere but can suffer a permeation leak via the diaphragm.*
- *A Teflon coated metal membrane can minimize permeation leaks in a diaphragm valve.*
- *Partial pressure is the driver of leakage, so leaks may occur even against a pressure differential.*
- *Advanced rotary or diaphragm valves feature leak interception grooves around or between the ports.*
- *A small flow of carrier gas may purge the leak interception grooves.*
- *To limit permeation leakage, some diaphragm valves also purge the space behind the diaphragm.*
- *The system designer may route the purge gas into a detector for automatic leak detection and alarm.*

Did you get it?

Self-assessment quiz: SAQ 08

Q1. Which common type of chromatographic valve is missing from the following list?
- **A.** Diaphragm valve
- **B.** Plunger valve
- **C.** Rotary valve
- **D.** Spool valve

Q2. What type of valve is only suitable for liquid sample injection?

Q3. What type of chromatographic valve was popular for gas sample injection and column switching, but is now obsolete?

Q4. Some slide valves have holes in the slider, but some do not. Explain the reasons for this difference.

Q5. What is the angle of rotation for a six-port rotary valve?

Q6. Diaphragm valves rely on pressure on the diaphragm to seal off the undesired flow paths. Exactly where is this pressure applied? There are two versions of the valve that apply the pressure at different locations. Explain.

Q7. Advanced rotary and diaphragm valves have leak interception grooves cut into the valve seat. Where would these grooves be located?

Check your SAQ answers with those given at the end of the book.

Student evaluation test: SET 08

Your instructor will provide answers to these questions.

S1. Which type of valve would use a "sample loop"?
Select the one correct answer:
- **A.** A gas sample injector valve
- **B.** A liquid sample injector valve
- **C.** A plunger valve
- **D.** A column switching valve

S2. Which of the valves listed below is *not* a linear valve?
Select all that comply:
- **A.** A spool (piston) valve
- **B.** A sliding-plate valve
- **C.** A rotary valve
- **D.** A plunger valve

S3. What is the angle of rotation in a ten-port rotary valve?
Your answer is a two-digit integer.

S4. Of the valves listed below, which one(s) are not appropriate for column switching?
Select all that comply:
- **A.** Diaphragm valve
- **B.** Slide valve
- **C.** Rotary valve
- **D.** Plunger valve

S5. Some diaphragm valves operate by direct air or gas pressure on the diaphragm, without internal pistons or plungers.
What are the PGC application(s) for such valves?
Select all that comply:

A. Gas sample injection
B. Liquid sample injection
C. Column switching
D. None of the above

S6. Which three of the categories listed below are enough to categorize all PGC sample injector valves?
For this question, ignore sample injection by solenoid valve assemblies. It was never any good, anyway.
Select three items:

A. Diaphragm valves
B. Linear valves
C. Plunger valves
D. Rotary valves
E. Sliding-plate valves
F. Spool valves

S7. Three major PGC manufacturers each adopted a specific type of chromatographic valve. What were the three rival valve technologies they adopted?
Your answer is a three-digit code, where:

- The first digit shows the type of valve mostly used by ABB.
- The second digit indicates the valve favored by Siemens.
- The third digit represents the valve adopted by Yokogawa.

Select each digit from the following key:

1. Solenoid valves
2. Diaphragm valves
3. Plunger valves
4. Rotary valves
5. Sliding-plate valves
6. Spool (piston) valves

For example, your answer might be 623 – or not!

S8. There's always the chance that a mechanical valve will leak. Assuming a very slow leak, which of the following leaks might affect the accuracy of the analytical measurements?
Be careful! Select only those that apply:

A. Sample gas or carrier gas leaking to the atmosphere.
B. Atmospheric gases leaking into the sample gas or carrier gas.
C. Valve actuation gas leaking into the sample gas or carrier gas.
D. Cross-port leakage.

S9. What is a sample splitter?
Select the one correct description:

A. The first column in a dual column system that splits the light components from the heavies and allows the second column to separate the light components.
B. A heated zone to selectively vaporize light components from an injected sample of a wide-boiling liquid.
C. A heated sample injector that catalytically cracks heavy molecules to yield lighter compounds that the columns can easily separate.
D. A flow divider that diverts most of the injected sample to vent, allowing only a small portion to enter the first column.

References

Cited

AFP (2007). *DV6-series positive port shut off diaphragm valve for chromatographic instruments, Product Brochure*. Thetford Mines, QC: Analytical Flow Products, division of Mécanique Analytique, Inc.

AFP (2008). *DV3-series positive port shut off diaphragm valve for analytical instrumentation*, Product Brochure. Thetford Mines, QC: Analytical Flow Products, division of Mécanique Analytique, Inc.

Annino, R., Curren, J. Jr., Kalinowski, R., Karas, E., Lindquist, R., and Prescott, R. (1976). Totally pneumatic gas chromatographic process stream analyzer. *Journal of Chromatography A* 126 (November 3, 1976): 301–331. https://doi.org/10.1016/S0021-9673(01)84082-6

ASDevices (2018). *μInProve ISV series GC valve*. Product Brochure: 1–10. Thetford Mines, QC: Analytical Sensing Devices, Ltd.

Bendix (1972). *Bendix C5 valve*. Product Brochure B527SB1172, (November): 1–4. Lewisburg, WV: Bendix Process Instruments Division.

Bendix (1973). *Bendix liquid sample valve*. Product Brochure B449SB273, (February): 1–4. Lewisburg, WV: Bendix Process Instruments Division.

Bendix (1975). *Sliding plate valves for chromatography*. Product Brochure B429SB1175, (November): 1–8. Lewisburg, WV: Bendix Process Instruments Division.

Broerman, A.B. (1964). *Pneumatic amplifier sampling valve for chromatographic analyzers*. US Patent 3140615A, granted July 14, 1964.

Clemons, J.M. (2016). Chromatography in process analysis. In: *Encyclopedia of Analytical Chemistry: Applications, Theory and Instrumentation*, (June 15): Wiley. https://doi.org/10.1002/9780470027318.a2107.pub3

Crum, W. (1961). The design and application of a small volume sliding plate valve for gas chromatograph. In: *National Symposium on Instrumental Methods of Analysis 1961*, 115–122. Pittsburgh, PA: Instrument Society of America.

Deans D.R. (1965). An improved technique for back-flushing gas chromatographic columns. *Journal of Chromatography A* 18, 477–481. https://doi.org/10.1016/S0021-9673(01)80403-9.

Deans D.R. (1968). A new technique for heart cutting in gas chromatography. *Chromatographia* 1 (1–2, January): 18–22. https://doi.org/10.1007/BF02259005.

Deming, P.L., Mehaffy, G.E., and Freeman, B.M. (1978). New valves in process gas chromatographs. In: *Analysis Instrumentation*, Vol. 16. (ed. F.D. Martin, W.H. Wagner, and T.J. Puzniak), 59–62.

Fellows, E.G. (1957). Chromatography analyzes gas and vapor products in the plant, *Control Engineering*, (July) 75–81.

Helms, C.C. and Claudy, H.N. (1958). The practical design of a vapor fractometer for automatic multicomponent analysis of process streams. In: *Gas Chromatography: Proceedings of the 1957 Analysis Division Symposium* (eds. V.J. Coates, H.J. Noebels, and I.S. Fagerson). New York, NY: Academic Press.

Helms, C.C. and Norem, S.A. (1957). A look at vapor-phase chromatography – it has excited the boys in the lab, but what about actual plant use? *Oil & Gas Journal* 55 (17): 146–149. Also presented at Pittsburgh Conference, March 7, 1957.

Hooimeijer, J., Kwantes, A., and van de Craats, F. (1958). The automatization of gas chromatography. In: *National Conference on Instrumental Methods of Analysis*, Volume 4. Published in 1959. Pittsburgh, PA: Instrument Society of America.

Karasec, F.W. and Ayers B.O. (1960). *A fast sampling valve for gas chromatography*, *ISA Journal* 7 (3): 70–71.

Karp, H.R. (1961). Industrial process chromatographs. *Control Engineering* (June): 87–100.

Matsuura, T. (1991). *Sample Valve*. Japanese patent: JP,03-051351,U1(1991), issued May 20, 1991.

Penther, C.J. and Hickling, J.W. (1961). New liquid-sampling valve extends usefulness of process chromatographs. *Oil & Gas Journal* (May 15), 130-133.

Servomex (1985). *Servomex process gas chromatographs*. Product Brochure 7981-3147, June. Crowborough, UK: Sybron Analytical.

Siemens (2003). *Model 50 diaphragm valve specification*, Technical Note 2017660-UI (REV XX), (November 26). Bartlesville, OK: Siemens.

Siemens (2007). *Model 50-SSO valve*, Technical Support Information No. 77c, (November): 1–11. Bartlesville, OK: SE&A Process Analytics.

Siemens (2011). *Process gas chromatographs – Maxum edition II*, Specifications and Data Sheet PA 01–2012. Karlsruhe, Germany: Siemens AG.

Spracklen, S.B. (1957). The development of gas chromatographs, *ISA Journal* (November), 514–517.

Turner, G.S. and Villalobos, R. (1961). Microsampling in process analysis. In: 3rd International Symposium on Gas Chromatography, ISA Analysis Instrumentation Division Symposium at Michigan State University 1961. Pittsburgh, PA: Instrument Society of America, 363–369.

Valco (2016). *Models DV-12 and DV-22 diaphragm valves*, Technical Note 605, September. Houston, TX: Valco Instrument Co. Inc.

Villalobos, R. (1975). Process gas chromatography. *Analytical Chemistry* 47 (11), September, 983A–1004A.

Wall, R. (1961). A sampling valve for gas chromatography. In: *National Symposium on Instrumental Methods of Analysis 1961*, 123–131. Pittsburgh, PA: Instrument Society of America.

Yokogawa (1966). *Chromatographic analysis*, Product Catalog 24-G1-1a (in Japanese), September. Larchmont, NY: Yokogawa Electric Works, Ltd.

Table

Figures

New technical terms

When first introduced, these new words and phrases were in bold type. Further information is available in the Glossary at the end of the book.

dead legs
diaphragm valve
electron capture detector
Hastelloy
Monel
partial pressure
permeation
piston valve
plunger valve
rotary valve
sample splitter
slide valve
solenoid valve
split ratio
spool valve

For information about chemical names, consult the Glossary.

9

Column systems

"When looking at a PGC for the first time, try to understand the column system. If you are new to process gas chromatography, this may not be an easy task, but you won't get far without this understanding. So be curious, find out what the columns are doing, and everything else will fall into place".

Two fundamental issues

The general elution problem

In Part One of the book, we outlined the function of the column and how it achieved the separation of injected substances. We looked at the patterns of retention on a single column and noted that the peaks gradually get wider. We also noticed that retention times increase rather rapidly for later peaks on the chromatogram.

The early practitioners of chromatography quickly realized that a column capable of resolving the early peaks would retain the later peaks far too long. In fact, they had a name for it: *the general elution problem* (Rubey 1991; Sewell 2000).

There are only two known solutions to the **general elution problem**:

- Use a single column, but gradually increase its temperature during the analysis to remove all injected components.
- Use column valves to divert groups of peaks into different columns, each optimized to separate some of the analytes.

These alternative solutions employ different analyzer hardware and tend to be mutually exclusive. The applications engineer must choose. But that decision must consider another mandate unique to process analysis; the **housekeeping rule**.

Process Gas Chromatographs: Fundamentals, Design and Implementation, First Edition. Tony Waters.

The housekeeping rule

In addition to solving the general elution problem, a PGC column system must take care of a vital housekeeping task:

- A PGC column system must remove all **heavies** (highly-retained substances) from the columns before injecting another sample.

Unlike the laboratory instrument, a PGC injects a new sample every few minutes, continuing for 365 days per year, without respite. At the end of each analysis, if even a trace amount of an injected substance were to remain in the columns, it would accumulate and would eventually form an unnavigable sea of waves on the chromatogram baseline, rendering the instrument unusable.

Thus, the **prime rule** of column design is to ensure the removal of all sample constituents before injecting another sample:

> *This is the prime rule of PGC column design: remove all injected substances from the columns before you inject another sample!*

In certain rare instances, a column designer may break the housekeeping rule, but it's always risky to do so. We can never be completely certain of the process stream composition, so the omission of the housekeeping function may lead to unexpected failures, particularly during process upsets.

Luckily, the two options for doing the housekeeping are the same two options already stated for solving the general elution problem:

- Ramp a single column to a high-enough temperature to remove all the heavies from the column.
- Use a multiple column system running at constant temperature to remove the heavies.

The temperature ramp solution

During the first five years of PGC development, temperature ramping was not an option in commercial instruments. When Beckman first introduced temperature programming to PGCs (Burnell and Said 1961), the process industries greeted the new technique with skepticism and saw it as too complex and costly to use for online process analysis. Thus, PGCs adopted temperature programming only for difficult applications with many analytes having a wide range of boiling points. A good example is the simulated distillation of hydrocarbons: that method was so successful it spawned two new standard methods (ASTM D2887; ASTM D7096).

Clearly, temperature-programmed PGCs can do measurements that isothermal PGCs can't do, and never will. We should use them for those applications. They include the estimation of process stream properties like **octane number**, **vapor pressure**, or **flash point**; and the aggregate measurement of whole classes of chemicals, such as the paraffins, isoparaffins, **naphthenes**, and **aromatics** (PINA).

Due to advances in chromatographic technology, temperature programming is now the *de facto* standard for laboratory chromatographs and has become more attractive for new process applications. Eventually, temperature programming may become the preferred technique for process gas chromatographs too, but we are not there yet. Later, we'll look at some developments in column heating and data processing that may lead to its more widespread adoption.

Yet there remain three housekeeping concerns about temperature programming that may ultimately inhibit its wider adoption in PGCs. Consider that:

- Additional analysis time will be necessary to ramp the column to a high enough temperature to remove the heavies, and then to cool the columns before starting a new analysis.
- The high temperature necessary to remove heavies may shorten the life of the column and any other exposed devices. It might also limit the use of the PGC in hazardous areas of the plant.
- At the end of the analysis, some uncertainty remains: there is no guarantee that the removal of heavies is complete.

A two-column **backflush** system doesn't expose us to any of those uncertainties and does the housekeeping easily and effectively, without raising the column temperature. These are strong incentives for continuing to use multiple column systems in online analyzers, even when a temperature-ramped capillary column would have superior separation power.

The multiple column solution

For process applications involving less than a dozen analytes, we see four reasons for preferring multiple columns over temperature programming. They are true for any kind of column and are likely to endure:

- Multiple columns can remove all injected substances from the columns before injecting another sample, without extending the analysis time.
- Multiple columns can quickly separate a minute concentration of analyte from a massive concentration of major component, a feat impossible to achieve on a single column.
- Multiple columns can easily separate some substance pairs that no single column can separate.
- Multiple columns can often achieve a faster analysis than is possible with a single column.

A multiple column arrangement comprises two or more columns and at least one column switching device; we call that arrangement a **column system**. Before describing the four main types of column system, let's briefly review what a column system must accomplish. Simply stated, a column system for online analysis must satisfy these demands:

- Complete the analysis as fast as possible.
- Separate and adequately resolve all analyte peaks from each other, and from the unmeasured substances in the sample.
- Do the housekeeping: remove all injected substances from the columns before injecting another sample.
- Protect a column from substances or temperatures that would damage it or would prevent it from performing the desired separation.
- Maintain the integrity of peaks by eliminating voids in the flow path and avoiding absorptive, reactive, or catalytic contact materials.
- Survive the injection of samples containing extreme concentrations, or unspecified substances, due to occasional process upsets.
- Fail to a safe condition upon loss of electric power or actuating gas pressure, thereby protecting its columns and detectors from overload.

The choice

Ultimately, the choice of whether to employ one temperature-ramped column or several isothermal columns should depend on the *number of measurements* required and the *desired analysis time*.

For the fast analysis of dozens of analytes, a single column with rapidly increasing temperature is unbeatable. But for measuring a few analytes, column heating and cooling might be an overly complex solution that is prone to downtime.

Before putting a high priority on the speed of analysis, consider the delays in your sampling system. The hold-up times that occur in process vessels and piping before sample extraction often extend to hours, so it's hardly likely that you'll need an analysis done in seconds. When the specified analysis time seems excessively fast compared with the process or sample system response time, try to negotiate more analysis time in exchange for a simpler column system that will be easier to maintain and ultimately more reliable.

Maybe, in the future, process gas chromatographs will use a combination of the two procedures. One can imagine a solution that blends a judicious application of programmed heating with an effective backflushing technique.

Delayed injection

In passing, it's interesting to note the difference between **analysis time** and **cycle time**. The analysis time is the time taken to separate and measure the analytes, whereas the cycle time is the time between successive sample injections. In most applications, there is little distinction between them, except, of course, that some peaks come out earlier than others and the PGC could make measurements available before the end of cycle. Once the PGC has measured the last peak, we usually try to avoid any wasted time before injecting the next sample. Yet there are some applications that require a fast response, but not a rapid repetition rate.

Several examples come to mind:

- If high-frequency updates are unnecessary for process monitoring, the analyzer might wait a predetermined time before running another analysis.
- In a batch process, the completion of a processing step might initiate an analysis run.
- In a continuous process, an alarm condition could automatically call for more frequent analyses.
- Initiation could be manual; a process might not need the analysis during normal operation, but operators can request an analysis any time they suspect an upset.

In each of these circumstances, the PGC would run a predetermined number of analyses and then stop. While waiting for the next command, the PGC would enter a standby mode, conserving resources and minimizing wear.

However, even in standby mode, the sample fluid should continue to flow at the same rates through the sample conditioning system and sample injector valve. It's usually a mistake to turn sample flows off and on.

Four types of column system

Recognizing the functions performed

Given that most PGCs use multiple columns, it's good for us to get familiar with the various column systems in use. Before proceeding, though, realize that all column valves used in PGCs have only two positions: deenergized (deactuated) or energized (actuated). Using these binary valves, a PGC can employ multiple columns to separate and measure analytes that no single column could separate. We shall see that they also reduce analysis time and keep the columns clean.

Since column systems are a fundamental function of a process chromatograph, it's essential that you understand them. The following paragraphs briefly explain what column systems do and how they work:

> *This chapter limits itself to reviewing the multiple column systems used by process gas chromatographs. The detailed theory and practice of column system design is beyond the scope of this introductory book.*

It would be unwise to attempt maintenance work on a PGC analyzer unit without first understanding the function of each column. Start by recognizing the column configuration it uses. All process gas chromatographs have one or more sample injectors, separating columns, and detectors, so observe the number of these devices used in your instrument. Many arrangements are possible, but most PGC applications will have one of the following recognizable characteristics:

- Type A: a single column.
- Type B: two or more columns, but only one detector.
- Type C: two or more detectors, but only one sample injector valve.
- Type D: multiple sample injector valves and multiple detectors.

Interestingly, these four categories follow the historical development of the PGC. Over many decades of development, technological advances have allowed PGC vendors to gradually relax the physical constraints on column system design. Each new product would accommodate more injectors, columns, and detectors, thus facilitating more complex applications and increasing the market for the product line. You will see this story unfold in the paragraphs that follow.

Type A: A single column

Figure 9.1 shows the simplest chromatographic configuration, comprising one sample injection valve, one column, and one detector. Naturally, the first prototype PGCs were like that.

Simple systems tend to be reliable and are easy to understand. Yet single-column systems are rare in process gas chromatographs because of

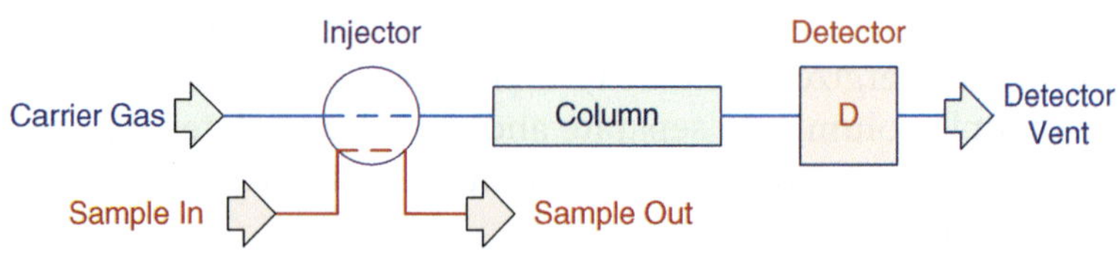

Typical arrangement of a single injector, single column, and single detector (for illustrative purposes only). All injected substances must pass through the detector.

Figure 9.1 Type A Column System.

the *general elution problem* and the *housekeeping rule* mentioned above. A fast analysis of multiple substances may be impossible on a single isothermal column. It could take weeks to elute all the peaks from a single injection.

We saw that temperature programming can overcome the doubling rule and will also remove heavies from the column if the final column temperature is high enough to elute all the injected substances. Then, a single capillary column can separate scores of peaks, even with samples of unpredictable composition. The single column also has an inherent simplicity of operation. Because of these attractive traits, temperature programming has become a standard technique of laboratory analysis.

In a newer application of the single-column approach, some entrepreneurial companies are now gaining experience with a radically different breed of process analyzer that uses innovative technology to mimic the success of high-speed columns in the laboratory. Using a combination of micromachined injectors, rapidly heated columns, and unique detectors, these novel analyzers can achieve fast multicomponent analyses. Originally positioned as portable analyzers, they are gradually encroaching on PGC markets. After repackaging for use in hazardous areas, they may foretell the future of process chromatography (Crandall *et al.*, 2014).

For the moment, though, this is too much technology for simple separations. When the process gas chromatograph needs to measure only a few components, it delivers superior performance and reliability by using two or more columns running at constant temperature. It's also less expensive that way.

Type B: Multiple columns, single detector

The twin mandates of good housekeeping and fast analysis make serial column systems essential in isothermal process gas chromatographs. Yet, that necessity is also one of the great strengths of gas chromatography as a process analyzer. When a PGC uses a multiple column system, it can separate and measure just about anything found in a fluids processing plant.

To accomplish the desired analysis, each PGC has a unique configuration derived by **application engineering**. An application engineer selects the necessary valves, columns, and detectors; configures the connections between them; specifies the operating conditions; and lists the sequence of timed events necessary to perform the analysis. When loaded into the PGC **control unit**, the latter program of instructions becomes the analytical **method** that controls the analysis, including the times to actuate and deactuate each column valve.

Type B column systems were universal in earlier models of PGC because those analyzers could support only one detector. Application engineers had to devise clever column arrangements to separate the analytes, vent the unmeasured components, and ensure that analyte peaks never arrived simultaneously at the detector. Typically, the first column would retain and

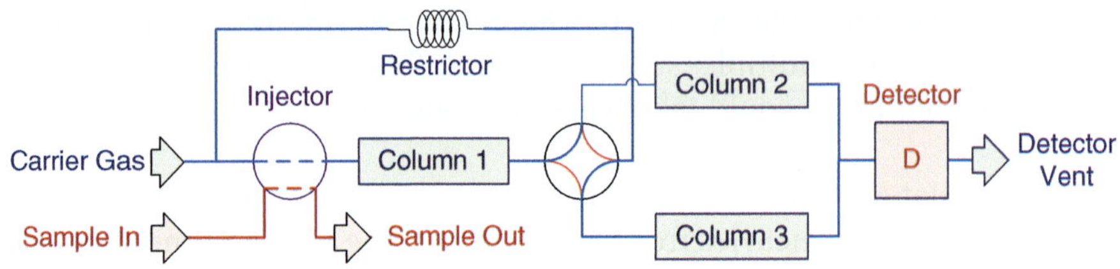

Figure 9.2 Type B Column System.

Typical arrangement of a single injector, multiple columns, and single detector. Shows a four-port rotary column valve (for illustrative purposes only). Other valves have equivalent function. The flow resistance of the restrictor is equivalent to Column 1. In the valve position shown (blue), peaks not strongly retained by Column 1 enter Column 2 for further separation. Then, upon 90° rotation of the column valve (red), later peaks from Column 1 enter Column 3. The column design must ensure that analyte peaks arrive separately at the detector.

then vent any unwanted component peaks, and the second column would separate a few analyte peaks.

Or, as in Figure 9.2, the column system might use two secondary columns, each providing its unique contribution to the overall separation.

These simple column functions are powerful tools for quickly separating a few peaks for measurement and are the foundation of process chromatography. In the next section, we briefly review typical column configurations for achieving common functions: you need to understand them.

The original PGC column systems worked well enough, as user expectations were modest, and the column systems were correspondingly simple. Over time, though, enhanced user aspirations and heightened cost constraints dictated that each analyzer should make more measurements. The application engineers had to get creative. The unfortunate consequence was a crop of excessively complex analyzers sporting beautiful column systems that were a challenge to maintain.

Clearly, a second detector would simplify the chromatography, reduce the skill level needed for maintenance, and improve reliability. At that time, however, the cost of several detectors and their electronics was prohibitive and would remain so until around 1980, when signal processing speeds finally became fast enough to time-share detectors. Then, why stop at two?

Type C: Multiple detectors, single injector

The need to manage complexity and reduce cost drove a further evolution in PGC technology that introduced the ability to house and continuously monitor two or more detectors.

The use of two or more detectors obviates the need to synchronize peak arrival times, so peaks from different columns can arrive concurrently in their separate detectors, thus simplifying maintenance and reducing the overall analysis time. Figure 9.3 illustrates a Type C column system and its improvement over a Type B.

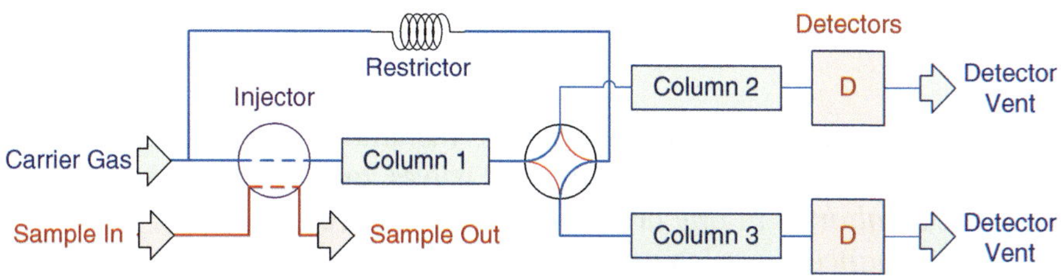

Typical arrangement of a single injector, multiple columns, and multiple detectors using a four-port rotary valve (for illustrative purposes only). Operation is similar to Figure 9.2, but peaks may now arrive concurrently in both detectors.

Figure 9.3 Type C Column System.

The use of a second detector has an interesting history. In addition to the twin advantages of simultaneous detection and reduced analysis time, an additional detector solves two perennial application problems: how to measure hydrogen and how to know when a peak has passed between columns.

In Chapter 6, we saw that the effective measurement of hydrogen requires argon or nitrogen as carrier gas, whereas other components like hydrocarbons are best measured with hydrogen or helium carrier. However, a conventional detector could not use two carrier gases. An early compromise using a mix of helium and hydrogen as carrier was unsuccessful, so creative application engineers used the reference sensor of the detector to measure the hydrogen. It's unlikely that the thermal conductivity detector (TCD) was equally stable with a different reference gas, but the compromise seemed to work; the hydrogen peak gave a strong signal that even had the same polarity as the other peaks.

It was a creative solution to an awkward problem, but a complex arrangement to overcome a hardware limitation never survives for long; development continues. The advent of multiple detectors provided an elegant alternative that eliminated much of the prior complexity. The measurement of hydrogen became a separate chromatographic system running on nitrogen carrier gas. This is our first instance of **parallel chromatography**; it's a Type D column system (see below) comprising two independent column systems each with its own sample injector, column system, and detector. Much easier to maintain.

The other demand for a second detector was the proven technique of monitoring the peaks passing between columns. A small uncalibrated **inter-column detector** can be useful for setting the valve timing. It's not always possible, but it's much easier to set the column valve timing when you can see the peaks leaving a column, as opposed to doing it by blind trial and error.

Once PGCs could host more than one detector, it was easy to install inter-column detectors. The new detectors were much smaller, more sensitive, and less costly than before; and fast digital processing techniques could easily capture multiple signals concurrently.

Unfortunately, though, once PGCs could support more than one detector, the continuous pressure to reduce cost resulted in the sale of complex analyzers that were difficult to maintain. One PGC manufacturer said it well:

> *Over 30 percent of today's process GC installations average two or more detectors, five to six analytical valves, and eight to ten analytical columns in one oven. The result is more complexity, more maintenance, more long-term cost, and less reliability.*
>
> (*Bostic and Clemons* 2009).

One way to improve maintainability is to break a complex application down into simple applications working in parallel. To do this, multiple detectors were not enough; new PGCs also had to support multiple sample injectors.

Type D: Multiple sample injectors

Modern PGCs can accommodate an almost unlimited number of valves, columns, and detectors. Paradoxically, this hardware versatility allows us to simplify the chromatography. We are no longer bound to the philosophy that all peaks must come from one sample injection and one detector. Instead, we can break a complex application down into several simple **applets**, each performing a portion of the desired analysis.

Figure 9.4 shows two applets; each has its own sample injector valve, column system, and detector. Several applets can run concurrently and independently, again taking advantage of parallel chromatography to simplify the analysis and reduce the total analysis time. However, while each applet is itself simple, a bunch of them in the same oven may at first glance look excessively complicated.

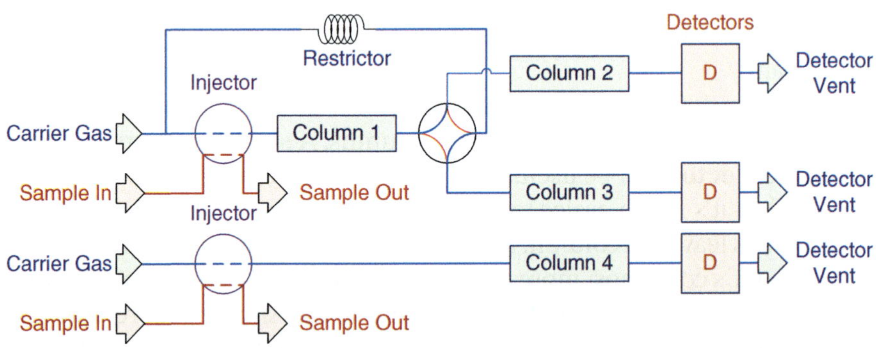

Typical arrangement of multiple injectors, multiple columns, and multiple detectors (for illustrative purposes only). The individual applets can be any of the Type A, Type B, or Type C column systems already considered.

Figure 9.4 Type D Column System.

> *Visual shock: at first sight, an oven full of valves and columns may look rather complex. Start by identifying the individual applets. Locate each sample injector, and then trace the columns, column valve (if any), and detector that it uses.*

It would be good if the manufacturers would color-code the separate applets. A small color sleeve on the tubing would suffice. Mostly, though, you just have to trace the tubing.

Be aware that applets with independent sample injector valves can be set up to inject the same sample or different samples. To discover which, trace the sample flow tubing from inlet to outlet.

- When two or more sample valves inject the same process sample, you have **concurrent separations**, each applet measuring a selected few of the desired analytes.
- When two or more sample valves inject different process samples, you have **concurrent streams**, each applet measuring a different process stream.

Concurrent separations are simple to understand and easy to maintain. The chromatography is easy when it separates only a few selected analytes. Also, the separations running in parallel take less analysis time.

Another variant of parallel chromatography is the duplicate analysis of the same sample stream. With simultaneous sample injections, this provides a concurrent reference analysis to assure the analysis is valid. Or, with staggered injections, it provides a more frequent update of process composition.

Concurrent stream analysis is also an advantage to the end user; it's like having several single-stream analyzers. Independent sample injectors eliminate stream switching, allow no possibility of cross-stream contamination, and consume no additional cycle time. It's best if the injected samples have similar composition, so the parallel column systems can be identical. Again, the resulting systems will be less complex and easier to maintain.

The supplier reaps some benefits too, as is clear from their sales literature:

> *With modular applications and parallel chromatography, application development can be virtually eliminated. Standard application modules and methods can be taken off-the-shelf and installed in the analyzer. Because each module is simple and developed for optimum performance, the result is not only faster delivery, but also a more reliable measurement.*
>
> (*AAI Siemens* 1999).

And they also save cost, since custom application engineering typically amounts to about 15 % of manufacturing cost.

Elemental column systems

Useful techniques

Here, in a simple way, we introduce some basic tools of the trade. This is not an exhaustive review of column system technology. That would be another book (see Mahler et al. 1995).

The basic analytical procedures introduced here are:

- Backflush sometimes called "stripping"
- Distribution sometimes called "dual column"
- Heartcut sometimes called "cutting"
- Trap and hold sometimes called "parking"

Recognize that most PGC analyzers use combinations of these elemental column systems. For instance, a heartcut also needs a backflush to do the housekeeping; it can't do it all by itself.

Backflush column system

Backflush is a Type B column system that removes all injected substances from the first column before injecting another sample. This housekeeping function is so valuable that it's included in nearly all PGC column systems.

Column arrangements that achieve backflush, may use six-, eight-, or ten-port valves, and often combine the backflush function with other functions. Figure 9.5 shows the popular ten-port valve system for sample injection and backflush. In this configuration, actuation of the column valve injects the gas sample and places the two columns in series. The valve stays in the inject position while the analyte peaks (A and B) pass through Column 1 and enter Column 2.

As soon as the analytes are completely out of Column 1, the PGC deactuates the column valve, thereby reversing the carrier flow in Column 1 to flush the heavies peaks (C and D) to vent. Figure 9.5 shows typical peak positions at that instant. From this point on, the column system is saving time by doing two jobs at once: removing the heavies peaks from Column 1 while completing the resolution of the analyte peaks on Column 2. Loss of power or actuating gas automatically puts the column system into this safe mode.

If the flow rate in Column 1 is the same in both valve positions, each heavies peak moves backwards through Column 1 at the same speed as it went in, so they all get back to the beginning of the column at the same time, *regardless of their retention time* on that column. Thus, the backflush technique regroups all peaks remaining in Column 1 as a single composite "heavies" peak.

> *Be sure to understand why the heavies peaks all come out of Column 1 at about the same time, regardless of their retention time in Column 1.*

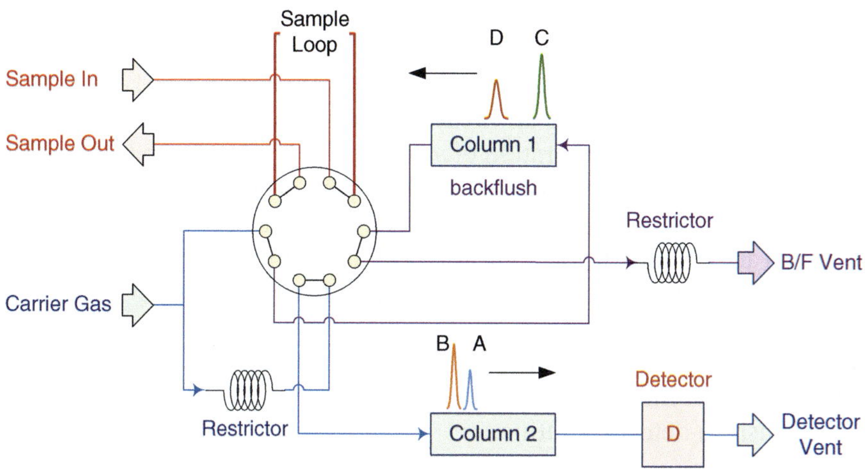

Showing the popular ten-port diaphragm valve in the deactuated state; other ten-port valves have equivalent function. The valve has two positions. In the rest condition shown, the system is backflushing Column 1 and refilling the sample loop. When actuated, each port connects to the opposite adjacent port. This connects the two columns in series and diverts the carrier gas through the sample loop to inject the gas sample into Column 1. The small peaks show the typical position of sample components just before and just after valve deactuation.

Figure 9.5 Ten-Port Gas Injector and Backflush System.

To guarantee the removal of heavies, most chromatographers set the backflush flow higher than the forward flow. This safety margin is acceptable, though it's usually unnecessary with good column design, since the backflush period (valve off duration) should always be longer than the inject period (valve on duration).

Note the use of flow restrictors to control column flow rates. In Figure 9.5, the first restrictor ensures that the detector flow is the same in each valve position. The second restrictor determines the backflush flow rate. These flow restrictors may be small needle valves or capillary tubing cut to the appropriate length. The latter method is cumbersome but ultimately more reliable.

The backflush technique is a wonderful asset to a PGC because it removes all the injected substances, even if we don't know what they are. Two operating modes are available: **backflush-to-vent** or **backflush-to-detector**.

The backflush-to-vent mode illustrated in Figure 9.5 discards the heavies peak and vents it to atmosphere. The alternative backflush-to-detector mode allows the PGC to measure the composite heavies peak, and there are two ways to make this measurement.

In an older PGC that has only one detector, the backflush vent flow tees into the flow from Column 2 and both flows mix together before entering the detector. In this arrangement, the backflush peak must arrive at the detector when no other peaks are eluting from Column 2.

A newer PGC may dedicate a second detector to separately measure the heavies peak in the backflush vent flow. The detection of the heavies is then independent of the main analysis, and peak timing is no longer an issue: both detectors can be measuring peaks concurrently.

The calibration of a composite heavies peak cannot be exact because it consists of multiple substances, some unknown, that have differing detector sensitivities. Nevertheless, the measurement is often useful and can become essential when normalizing all measurements to sum to 100 %. For more information about the calibration of a composite peak, refer to Chapter 14.

Another use of backflush is to assist in a difficult separation. A short-duration backflush delays a few analyte peaks, briefly moving them backward in Column 1 and then forward again into the next column. Of course, the backflush valve must be separate from the injector valve. This "stuttering" technique allows some time for the peaks in Column 2 to move ahead of those briefly delayed in Column 1. Once all the analyte peaks have eluted from Column 1, the analyzer performs a complete backflush-to-vent or backflush-to-detector routine.

Distribution column system

A **distribution** system uses dual columns operating in parallel and may separate the analytes in less time than is achievable on a single column. This arrangement was popular in the earliest PGCs and is still useful today.

Figure 9.6 shows a typical Type C distribution system. After Column 1 achieves a partial separation of the sample components, the column valve directs some of the peaks into Column 2 and some of the peaks into Column 3. This distributes the components between the dual columns for parallel separation, which is often faster and more effective than serial separation.

To clarify, although the concurrent separations occurring in a distribution column system occur in parallel columns, this is not *parallel chromatography*. We reserve the latter moniker for two or more fully independent applets each with their own sample injector, column system, and detector.

In a distribution column system, each column must separate the analyte peaks that will pass through it, so each column might be a different length

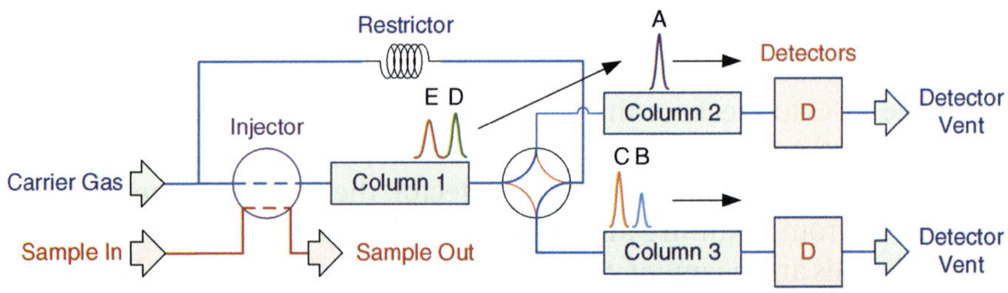

Showing a four-port rotary column valve in its rest condition (blue); other valves have equivalent function. After sample injection, some early peaks (e.g., A) pass through Column 1 and enter Column 2. Then, 90° rotation of the valve (red) allows selected peaks (e.g., B and C) to flow into Column 3. Once the selected peaks are in Column 3, the valve reverts to its rest position allowing later peaks (e.g., D and E) to enter Column 2.

Figure 9.6 Distribution Column System.

and use a different stationary phase. Therefore, a real system might need additional flow restrictors to ensure the same flow rate in each position of the column valve. The flow restrictor shown in Figure 9.6 has the same flow resistance as Column 1. Some applications use a flow restrictor in place of Column 2 or Column 3, giving selected peaks a shortcut to the detector without further separation.

Of necessity, many older PGCs used a Type B distribution column system with a single time-shared detector. With that timing constraint, the lengths of Column 2 and Column 3 became critical. They had to ensure that peaks would arrive in the detector at precisely scheduled times. Although such systems can work very well, the ongoing care of a column system that relies on stable retention times can be challenging and might demand a higher maintenance skill level. It follows that any column system using a time-shared detector may suffer low reliability.

In most applications, the system would also need a backflush function to remove heavies from Column 1.

Heartcut column system

When low concentration peaks follow a high concentration peak, it often turns out to be impossible to measure the small peaks because they disappear in the large **peak tail** left behind by the passage of the **major component**.

For part-per-million measurements, tailing of the major component is not a fault condition; it's an inevitable consequence of high detector sensitivity and large sample size.

The **heartcut** column system is the only way to measure those small peaks. It's a powerful technique, often capable of separating peaks that are a million-to-one in size disparity, yet it uses the simple valve system shown in Figure 9.7.

During sample injection, the column valve is in its rest condition, and Column 1 flows to vent via flow restrictor R2. Column 1 holds back the analyte peaks and allows most of the major component to quickly flow to vent. The column valve briefly operates each time an analyte peak is ready to leave Column 1 and enter Column 2. As soon as the analyte is safely in Column 2, the valve deenergizes.

Each heartcut also captures a small remnant of the major component tail. That's a job for Column 2: it separates the analyte peak(s) away from the remnant peak.

Figure 9.8 illustrates a typical heartcut chromatogram. In this chromatogram, the control unit has attenuated the remnant peak so that it remains visible on scale, and it's easy to see that it retains the shape of the major peak. Frequently, however, remnant peaks are quite large and will fly offscale.

A simple heartcut to measure one or two components is not too difficult to set up and maintain, although the on and off times of the column valve must often be set by blind trial and error. It may seem like an intercolumn

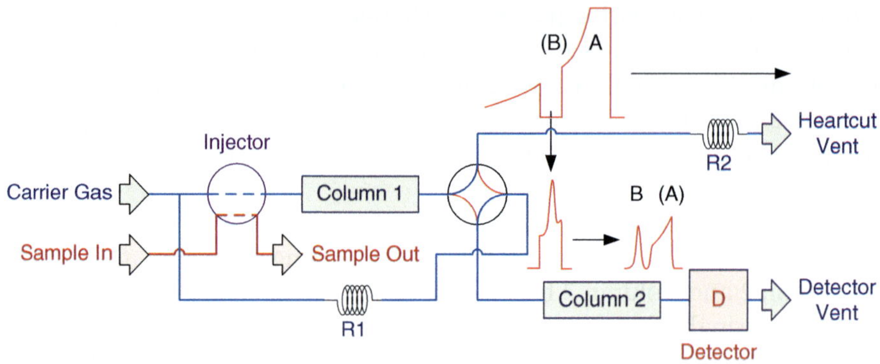

Figure 9.7 Heartcut Column System.

Showing a four-port rotary valve in its rest position (blue); other valves have equivalent function. After sample injection, most of the major component (A) flows to vent. Just before an analyte peak is due to exit Column 1, a short-duration 90° valve rotation (red) allows that peak to enter Column 2, together with a small slice from the unseparated tail of the major component. Once the analyte is fully into Column 2 the valve reverts to its rest state (blue), and Column 2 separates the small analyte peak (B) from the residual portion of major component (A).

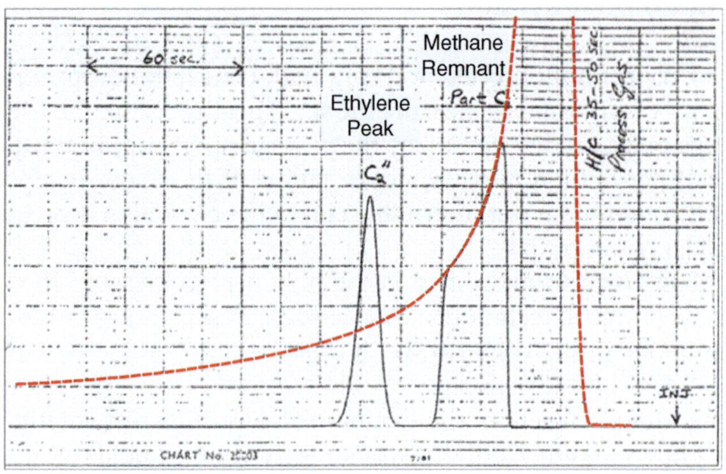

Figure 9.8 A Real Heartcut Chromatogram. *Source*: Author's collection.

This older chromatogram illustrates heartcut well (time progresses right to left). The 700 ppm ethylene peak enters Column 2 together with a small portion of the methane tail. Column 2 easily separates the ethylene peak for measurement, with a nice flat baseline on both sides of the peak. The control unit has attenuated the methane remnant to bring it onscale, and you can see that its top surface retains the original curvature of the methane peak tail, superimposed here in red.

detector would be helpful for visually sighting the valve timing, but it's not so: the separation at the end of Column 1 is rarely enough to see the small analyte peaks.

Multiple heartcuts are far more difficult to maintain. Each heartcut valve action catches one or more analyte peaks, but also transfers a slice of the major tail into Column 2, and all these slices turn up on the chromatogram as remnant peaks. It's sometimes difficult to avoid interference between the flock of remnants and the true analyte peaks.

Most heartcut systems also include a backflush for Column 1 that starts as soon as the heartcut is complete. But some older systems omitted the backflush in an attempt to save money or simplify maintenance. The logic given was that the first column was very short (sometimes only half a meter) and the heartcut valve action was very early in the analysis cycle. Therefore, the first column had a long forward flush to vent, which would remove any late components before the next injection. That might be true for some fairly pure process samples … most of the time!

Let's say it again:

> *The savings gained by running without backflush is not worth the risk. Never run an isothermal PGC without backflushing the first column.*

Trap-and-hold column system

This unique technique—now rarely used—is also known as trap-and-release or trap-and-bypass and allows a complex analysis using a single detector.

The **trap-and-hold** column system immobilizes selected peaks while other peaks separate and clear the columns. It then releases the trapped peaks for further separation and measurement.

Refer to Figure 9.9. When the column valve operates, Column 2 traps selected analyte peaks while the other peaks elute from Column 1 into Column 3. When released, the trapped peaks enter Column 3 for final separation.

Most complete systems also include a backflush for Column 1 that starts as soon as all of the analyte peaks have exited the first column.

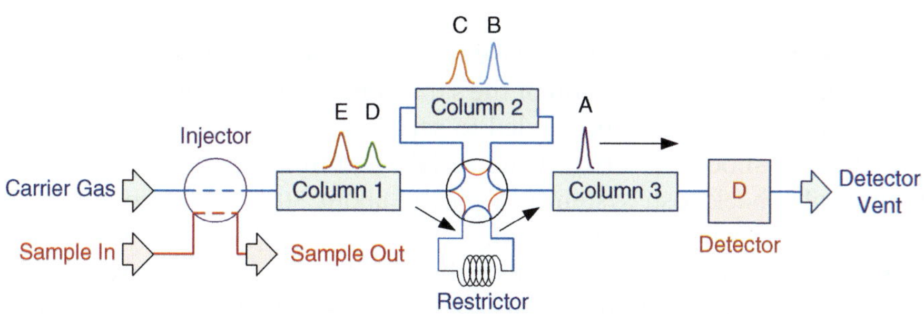

A six-port rotary valve is in its rest state (blue) puts all three columns in series. Other valves have similar function. The flow resistance of the restrictor is equivalent to Column 2. A valve rotation of 60° (red) seals Column 2, trapping selected peaks (e.g., B and C), and connecting Column 1 to Column 3 to allow later peaks (e.g., D and E) to bypass Column 2. When the column valve returns to its rest state (blue) the trapped peaks enter Column 3 for further separation before reaching the detector.

Figure 9.9 Trap-and-Hold Column System.

Trap-and-hold can sometimes act as a heartcut, and at other times it just delays a few components while the other peaks are clearing Column 3. Another valve configuration can backflush the components out of Column 2 instead of trapping them, which is a technique for regrouping bunches of components into single peaks, such as total C_4, total C_5, total C_6, etc. (Villalobos and Turner 1963).

While the trap-and-hold technique was useful for expanding the application of older PGCs having limited column valve capability, it has some disadvantages:

- The trapped component peaks gradually become wider due to longitudinal diffusion of their molecules, thereby reducing the resolution of adjacent peaks.
- The trapped peaks take longer to elute, increasing the analysis time.
- Switching the trap column in and out of circuit may cause pressure and flow upsets that disturb the detector baseline.

Although parallel column systems have generally superseded trap-and-hold, you may encounter some older systems in the field that are still using the technique.

The real power

We have seen that column switching techniques can reduce analysis time, do the housekeeping, and separate extremely small peaks for measurement. Table 9.1 summarizes these splendid features. Yet valve configurations are just the tools we use; the real power of a column system is in the columns themselves.

All isothermal column systems have one feature in common; they all employ two or more columns in series.

If the serial columns all contain the same stationary phase, their performance is predictable. All peaks have the same retention time and width as they would on a single column of the same length, so system design is easy. The total length of column needed to separate the analyte peaks is determined, then allocated between the serial columns. For instance, the first column of a backflush system is typically about 30 % of the total length, leaving 70 % for the second column.

But to release the real power of the serial column system two of the columns must use different stationary phases. Then, the first column achieves a partial separation of analytes from other components and passes those partly separated analytes to a different column for final separation.

In a backflush system, for instance, the first column is often a non-polar silicone oil that separates in boiling point order. This column will keep all

Table 9.1 Summary of Column Systems.

Elemental Column System	Type	Valve Ports	Detector Used	Usage
Single column	A	None	1	Only for temperature ramping.
Backflush-to-vent	B	6, 8, or 10	1	Removes all unwanted heavies peaks.
Backflush-to-detector	B	6, 8, or 10	Shared	Regroups all heavies for measurement.
Distribution	C B	4 or more	2 Shared	Divides analytes into two groups and separates each group on a different column.
Heartcut	C B	4	2 1	Separates ppm peaks from a large major component peak.
Trap-and-hold	B	6	1	Holds selected peaks, then releases them for further separation.
Parallel chromatography	D	Per applet	2 or more	Independently injects, separates, and measures two or more samples concurrently.

the C_4 hydrocarbons close together while holding back the C_5+ hydrocarbons. The bunch of C_4 peaks then pass into Column 2 for separation by a more polar liquid phase while the C_5+ peaks are being backflushed. The wide separation of C_4 peaks in the second column is possible only because the C_5+ peaks are no longer there.

Though hundreds of stationary phases are available, most application engineers limit their choice to about one dozen, which they get to know well. Even so, over one hundred stationary phase combinations are possible in a column system having two columns. This is the true power of the serial column system.

Endnote

PGCs have used two liquid phases to perform complex separations since the early days of the technique (Baker and Zinn 1961). More recently, the technique was rediscovered by academia and adapted to capillary columns for the laboratory analysis of samples containing thousands of analytes. By tenuous analogy with a laboratory method called **planar chromatography**, these researchers tagged the serial-column technique **multidimensional gas chromatography** (Liu and Phillips 1991). This fancy labeling doesn't seem to have improved the technique at all.

Knowledge Gained

- *General elution problem: A column that separates early peaks takes too long to elute later peaks.*
- *Housekeeping Rule: Remove all injected substances before injecting another sample.*
- *There are two ways to solve the general elution problem and satisfy the housekeeping rule.*
- *Temperature programming can reduce analysis time and drive heavies from a single column.*
- *Multiple columns can reduce analysis time and can backflush heavies from the first column.*
- *Older PGCs didn't much use temperature ramping because of the additional complexity involved.*
- *Temperature ramping is successful in PGCs that analyze multiple analytes in wide-boiling mixtures*
- *Temperature ramping has created new ways to measure physical properties like Boiling Range.*
- *The percentage of PGCs using temperature programmed capillary columns is likely to increase.*
- *As yet, most PGCs use multiple column systems running at constant temperature.*
- *Multiple column systems can remove heavies without extending the analysis time.*
- *Multiple columns can separate and measure peaks having a size disparity of one-million-to-one.*
- *Multiple columns can sometimes separate analyte pairs that a single column can't resolve.*
- *If fast repetition is unnecessary, a PGC can wait for an event or time, before running another analysis.*
- *Most PGCs used with temperature programming use a single capillary column.*
- *Application engineers design custom combinations of columns to rapidly perform desired separations.*
- *Complex applications are easier to design and maintain with the option of a second detector.*
- *With multiple detectors, some separations can run in parallel columns.*
- *With multiple injectors, complete analytical applets can run simultaneously and independently.*
- *A PGC applet can measure hydrogen using nitrogen carrier while other applets use a different carrier.*
- *Two or more applets injecting the same sample are each measuring some of the analytes concurrently.*
- *Concurrent analysis makes each separation faster, and easier to set up and maintain.*
- *Two or more applets injecting different samples are measuring different process streams concurrently.*
- *Concurrent stream analysis eliminates cross-stream contamination and needs no extra cycle time.*
- *Backflush-to-vent removes all heavies from the columns without extending the analysis time.*
- *Backflush-to-detector routes the regrouped backflush peak into a detector for measurement.*

- *A brief backflush will delay selected components while other peaks move ahead in the next column.*
- *Distribution allocates analytes to parallel columns for better resolution and reduced analysis time.*
- *Heartcut captures and measures a few small peaks riding on the tail of a large peak.*
- *Trap-and-hold stores a few peaks while others elute, then releases them for further separation.*
- *When a single stationary phase can separate the analytes, it leads to a nice simple column system.*
- *Columns using different stationary phases are a powerful tool for achieving difficult separations.*

Did you get it?

Self-assessment quiz: SAQ 09

Q1. What two mutually exclusive PGC techniques can solve the general elution problem?

Q2. Which standard column system function is missing from the following list? Give the name of the missing column function:

- Backflush-to-vent
- Backflush-to-measure
- Heartcut
- Trap-and-hold

Q3. Why is the ten-port valve so commonly used in PGC column systems?

Q4. How can we be sure that a backflush column system will remove every injected component from the columns when we don't know the identity of all the components present in the sample?

Q5. For a distribution column system, why is it better to use two detectors rather than one detector?

Q6. Is it necessary to use a heartcut column system to measure a ppm peak that elutes before the major component peak?

Q7. What is the purpose of the flow restrictor in the trap-and-hold column system shown in Figure 9.9?

Check your SAQ answers with those given at the end of the book.

Student evaluation test: SET 09

Your instructor will provide answers to these questions.

S1. In gas chromatography, the general elution problem relates to isothermal analysis. Which one of the below-listed statements is the best exposition of the general elution problem?
Select the one best option:

A. It's impossible to resolve all the peaks on a chromatogram by using only one column.

B. A column that resolves early peaks on the chromatogram will strongly retain later peaks and will thereby cause a long analysis time.

C. A column that resolves a pair of peaks on the chromatogram cannot resolve another pair of peaks without spoiling the resolution of the first pair.

D. The settings for optimum resolution are different for each pair of peaks on the chromatogram; therefore it's impossible to set the column conditions to optimize all the desired analyte resolutions.

S2. Which one of the below-listed statements is the best exposition of the Housekeeping Rule?
Select the one best option:

A. The column system must prevent any injected substance from reaching a column or detector that it might deactivate or damage.

B. The column system must remove every injected substance from the columns before another sample injection occurs.

C. The heartcut purging system must sweep away the long trail of molecules that follow a major component peak.

D. The columns must survive the injection of the wildly different sample compositions that may occur during process upsets.

S3. What is the best way to describe a PGC applet?
Select the best description:

A. An applet is a simple PGC column system that doesn't need a separate column valve.

B. An applet is a completely functional chromatography system with injector, column system, and detector, that separates and measures a few (but not all) of the desired analytes.

C. An applet is an elemental column system that achieves a certain type of separation; when combined with other applets it provided a full analysis of the injected sample.

D. An applet is a computer program that resides in the PGC Controller and uses the analysis data to calculate a physical property of the sample, such as its boiling range, Reid vapor pressure, or heating value.

S4. If well-designed, which of the column systems listed below can guarantee an effective housekeeping function?
To receive a score, you must select all the effective systems and none of the ineffective systems:

A. Single column with temperature ramp
B. Backflush-to-vent
C. Backflush-to-detector
D. Distribution
E. Heartcut

S5. If well-designed, which of the column systems listed below can measure a very small impurity peak – say, about 10 ppm – eluting after a 99.999 % major component?
To receive a score, you must select all the effective systems and none of the ineffective systems:

A. Single column with temperature ramp
B. Backflush-to-vent

C. Backflush-to-detector
D. Distribution
E. Heartcut

S6. If well-designed, which of the column systems listed below can regroup selected components into a single composite peak for an aggregate measurement?
To receive a score, you must select all the effective systems and none of the ineffective systems:

A. Single column with temperature ramp
B. Backflush-to-vent
C. Backflush-to-detector
D. Distribution
E. Heartcut

S7. Parallel chromatography is a useful technique in a PGC. Which of the following statements about parallel chromatography is/are correct?
To receive a score, you must select all of the correct statements and none of the incorrect statements:

A. It may concurrently perform the separation and measurement of different analytes in the same process stream.
B. It may concurrently perform the separation and measurement of the same analytes in the same process stream.
C. It may concurrently analyze the calibration sample and the process stream, eliminating the downtime for calibration.
D. It may concurrently perform the separation and measurement of the same analytes in different process streams.

S8. Temperature programming is a useful technique in gas chromatography. Which of the statements listed below about temperature programming is/are correct?
To receive a score, you must select all of the correct statements and none of the incorrect statements:

A. It solves the general elution problem by reducing the retention time of later peaks, thereby reducing the overall analysis time.
B. In addition to the new analysis time, it requires more cycle time to cool the column back to its starting temperature.
C. It reduces the peak width of later peaks, thereby improving their resolution.
D. It is certain to remove all heavies from the column before injecting another sample.

S9. Refer to Figure 9.5, which shows a backflush column system. Imagine that the valve was deactuated two minutes after sample injection. Peaks C and D are now being backflushed to vent.
Assume that the backflush vent flow rate is equal to the column flow rate, as measured at the detector vent.
Assume that Peaks A and B move quickly and are completely out of Column 2 within two minutes after valve deactivation.
Estimate the shortest practical cycle time for this application.
Express your answer in integer minutes.

References

Cited

AAI Siemens (1999). *Advance Maxum™ The First "Plug and Play" Gas Chromatograph System*. Product Brochure (16 pp.). Bartlesville, OK: Siemens Applied Automation, Inc.

ASTM D2887-18 (2018). *Standard Test Method for Boiling Range Distribution of Petroleum Fractions by Gas Chromatography*. West Conshohocken, PA: ASTM.

ASTM D7096-16 (2016). *Standard Test Method for Determination of the Boiling Range Distribution of Gasoline by Wide-Bore Capillary Gas Chromatography*. West Conshohocken, PA: ASTM.

Baker, W.J. and Zinn, T.L. (1961). *Multiple columns in chromatography,* Control Engineering 8 (January): 77–81.

Bostic, S.M. and Clemons, J.M. (2009). 50 Years: Online gas chromatograph. *Process and Control Engineering* (February 12).

Burnell, M.R. and Said, A.S. (1961). Optimized temperature programming in gas chromatography. In: *Progress in Industrial Gas Chromatography,* Vol. 1, 19–30. New York, NY: Plenum Press.

Crandall, J., Roques, N., and Bostic, S. (2014). Rethinking the anatomy of a gas chromatograph. In: *Proceedings of the ISA Analysis Division Symposium,* Baton Rouge, Louisiana (May 5, 2014). *Research Triangle Park*, NC: International Society of Automation.

Liu, Z. and Phillips, J.B. (1991). Comprehensive two-dimensional gas chromatography using an on-column thermal modulator interface. *Journal of Chromatographic Science* 29 (6): 227–231.

Mahler, H., Maurer, T., and Mueller, F. (1995). Multi-column systems in gas chromatography. In: *Chromatography in the Petroleum Industry* (ed. E.R. Adlard). *Journal of Chromatography Library*, Vol. 56, 231–268. Amsterdam, Netherlands: Elsevier.

Rubey, W.A. (1991). A different operational mode for addressing the general elution problem in rapid analysis gas chromatography. *J. High Resolution Chromatography* 14, 542–548. doi:10.1002/jhrc.1240140806

Sewell, P.A. (2000). Gas chromatography: Theory of gas chromatography. In: *Handbook of Methods and Instrumentation in Separation Science,* Vol. 1 (ed. I.D. Wilson and C.F. Poole). London, UK: Elsevier.

Villalobos, R. and Turner, G.S. (1963). The role of column backflushing in gas chromatography: II. Multiple column systems. In: *Gas Chromatography* (ed. L. Fowler). New York, NY: Academic Press.

Table

Figures

9.4 Type D Column System
9.5 Ten-Port Gas Injector and Backflush System
9.6 Distribution Column System
9.7 Heartcut Column System
9.8 A Real Heartcut Chromatogram
9.9 Trap-and-Hold Column System

New technical terms

When first introduced, these new words and phrases are in bold font. For further information consult the Glossary at the end of the book.

analysis time
applet application engineering
aromatics
backflush
backflush-to-detector
backflush-to-vent
column system
concurrent separations
concurrent streams
control unit
cycle time
distribution
flash point
general elution problem
heartcut
heavies
housekeeping rule
intercolumn detector
major component
method
multidimensional chromatography
naphthenes
octane number
parallel chromatography
peak tail
planar chromatography
prime rule
trap-and-hold
vapor pressure

For information about chemical names, refer to the Glossary.

10

Detectors

"Given an effective column system design, the analytes elute from the column one by one, adequately separated from each other. The PGC detector has a simple task: measure analyte molecules in the presence of carrier gas molecules; nothing else is present".

Introduction

As a method of analysis, gas chromatography has a unique advantage: a single detector can measure several different analytes without interference from other components in the sample. Unlike other process analyzers, a gas chromatograph neither needs nor wants a different type of detector for each analyte.

Yet PGCs have used many different detectors.

The detector is central to how a gas chromatograph becomes an analyzer. Without a detector, chromatography would be just a separation technique.

Consequently, several authors have written about detectors. Although slightly dated, you may find Scott (2003) a useful and easily accessible review of detector types. Technical reviews that relate more to process gas chromatographs include those of Clemons (2016, 11–14), Annino and Villalobos (1992, 209–247), and Guiochon and Guillemin (1988, 393–480).

Types of detector

Two measured variables

The two most popular PGC detectors are the **thermal conductivity** detector (TCD) and the flame ionization detector (FID). They couldn't be more different in the techniques they use to acquire the measurement. Yet, more than that, these two detectors illustrate two fundamental properties of all chromatograph detectors.

Process Gas Chromatographs: Fundamentals, Design and Implementation, First Edition. Tony Waters.

All detectors quantify the analyte molecules in one of two ways:

- Their *concentration* in the carrier gas in mol/m^3
- Their *rate of arrival* at the detector in mol/s

Since all detectors fit into one of these categories, it's useful to know which one you are working with when calibrating or troubleshooting a PGC.

Concentration detectors

The thermal conductivity detector (TCD) is a **concentration detector**. It measures the concentration of analyte molecules in the carrier gas passing through it. Since the detector doesn't harm the analyte molecules, it can continue to measure them indefinitely.

> *Be sure to understand that the concentration of analyte in the detector is not the same as the concentration of analyte in the injected sample.*

In principle, the measurement signal from a concentration detector is unaffected by the carrier gas flow rate. Given a constant concentration of analyte, it would continue to output the same measurement signal at any flow rate – even if the carrier gas were to stop flowing completely.

In practice, no detector is perfect, and the response from most concentration detectors depends on the flow rate passing through them. Mostly, they become a little more sensitive at lower flow rates.

Of course, a PGC detector doesn't see a constant concentration of analyte. As a peak elutes from the column, the analyte concentration in the carrier gas rises from zero to a maximum value and then drops back to zero again. The analyte measurement may be based on the peak height or the peak area.

When measuring peak height, a concentration detector will faithfully measure the concentration of analyte in the detector at the peak apex. However, the apex concentration itself might change with a change in carrier gas flow rate; then, the peak height measurement would follow that change. For example, at a reduced carrier gas flow rate the peak would be wider, with a lower concentration at the apex. Naturally, the chromatogram would display the lower peak height.

It gets more interesting, when you're measuring peak area. At a lower flow rate, one might think that the peak area should be unchanged – that the increase in peak width would exactly compensate for its loss of height – but the anticipated compensation doesn't work for a concentration detector. Although the wider peak has a lower apex concentration, it spends more time in the detector and accumulates a larger measured area. Should you wish to check it out, the SCI-FILE that follows lays out the mathematical evidence for this conclusion.

In summary, a concentration detector measures the true height of a peak (which depends on the column performance at different flow rates), whereas the peak area measured by the same detector changes in inverse proportion to the flow rate. Thus, the calibration of the detector is valid only at one flow rate. Therefore, it's important to maintain a constant flow rate when measuring peak area with a concentration detector.

In applications using programmed temperature or programmed pressure, the carrier gas flow rate is not constant throughout the analysis, so a concentration detector is not the best choice.

Rate-of-arrival detectors

The flame ionization detector (FID) measures the arrival rate of analyte molecules into the detector. The act of measurement destroys the analyte molecules, so the detector cannot produce a continuous signal unless it receives a continuous supply of new molecules.

Laboratory workers call this kind of detector a **mass-flow detector**, but talking about the detection of mass can lead to confusion. A process gas chromatograph injects a volume sample and inherently measures volume percent rather than mass or weight percent. Refer to Chapter 7 for a more detailed discussion about calibration in volume percent or weight percent.

To avoid confusion, it's better to think of these detectors as molecule counters. They respond to the number of analyte molecules now present, rather than their concentration. Therefore, for process chromatography, we prefer to emphasize analyte arrival rate in moles per second instead of analyte mass flow in grams per second.

In contrast to a concentration-sensing detector, the signal from a **rate-of-arrival detector** varies with the carrier flow rate. Given a constant concentration of analyte, the detector would output a signal directly proportional to the incoming flow rate. That signal would drop to zero if the carrier gas were to stop flowing entirely.

To evaluate the effect of a flow change, consider for example an increase in the flow rate of the carrier gas. For a given chromatogram peak, the arrival rate of analyte molecules into the detector would then increase, and the detector signal would also increase, indicating the new height at the apex of the peak. The taller peaks would be easier to measure and would improve the limit of detection, as long as the **signal noise** level didn't also increase proportionally.

The effect of flow rate is different when measuring peak area. A flame ionization detector responding to the rate of arrival of analyte molecules is effectively counting the molecules – each arriving molecule contributes to the total integrated response. Therefore, the measured peak area is unaffected by a change of carrier gas flow rate. This is a strong advantage of rate-of-arrival detectors: their calibration by peak-area does not depend on maintaining a constant flow rate.

In summary, a rate-of-arrival detector measures the true height of a peak, which will vary in proportion to the flow rate, whereas a measurement of peak area from the same detector is unaffected by the flow rate.

Multiple detectors

In the early PGCs, it was not practical to have multiple detectors, since that would require duplication of the entire analog signal-processing circuitry. Therefore, most PGCs had one detector, and there was no provision to add a second one. Application engineers had the tough task of ensuring each analyte peak would pass through the detector at a different time. They often met this challenge with delightfully original column designs that worked well in the vendor's factory but were often too complex to maintain onsite.

The development of fast microprocessors that could time-share the digital signal processing changed everything. Suddenly, it was possible to add a detector to separately measure a few analytes at low additional cost. After that, it was a small step to add a second sample injector valve and break an analysis into two or more separate parts. Today, a PGC can analyze a complex mixture using two or more independent applets, each with its own injector valve, simplified column system, and dedicated detector.

The manufacturers call this technique parallel chromatography.

Signal capture

The signal from a sensitive detector is often delicate, being just a few microvolts or nanoamps. The wire connecting the detector to the first stage of amplification can pick up stray voltage signals synchronous with plant power, radio transmissions, or random switching of electrical equipment.

In particular, the signal from an ionization or photometric detector is usually a small flow of electrons, a minute electric current often measured in nanoamps. Such signals require amplifiers with high input impedance that are extremely sensitive to induced voltages. Early FID amplifiers were so sensitive they would detect your presence up to two meters away!

Electronic amplifiers have greatly improved and are now close to the detector, so signal noise from the plant environment is not the problem it once was. Nevertheless, you should be aware of the potential for induced noise.

Induced noise is the chief drawback of analog signal transmission. To minimize pick-up, ensure that the detector conductors comply with all shielding and grounding specifications. Prevention is always easier than cure.

The ultimate signal capture philosophy now used by some PGCs is to digitize the signal right at the detector and transmit the processed detector response to the microcomputer by a short serial-data link.

The SCI-FILE *On Detectors* that follows is optional reading. It gives an overview of detector performance evaluation.

SCI-FILE: *On Detectors*

The reference section at the end of the chapter defines the symbols used here, and their units of measure.

Signal noise

Signal noise is the only real limit to detector sensitivity. In the absence of noise, the electronic signal processors could amplify the smallest detector response and measure an infinitesimal analyte concentration. Thus, most advances in metrology come from reducing signal noise or by finding ways to mask it. Signal-to-noise ratio is the key performance parameter.

Noise is the random variation in a signal not associated with the measurand, and unfortunately present in every measurement. In a PGC detector, these perturbations may come from actual detected events like flow eddies, pressure pulses, contamination, or fine solids in the detector; from induced electric fields or thermal variations in the environment; or from spurious fluctuations within electronic devices.

Two ways to evaluate the noise level in a signal are by peak-to-peak measurement or by statistical evaluation. The peak-to-peak method simply measures the overall band of the variations – it's the way most people would evaluate the noise just by looking at it. Modern PGCs use the statistical method discussed in Chapter 14. By that estimation, the definition of noise level is four times the standard deviation.

Regulatory agencies have found many ways to define acceptable performance, and you should become familiar with those that apply to your industry (refer to Miller 2005, 1–34).

It's generally accepted that for reliable detection a peak must be at least twice the noise level. Environmental standards often specify a minimum detector response of three times the noise level for reliable detection, and ten times the noise level for quantitative measurement.

Process analysis is not so much worried about limits of detection and more concerned about the effect of noise on the accuracy of peak integration.

Chapter 14 describes how analog or digital signal filters can block high-frequency noise, and how an automatic zero can cancel slow baseline drift. But any attempt to remove noise that has frequency characteristics similar to a chromatogram peak is certain to degrade the measurements. Instead of electronic filtering, the PGC must have a column system design that minimizes the baseline disturbances and spurious peaks sometimes caused by column valve switching, and ensures they occur well away from the analyte peaks.

Speed of response

A PGC detector must respond fast enough to reliably follow the rate of change of analyte molecules leaving the column. Since no detector can be instantaneous, all detectors distort the peaks a little. Figure 10.1 shows how that distortion can become severe when the detector can't keep up with the rise and fall of the peak.

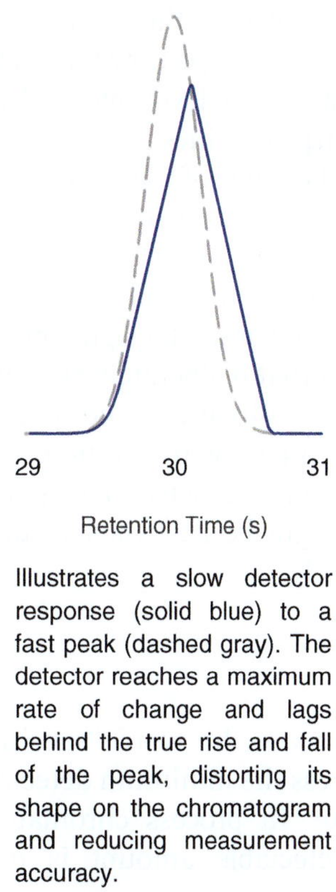

Figure 10.1 Peak Distortion by Slow Detector Response.

Narrow-bore columns generate really fast peaks, and a detector that works well with packed columns may be useless for capillary columns.

Some detectors are much faster than others. The FID has always had the advantage of low internal volume and fast response. More recently, the TCD has almost caught up. Advanced TCDs for process gas chromatographs now have internal volumes and response speeds approaching those of an FID.

The **time constant** (τ) of a detector is the time it takes to output 63 % of the value of a step change applied to its inlet port. After three time constants, the output will reach about 95 % of the inlet value.

The **response time** of a detector is an undefined term. Ideally, it should be four or five time constants (about 98 % or 99 % response), although some authors use *response time* and *time constant* interchangeably.

The time constant of a detector is the sum of several delays due to flow dynamics, rate of sensor change, and electronic or digital signal dampening.

The flow dynamics relate to the internal volume of the detector, the shape of the gas-flow paths, and the carrier gas flow rate. A fast detector will have low internal volume and a flow path with no unpurged cavities.

The rate of change of the sensing element may be a key contribution to the time constant. For instance, a thermal sensor may take a significant time to reach a new temperature.

The electronic or digital signal processing may include filters that deliberately slow the response to eliminate high-frequency noise. Always apply these signal filters with caution, as they can severely dampen the detector response. For fast analysis, it might be necessary to disable signal filtering entirely.

Sensitivity

Some authors use the word "sensitivity" incorrectly to express the minimum detectable amount of an analyte. In process chromatography, the minimum detectable amount is not a useful parameter; we are much more interested in the minimum reliable measurement range.

The **sensitivity** of a detector is the signal it produces per unit quantity of analyte. Therefore, the sensitivity of a concentration detector like a TCD has different units than the sensitivity of a rate-sensitive detector like an FID.

Concentration detectors

By definition, the sensitivity of a concentration detector is the change in signal output per unit change in the *concentration* of analyte in the detector. Therefore, the sensitivity (S_c) of a detector represents its output signal (E) in volts per unit concentration (c) in moles per cubic meter:

$$S_c = \frac{E}{c} \quad (10.1)$$

Thus:

$$E = S_c \cdot c \quad (10.2)$$

The above equation expresses the instantaneous response of the detector to the concentration of analyte molecules in the carrier gas. Ideally, the detector signal would indicate the correct concentration, regardless of the carrier gas flow rate.

Since most PGCs measure peak area, it's useful to see how the detector sensitivity relates to peak area. To get the peak area (A_c), multiply the average analyte concentration during peak integration ($\bar{c}$) by the detector sensitivity and by the integration time period (t_i):

$$A_c = S_c \cdot \bar{c} \cdot t_i \quad (10.3)$$

Furthermore, the average concentration is the quantity of analyte injected (M), divided by the carrier gas volume (V) that passes through the detector during the integration period:

$$\bar{c} = \frac{M}{V} \quad (10.4)$$

And, the carrier gas volume is equal to its flow rate ($\dot{V}$) multiplied by the integration time:

$$V = \dot{V} \cdot t_i \quad (10.5)$$

Combining Equations 10.3, 10.4, and 10.5 gives the peak area (A_c) in volt-seconds:

$$A_c = S_c \cdot \frac{M}{\dot{V}} \quad (10.6)$$

Equation (10.6) reveals two important characteristics of a concentration detector like the TCD: the peak area is directly proportional to the injected quantity of analyte and inversely proportional to the carrier gas flow rate.

Therefore, when measuring peak area, the calibration accuracy of a concentration detector depends on maintaining a constant carrier gas flow rate.

Rate-of-arrival detectors

The sensitivity (S_r) of a rate-of-arrival detector is the output signal (I) in amperes generated in response to a unit change in analyte arrival rate ($\dot{M}$) in moles per second:

$$S_r = \frac{I}{\dot{M}} \quad (10.7)$$

Thus, the instantaneous detector signal is proportional to the analyte arrival rate:

$$I = S_r \cdot \dot{M} \quad (10.8)$$

Clearly, the analyte arrival rate is proportional to the carrier gas flow rate, so the detector response varies with carrier flow. It measures the true height of a peak.

During the integration of a peak, let the average molecule arrival rate be $\overline{n}$ mol/s. From Equation 10.8, the average detector response ($\overline{i}$) is then:

$$\overline{i} = S_r \cdot \overline{n} \quad (10.9)$$

And the total peak area (A_r) is:

$$A_r = S_r \cdot \overline{n} \cdot t_i \quad (10.10)$$

In Equation 10.10, the product of the average arrival rate ($\overline{n}$) and the time of integration (t_i) gives the total quantity (M) of analyte injected in moles. Therefore:

$$A_r = S_r \cdot M \quad (10.11)$$

Equation 10.11 illustrates the characteristics of a rate-of-arrival detector like an FID: the peak area is directly proportional to the quantity of analyte injected and is unaffected by peak shape, retention time, or carrier gas flow rate.

Note: this SCI-FILE could have derived the sensitivity equations more succinctly using the calculus, but we promised not to do that.

Thermal conductivity detector

TCD application

The thermal conductivity detector (TCD) is a concentration-sensing detector that responds to the concentration of analyte molecules in the carrier gas as they elute from a column. It outputs a signal that is linear with analyte concentration in mol/m^3.

The thermal conductivity detector predates the invention of chromatography, so it was the detector of choice in the very first process gas chromatographs. In those days it was known as a **katharometer**. Today, this plucky detector remains a firm favorite of PGC users because it will detect any component – mostly with adequate sensitivity and stability – and is easy to install and cheap to operate, needing no support gases.

The TCD is a universal detector; it will run on any carrier gas and measure any analyte peak, including the **noble gases** and the various

inorganic compounds of hydrogen, carbon, nitrogen, oxygen, sulfur, and halogens.

The TCD doesn't burn the sample, so it's particularly useful for analyzing samples that would burn to yield a corrosive vent gas, such those containing chlorine.

Since it doesn't harm the peaks, a TCD often serves as an **intercolumn detector**, installed between columns to assist with the programming of valve switching times. It's even possible to install another detector directly downstream of a TCD to measure a specific component at high sensitivity, such as a flame photometric detector to measure trace sulfur compounds.

In earlier times, the absence of a live flame was an attractive safety feature of a TCD, although the safety of flame detectors is no longer in question. Some PGC manufacturers have taken the safety advantage one stage further by producing a thermal conductivity detector certified as **intrinsically safe** for use in hazardous areas.

In a process gas chromatograph, the TCD can operate at temperatures up to about 180 °C and usually resides in the column oven to minimize the connection volume and take advantage of the oven's tight temperature control. The thermal mass of the detector provides additional temperature stability, typically maintaining the detector body at ten times less variation than the oven itself.

TCD basic function

The TCD detects the difference in thermal conductivity between pure carrier gas and carrier gas containing some analyte molecules.

The detector body is an aluminum or stainless steel block with narrow passages drilled for the carrier gas flow. As the carrier gas passes through these passages it encounters one or two electrically heated sensing elements, either resistive wires or **thermistor** beads.

When a peak emerges from the column, the component molecules conduct more or less heat away from the elements, changing their temperature and also their electrical resistance. The electronic circuits convert the resistance change to a voltage that becomes the chromatogram signal.

The sensitivity of a TCD improves when the carrier gas and the analyte gas have a large difference in their ability to conduct heat. The thermal conductivity of hydrogen and helium is much higher than for other gases; about 7 and 5.8 times higher than nitrogen, respectively. Therefore, these two gases are by far the most common carriers used with a TCD.

After passing over the heated element, the gas flow continues to an exit port and vents to atmosphere. No chemical reaction occurs in the detector; brief contact with the heated elements has no effect on the carrier gas or analyte molecules.

TCD detection principle

To understand TCD function, consider a single resistive element with a constant voltage across it. Figure 10.2 illustrates this situation. The applied voltage causes a current to flow, and the electrical energy heats the element, raising its temperature. Then, as soon as the element becomes hotter than its surroundings, it starts to lose energy by four familiar mechanisms of heat transfer (Schupp 1968, 138):

- Conduction of heat from one gas molecule to the next one.
- Absorption of heat by the gas, raising its temperature.
- Radiation of heat to nearby colder surfaces.
- Conduction of heat along the connecting wires.

The design of a detector ensures that thermal conduction by the gas is by far the largest contributor to the heat loss. The effect of heat absorption (often called convection) is small and depends on the **specific heat** of the carrier gas. The heat lost by radiation and conduction along the connecting wires is unaffected by gas composition and is also minor, amounting to about 10 % of the total heat loss with nitrogen carrier and only 2 % with helium carrier (Maeda and Kawamura 1981).

The element temperature quickly rises until the total rate of energy loss from all four of the above mechanisms is exactly equal to the electrical power input. From that moment, the element temperature remains constant. Given a constant power input, it can change only if one of the heat-loss mechanisms changes.

The constant temperature so attained is the baseline condition of the TCD and occurs when a constant flow of pure carrier gas is passing through the detector. At that time, the detector outputs a constant millivolt signal. The PGC control unit adjusts this baseline signal to the desired zero position on the chromatogram display. Most chromatographers set the baseline to just above zero so they can see and evaluate any negative excursions.

When an analyte peak elutes from the column and enters the detector, it affects only two of the four heat loss mechanisms:

- The analyte gas conducts more or less heat than the carrier gas does; its thermal conductivity differs.
- The analyte gas absorbs more or less heat than the carrier gas does; its specific heat differs.

The arrival of component molecules has changed the rate of heat loss, so the element is now shedding more or less heat energy than it did before.

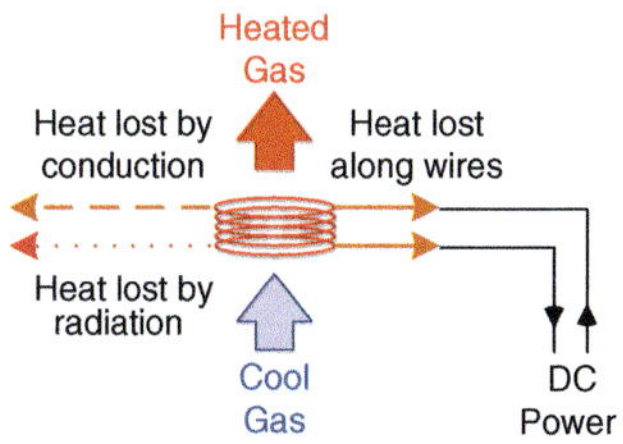

Illustrating the four heat-loss pathways from a heated element in a gas stream: thermal conduction to the surrounding walls, heating the gas (convection), infrared radiation to the walls, and conduction along the connecting wires. The element temperature becomes constant when the total heat loss equals the energy gained from the electrical power supply.

Figure 10.2 Principle of the Thermal Conductivity Detector.

As a result, its temperature rapidly adjusts to maintain the delicate balance between the rate of heat loss and the electrical power input. A higher rate of heat loss cools the element, whereas a lower rate of heat loss warms it to a higher temperature.

Specifically:

- When the carrier gas is hydrogen or helium, its high thermal conductivity keeps the thermal elements cool, since a low temperature differential is enough to conduct the heat away. When an analyte peak enters the detector, the element gets hotter, reaching a maximum temperature at the peak apex.
- When the carrier is another gas – like nitrogen, for instance – the element runs hot and cools down when more-conductive component molecules enter the detector.

Note that the TCD responds to a change in the *thermal conductivity* or *specific heat* of the gas passing through it. In practice, TCDs designs maximize the thermal conductivity response and minimize the specific heat response.

> *One should never forget the effect of* ***specific heat****. It's the reason that the detector response doesn't exactly match the thermal conductivity of the analyte gases and is why the detector is sensitive to flow rate. What's more, it can cause some strange baseline behavior.*

Even when no analytes are present, the temperature of the thermal element depends on several controlled variables; these include the electrical power input, the temperature of the metal block, and the purity of the carrier gas.

To compensate for any fluctuation in these variables, TCDs use a reference sensor that is identical to the measurement sensor but receives only a constant flow of carrier gas. The reference sensor carries the same electric current, feels the same detector block temperature, and touches the same carrier gas. Therefore, any variation in supply voltage, detector block temperature, or gas purity affects both elements equally.

As noted below, the electronic signal from a TCD may derive from sensor voltage, sensor current, or sensor resistance, but it's always founded on the difference between the measurement element and the reference element.

TCD thermal elements

The thermal elements in a TCD act in a dual role: they detect the quantity of analyte molecules present in the column eluent by changing their temperature, and they also provide a measurement of that temperature by a matching change in their electrical resistance. Therefore, for good sensitivity, an element should have a high temperature coefficient of resistance.

In a traditional **hot-wire detector**, most thermal elements are coiled wires of tungsten, platinum, or tungsten-rhenium alloy. Of these, pure tungsten is often the preferred wire material because it has the greatest temperature coefficient and maintains high tensile strength at elevated temperature. In certain applications, thermal elements made of platinum or tungsten-rhenium alloy improve corrosion resistance or facilitate higher wire temperatures.

Figure 10.3 shows another version of the TCD that uses thermistor sensors. A thermistor is a solid-state electronic resistor that exhibits a large sensitivity to temperature change. Its resistive component is a small metal-oxide bead encapsulated in glass to protect the sensor. Thermistors provide the highest detection sensitivity, exhibiting a large *reduction* in resistance for only a small increase in temperature.

However, thermistor elements lose sensitivity at elevated temperature and are rarely effective above 120 °C. The hot-wire detector is a better choice for high-temperature applications.

There was some concern that thermistors might not be reliable with hydrogen carrier gas because the hydrogen could attack the metal oxide

This intrinsically-safe TCD comprises one or two separate four-element detector assemblies. The photograph shows one detector removed for maintenance without disconnecting the tubing. Each detector module can house two measuring and two reference thermistors or three measuring and one common reference thermistor.

Figure 10.3 Intrinsically-Safe Thermal Conductivity Detector. *Source: Siemens Analytical Products and Solutions. Reproduced with permission.*

beads if their glass coating cracked, leading to early failure (Annino and Villalobos 1992, 219). However, other researchers found no cause for concern (Conlan and Szonntagh 1968; Hill 1969, 39). PGC experience over many years has also found that glass-coated thermistors are reliable in hydrogen carrier. As noted below, electronic circuits now keep the thermistor bead at a low constant temperature, avoiding the thermal stress on its glass coating that would otherwise occur during rapid temperature change.

TCD electrical arrangement

The measured variable in a basic TCD is the temperature of the sensing element. To track temperature change, it's convenient to monitor the electrical resistance of the element itself, since its resistance changes with temperature.

The traditional way to measure electrical resistance uses a **Wheatstone bridge** circuit. As shown in Figure 10.4, this circuit has four resistive elements in a series-parallel arrangement. If the elements all have the same resistance, the voltage across the output terminals is zero, but if one element then experiences a small change in resistance, the bridge circuit gives a significant millivolt output signal.

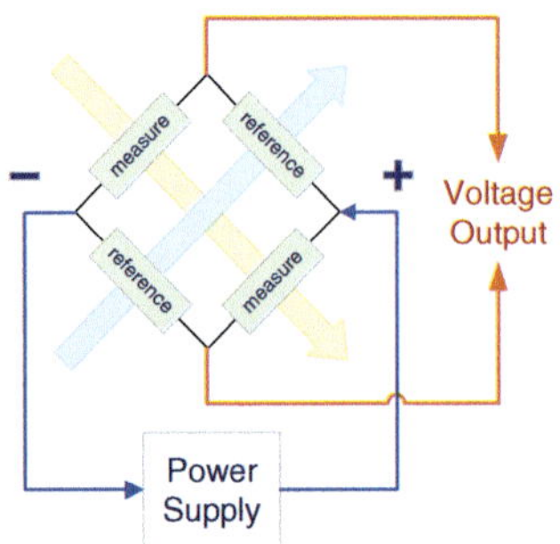

The electrical resistance of the four thermal elements is about the same, resulting in an output voltage of zero millivolts. When the flow of carrier gas over the measuring elements (yellow arrow) contains other molecules, the temperature of those elements changes, causing a proportional change in element resistance and a different bridge output voltage.

Figure 10.4 Typical Wheatstone Bridge Circuit.

Most TCD bridge circuits work on direct current. A notable exception was the 1962 Foxboro Model 91-200 that used 1000 Hz excitation to facilitate amplification and reduce signal noise.

It's evident that two sensing elements in opposite arms of the bridge circuit will generate twice the output signal, so many TCDs adopted four thermal elements, as shown in Figure 10.4, two actively sensing the column eluent and two exposed only to the carrier gas.

TCD electrical improvements

The first TCDs operated with a constant bridge voltage and generated a millivolt output direct to a recorder. This mode of operation has several disadvantages, the worst of which is detector burnout whenever the carrier gas stops flowing.

When a peak arrives in the detector, the element temperature changes, and so does its electrical resistance. That means the current has also changed, so the detector is not operating with constant input power.

For example, when the carrier gas is hydrogen or helium, the element temperature rises when it detects a peak.

Then, if the sensing element is a hot wire, its electrical resistance increases as its temperature rises, reducing the power in the circuit. The power reduction automatically diminishes the temperature change, thereby limiting potential detector sensitivity. On the other hand, the power reduction leads to less chance of an element burnout when a large peak passes through the detector.

Alternatively, if the sensing element is a thermistor, its resistance falls rapidly at higher temperatures, putting more power through the circuit and risking burnout. Therefore, a thermistor circuit should not operate on constant voltage.

To avoid these issues, some detectors have employed a constant-current power supply. With hydrogen or helium carrier gas, an analyte peak causes the temperature of a hot-wire element to rise, increasing its resistance. Then, to maintain constant current, the power supply increases the applied voltage. The additional power drives the element to higher peak temperatures, developing more signal at the expense of reduced sensor lifetime.

With thermistor elements, the constant current results in less power dissipation when sensing a peak, thereby protecting the elements from overheating.

The constant-current philosophy soon gave way to the direct control of element temperature. By monitoring the element resistance, an electronic controller can adjust the applied voltage to hold the resistance constant, thereby holding the element at constant temperature.

Consider a hot-wire TCD element running on hydrogen or helium carrier gas with direct control of element temperature. When a peak enters the detector, the gas conductivity falls, and the element temperature starts to rise. The controller senses the incipient increase of resistance and reduces the applied voltage to hold the resistance constant, thereby maintaining a constant temperature. The applied voltage or current then becomes the measured variable.

Constant-temperature detection has several advantages over the classical modes of operation. A key benefit is its faster speed of response. Temperature change always takes time, which means the output signal from a constant-voltage or constant-current detector always lags the true peak. If the peak is narrow and changes rapidly it may exceed the maximum response speed of the detector, resulting in a distorted and inaccurate chromatogram peak like the one shown earlier in Figure 10.1. A constant-temperature TCD doesn't have the thermal lag of a classical bridge circuit and can respond much faster. It will more faithfully follow a fast peak.

From a practical viewpoint, the nicest thing about the constant-temperature TCD is that its elements don't burn out, even if the carrier gas stops flowing.

TCD performance enhancement

In the past, a thermal conductivity detector had two major disadvantages:

- A baseline noise level that prevented its use for low ppm measurements.
- An unfortunate tendency to respond to the flow and pressure surges created by valve switching.

A low-density carrier gas – helium or hydrogen – minimizes flow upsets and tends to alleviate the latter nuisance. Nitrogen or argon carrier gas creates much larger flow surges.

The design of a TCD is always a compromise between two opposing principles:

- Locating the thermal elements in the direct path of the carrier gas flow provides a fast response, but this geometry suffers from signal noise and a high sensitivity to flow surges.
- Locating the thermal elements in a diffusion pocket where they are not subject to direct impact by the carrier flow generates less noise and flow sensitivity, but this geometry causes slow response.

Over the years, detector manufacturers have offered various **semi-diffusion** flow geometries to explore different trade-offs between the two extremes. Then a bigger challenge appeared. Due to the new capillary columns, it was necessary to reduce the detector internal volume to a size compatible with narrow-bore columns running at low flow rates. A large four-element detector that works well with packed columns can't follow the fast peaks eluting from a low-flow capillary column.

For detecting fast peaks, the classical TCD has two deficiencies that could cause peak broadening:

- A four-element detector often has an excessive internal volume, and
- The signals from two sensors in series are inherently asynchronous.

As column diameters shrank, the TCD evolved, at first reverting to one active and one reference element, and then to dual active elements in parallel, and finally to one reference sensor serving multiple active sensors.

Larger PGCs now have one or two TCD blocks, each containing up to eight sensor elements that may each act in a measurement or reference role.

Each step of this evolution reduced internal volume and increased sensitivity. TCDs are much smaller now. Their internal volume has reduced more than tenfold, from about 300 μL to less than 30 μL.

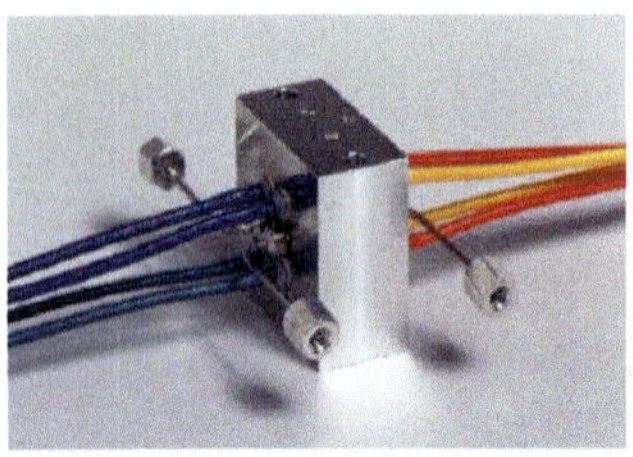

This process TCD can measure low-ppm peaks. It uses hot-wire elements in narrow bores that total only 11 μL in volume. The two measuring elements are in parallel flow paths to eliminate serial delay, as are the two reference elements.

Figure 10.5 High-Sensitivity Thermal Conductivity Detector. *Source: Yokogawa Electric Corporation. Reproduced with permission.*

These exciting developments have greatly improved the performance of TCDs for process gas chromatographs. A good example is the detector pictured in Figure 10.5, which has a volume of about 11 μL. These masterpieces of engineering will measure less than 10 ppm of many analytes – including hydrogen sulfide – and exhibit rapid response.

The response of a concentration detector depends on the dilution of the injected sample in the carrier gas flow. Due to the adoption of narrower columns, the volume of injected sample has much decreased, which, taken alone, would reduce the detector sensitivity. However, the narrower columns also run with much reduced carrier flow, which tends to restore the detector response.

A rate-sensitive detector doesn't have this advantage. It responds to the injected quantity of sample molecules, not their concentration in the detector. As sample volumes diminish, the rate-sensitive detector is at a disadvantage. This suggests that a low-volume micro-TCD will eventually surpass the

Table 10.1 Key Features of the Thermal Conductivity Detector.

Upside:	**Advantages:**
Detects concentration:	Gives larger peak areas at lower column flow rate.
Universal application:	Can measure almost any analyte in any carrier gas.
Linear response:	For about four orders of magnitude (10^4).
Modest sensitivity:	Achieves 0–50 ppm measurement range, or better.
Doesn't burn the sample:	No corrosive vent gases.
	Often used as an intercolumn detector.
Can be intrinsically safe:	Convenient for use in hazardous areas.
Requires no support gases:	Low ownership cost.
Downside:	**Disadvantages:**
Sensitivity for trace analysis:	Older TCDs had poor sensitivity, but new designs have improved – can now measure <10 ppm.
Sensitive to temperature change:	Requires good temperature control, usually mounted in column oven. Unsuitable for temperature programming.
Sensitive to flow change:	Older models may respond to flow disturbances due to valve switching.
Sensitive to pressure change:	Older models may respond to vent pressure change and pressure waves caused by valve switching.
Response varies with carrier gas flow rate:	Calibration depends on constant column flow rate. Unsuitable for pressure/flow programming.
Limited life of elements:	Thermal elements may burn out or change their characteristics and must be replaced.
	Reduced element life with hydrogen carrier gas.

performance of the FID for low-flow capillary columns (Annino and Villalobos 1992, 234).

This prediction is becoming true. The best TCDs already rival the performance of the FID. And are intrinsically safe.

Table 10.1 provides a summary of the performance characteristics of a thermal conductivity detector.

Flame ionization detector

FID application

The flame ionization detector (FID) is a rate-of-arrival detector that responds to **organic compounds** and ignores other peaks. It is very popular for the analysis of hydrocarbon mixtures.

The detector outputs a signal that is linear with the instantaneous flow of analyte into the detector in mol/s. Variations in pressure, temperature, or carrier flow rate have little effect on this detector – it's a very stable device.

The FID has an extraordinary linear range – about six **orders of magnitude**. To put that into perspective, one could calibrate a 0–100 vol. %

A PGC flame ionization detector installed in a flameproof enclosure.

Figure 10.6 PGC Flame Ionization Detector.
Source: ABB Process Analytics. Reproduced with permission.

measurement range and then amplify it exactly one million times to get a 0–1 ppm range, without adjusting the calibration.

A flame ionization detector designed for use in a process gas chromatograph has several safety features that may not be necessary in a laboratory instrument. As depicted in Figure 10.6, the FID may reside in a **flameproof enclosure**, rated for use in a specified **hazardous area**. Alternatively, the outer casing of the detector is itself certified for use in the hazardous area, without further protection.

FID detection principle

To detect the analytes, the flame ionization detector uses the temperature of a small oxygen-rich hydrogen flame to fragment the carbon-hydrogen bonds that are present in the molecules of all organic compounds, forming positive **ions** and free electrons within the flame. A polarized electrode collects the ions, creating a small electron flow to a sensitive amplifier that outputs a millivolt signal.

Figure 10.7 illustrates a typical FID configuration. A mixture of column eluent gas and hydrogen flow to the burner jet, where the hydrogen burns continuously. To achieve the highest signal-to-noise ratio, the hydrogen fuel gas should have a purity of 99.999 % and a hydrocarbon content of less than 2 ppm.

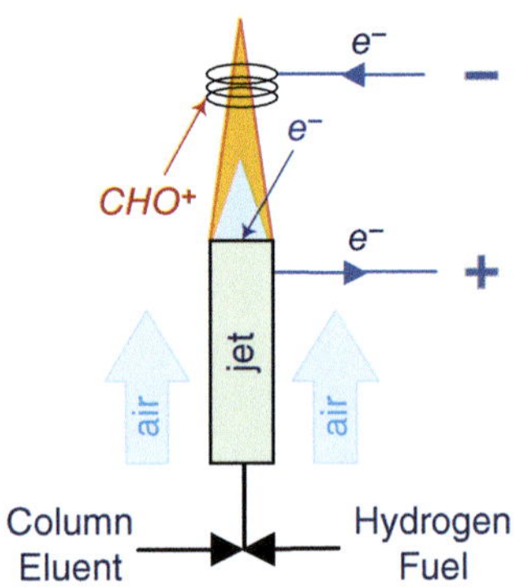

The high temperature of the flame ionizes organic compounds, releasing free electrons that are captured by the postively charged burner jet. The electrons flow through the external measurement circuit to reach the collector where they neutralize positive carbon ions. The external current flow is a measure of the rate of arrival of carbon atoms into the flame.

Figure 10.7 Principle of the Flame Ionization Detector.

To provide excess oxygen for combustion, purified air flows through the detector at about ten times the flow rate of the hydrogen fuel. The air must be dry and devoid of particles and hydrocarbons. When using instrument air, it's purified by passing through a heated catalyst tube that removes all hydrocarbons and other combustible compounds.

Typical flow rates are about 30 mL/min of carrier gas, 30–40 mL/min of hydrogen fuel, and 300 mL/min of pure air (Clemons 2016, 12).

The carrier gas transports peaks from the column directly into the flame where the gas temperature is about 2,000 °C, hot enough to fracture any hydrocarbon molecules and form fragment ions. It's thought that the fragmentation creates CHO^+ **cations** and free electrons in the flame.

A collector electrode in the flame held at about –300 V dc relative to the flame jet creates a strong electrostatic field that drives electrons to the jet and cations to the collector. The shape of the collector electrode is unimportant – it can be a coil, plate, or cylinder – but the collector voltage should be negative, so it draws carbon ions away from the jet to avoid a buildup of carbon deposits there.

A high-impedance amplifier measures the electron flow in the external circuit and creates a corresponding chromatogram signal.

An internal electrode or coil allows remote ignition, and a temperature sensor affirms that the flame is burning or signals a flameout condition if not. Upon receiving a flameout signal, the PGC software can attempt automatic reignition for a specified number of times before issuing a flameout alarm notification.

To manually test whether a flame is present, first check and confirm that it's safe to disconnect the detector vent tubing. If so, disconnect the vent tube and hold a mirror or glass close to the detector vent. A misty condensation on the mirror confirms the flame is burning.

Caution: Remember, there may be a live flame inside the detector. Don't disconnect the detector vent tubing unless it's safe to do so.

FID makeup gases

Since the FID is a rate-sensitive detector, its response should be unaffected by adding a makeup gas to the column eluent gas prior to detection. The makeup gas has no effect on the arrival rate of analyte molecules into the flame, but it does affect flame temperature, causing a minor change of sensitivity. We shall see that the makeup gas may be nitrogen or hydrogen fuel.

The ability to add makeup gas allows the FID to operate on any carrier gas. If the carrier gas is an inert gas like nitrogen or helium, pure hydrogen fuel mixes with the column eluent gas before they flow together into the jet. If the carrier gas is hydrogen, it's common to mix nitrogen with the column effluent instead. Or, for simplicity, the carrier may itself fuel the flame, sometimes with the help of some additional makeup hydrogen to provide enough fuel.

The FID is highly suited to working with capillary columns. Its internal volume in the carrier flow path is exceedingly small, and the possibility of adding makeup gas before detection facilitates a fast response even with the very low flow of carrier gas – sometimes as low as 1 mL/min – preferred by capillary columns.

FID sensitivity

The FID outputs an electronic signal proportional to the momentary population of carbon ions in the flame, which in turn depends on the rate of arrival of carbon atoms therein. The hydrocarbon molecules measured by a PGC may contain any number of carbon atoms from one to more than twenty. For instance, a molecule of pentane [C_5H_{12}] contains five times as many carbon atoms as a molecule of methane [CH_4]. Since equal volumes of low-pressure gases contain the same number of molecules, the FID response to a volume of pentane is five times its response to the same volume of methane.

The increased sensitivity for larger molecules is a big advantage of the FID. You may remember that the peaks for larger molecules tend to appear later on the chromatogram and are wider and lower than earlier peaks. The FID boosts the signal for these later peaks making them taller and easier to measure.

It might be surprising to learn that the FID ionization is not very efficient and only a small percentage of the molecules form ions. Yet the detector is quite sensitive, easily measuring a 1 ppm peak. Its high sensitivity is due to the low noise content of the signal (<10^{-13}A), which enhances the signal-to-noise ratio and allows abundant amplification.

Yet some noise is always present in the output signal due to traces of ionizable impurities in gas supplies, electromagnetically induced interference, and random fluctuations in electronic components. The processor can easily filter out the high-frequency noise, but some signal noise remains, and this remaining noise determines the ultimate sensitivity of the detector. For a particular instrument, sensitivity also depends on the various gas flow rates, the position and the shape of the electrodes, and the flame jet exit diameter (Hinshaw 2015).

To maximize the production of ions – and the detector sensitivity – the FID setup procedure may optimize the flame temperature by adjusting the mixing ratio of hydrogen fuel and an inert gas, such as nitrogen or helium. The mixing ratio is not critical but should stay reasonably constant to preserve the calibration. In practice, the FID sensitivity depends on all three gas flows – carrier gas, hydrogen fuel, and combustion air – so always tune a detector at the carrier flow it will receive during analysis.

FID vent arrangements

Although the FID is not very temperature sensitive, it's good to remember that the hydrogen flame produces water vapor, so the detector must be hot enough to avoid the internal condensation of water. Most PGCs keep the FID at 125 °C. Some detectors vent downwards to facilitate the drainage of condensate, but it's still important to avoid condensation entirely.

The FID must vent to atmosphere, not to flare, and the vent line may also need purging or heating to avoid condensation. A line temperature difference (ΔT) of 20 K above the anticipated water dew point is usually adequate. If using a vertical vent header, don't forget to fit a low-point drain and seal pot.

If working in a potentially hazardous area of the plant, the FID vent line should include an adequate **flame arrestor** to eliminate the risk of a flame front reaching the atmosphere, especially during flame ignition procedures. Some proprietary detectors include a flame arrestor as standard equipment in the detector vent pathway. If so, never remove it.

FID methanator

The FID doesn't respond to carbon monoxide and carbon dioxide, so a PGC often will use a thermal conductivity detector to measure them.

But a TCD is not sufficiently sensitive to measure low ppm amounts of these gases. Therefore, to measure ppm concentrations of carbon monoxide and carbon dioxide – and occasionally formaldehyde – a PGC will separate the analytes and then convert them to methane. A flame ionization detector easily measures the ppm methane peaks.

To convert the peaks to methane, they are reacted with the hydrogen fuel. Before reaching the detector, the mixed fuel and carrier gases flow through a heated **methanator** tube packed with a nickel catalyst that facilitates the reaction with hydrogen. The chemical reaction converts the peaks to methane and water.

For example, here is the chemical equation for the reaction of carbon monoxide and hydrogen:

$$CO + 3H_2 = CH_4 + H_2O$$

You might like to write similar equations for the hydrogenation of carbon dioxide [CO_2] and formaldehyde [CH_2O]. They also convert to methane and water.

The flame ionization detector responds to the methane and ignores the water vapor. The calibration procedure uses a certified standard

Table 10.2 Key Features of the Flame Ionization Detector.

Upside	Advantages
Detects rate of arrival:	Gives taller peaks at higher column flow rate.
Peak area is not flow sensitive:	Doesn't require constant column flow rate. Good for temperature or pressure programming.
Responds only to organic compounds:	Easier to measure low concentrations in non-hydrocarbon samples such as air.
Responds to carbon atoms:	More sensitive to later peaks on the chromatogram.
Linear response:	For about six orders of magnitude (10^6).
Good sensitivity:	Sensitivity for 0–10 ppm measurement range, or better.
True zero:	Unlike TCD, gives zero response in absence of peak.
Can measure ppm of CO and CO_2:	By catalytic conversion to methane.
Very low internal volume:	Suitable for working with capillary columns.
Low temperature sensitivity:	Doesn't need close temperature control – can reside in a separate detector enclosure.
Low pressure sensitivity:	Doesn't respond to vent pressure variations. Smoother baseline during column switching.
Simplicity of design:	Reliable; few modes of failure.
Destroys the sample:	Doesn't vent toxic substances into the environment.
Downside	**Disadvantages**
Doesn't respond to inorganic gases:	Limits FID applications.
High-impedance electron source:	Susceptible to induced signals. Older systems suffered spurious variations of chromatogram signal.
Requires support gases:	Expense and inconvenience of supplying hydrocarbon-free gases; carrier, hydrogen, and pure air.
Additional equipment:	May need purifying devices for support gases.
Potential safety risk:	Cannot be intrinsically safe. Precautions needed due to live hydrogen flame and high voltage.
Burns the sample:	Vent system must prevent condensation of water vapor. Can sometimes produce corrosive vent gases.
Potential for flameout:	An additional mode of failure but mitigated by alarm notification and automatic reignition.

gas containing known amounts of the original analyte gases. Thus, the calibration of peak area includes the efficiency of converting each analyte to methane.

Table 10.2 provides a summary of the performance characteristics of a flame ionization detector.

Flame photometric detector

FPD application

The flame photometric detector finds far more frequent application in process analysis than in laboratory analysis, and it's not hard to see why. As used in a PGC, the detector is selective to sulfur; it doesn't measure anything else.

Like the FID, the flame photometric detector burns the sample components in a hydrogen flame. Instead of collecting ions in the flame, though, it examines the light emitted by the flame, using the principles of **flame photometry** to measure selected atoms present in the column eluent. FPDs tuned for process analysis typically respond only to sulfur atoms.

The process FPD is a rate-of-arrival detector. It outputs a signal that represents the instantaneous population of sulfur atoms in the flame.

The measurement of sulfur has become very important to process plants. Due to regulations that limit the discharge of sulfur compounds into the environment, processing plants need to measure the sulfur content of their raw materials, their products, and their wastes. The FPD is well-suited to this task. Its preferential response to sulfur and high sensitivity make it perfect for measuring low levels of sulfur compounds in hydrocarbon fluids or process waste streams.

PGCs with flame photometric detectors are also employed to protect process catalysts that cannot tolerate the presence of sulfur in their hydrocarbon feed. The catalyst charge in a process reactor is a costly investment worthy of special care (Clemons 2016).

Figure 10.8 shows a typical flame photometric detector for a PGC. The detector is exceptionally sensitive and can measure parts-per-billion amounts of sulfur.

Showing the burner unit of the FPD. The burner unit connects to the photomultiplier by a tubular light guide. The photomultiplier subassembly typically mounts on the outer wall of the enclosure, close to the signal amplifier.

Figure 10.8 Typical PGC Flame Photometric Detector. *Source: ABB Process Analytics. Reproduced with permission.*

FPD detection principle

Figure 10.9 illustrates the basic function of an FPD.

Under the conditions they experience in the flame, sulfur or phosphorus atoms **fluoresce**, emitting light at wavelengths unique to sulfur atoms (centered around violet at 394 nm) or phosphorus atoms (centered around green at 526 nm).

In a process FPD set up to measure sulfur atoms, the light from the flame travels along a fiberoptic cable or light pipe to a measuring device. There, it confronts a fixed **optical filter** selected to pass a band of wavelengths centered at 394 nm and to block other wavelengths.

The filtered light impinges on a **photomultiplier**, a sensitive electronic device that produces a shower of electrons for each photon of light it sees. Electronic circuits amplify the electron flow to give a strong measurement signal for sulfur.

Notice that an FPD responds to atoms, not molecules. To measure the individual sulfur species, the columns must separate them before detection. If an indication of total sulfur content is your goal, a complete separation may not be necessary, but for the highest accuracy it is still advisable.

It's generally thought that the detector doesn't respond to hydrocarbons. In practice, the hydrocarbon response is not zero, but about one ten-thousandth of the sulfur response (Annino and Villalobos 1992, 243). While it's true that a small hydrocarbon peak may go unnoticed by the detector, a large one will upset the chromatogram baseline. The column system must separate and vent any large hydrocarbon peaks before the analyte peaks reach the detector.

Unfortunately, the detector's extreme sensitivity may result in a level of baseline noise that is enough to interfere with low ppm measurement. To overcome this difficulty, a PGC may continuously dose the column eluent with a small but constant addition of a sulfur dopant gas, thereby elevating the baseline and suppressing the noise. Then, when the FPD detects a sulfur peak, it more easily responds to the additional sulfur atoms in the flame.

Like the FID, the FPD sports a flameout alarm and a remotely operated ignition device. One must also keep the detector and its vent system warm enough to preclude moisture condensation within.

Unlike the FID, the FPD flame is short of oxygen, which provides an environment conducive to sulfur fluorescence. Reducing the combustion air flow increases the detector sensitivity but can sometimes make it difficult to light the flame. If so, increasing the combustion air flow may help to ignite the flame, but don't forget to revert to the previous flow setting once the flame is burning. Microprocessor PGCs may increase the air flow automatically during ignition.

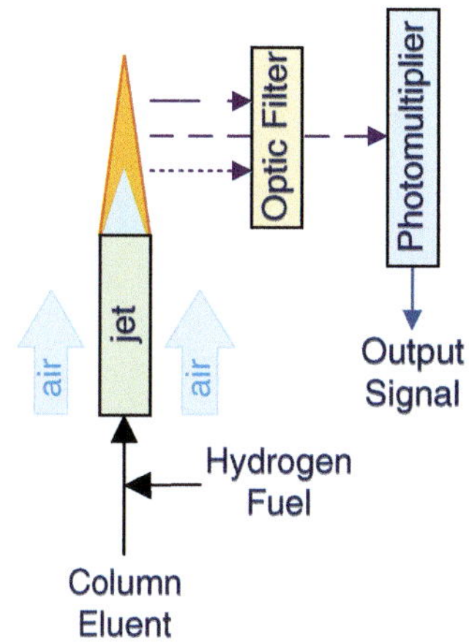

When sulfur compounds burn in the flame, the sulfur atoms emit radiation. The light from the flame falls onto an optical filter that passes only the sulfur wavelengths centered at 394 nm. The filtered light impinges on a sensitive photomultiplier to generate a measurement signal.

Figure 10.9 Principle of the Flame Photometric Detector.

FPD concerns

The FPD is a more temperamental detector than the TCD or FID, mostly due to its exceptional sensitivity. It requires more care and attention:

- All gases used by the detector must be scrupulously clean, and never in contact with polymeric materials.
- All sample-gas passageways must be silicon-coated stainless steel.
- The flow rates of carrier gas, hydrogen, and pure air must be constant because any change in flow ratios will affect the calibration.
- The light path must remain sealed; no ambient light must ever reach the sensitive photomultiplier.
- The detector and vent tubing must be hot enough avoid condensation of water.

The emission of photons is a consequence of electron activation within the sulfur atoms, and it would be nice if that radiation were unaffected by other atoms in the molecule. Unfortunately, we're not that lucky; it's found that the detector response to sulfur varies with molecular composition, so a separate calibration for each molecular species is advisable.

What's more, the detector output is nonlinear with concentration. In theory, the fluorescence centered at 394 nm comes from excited S_2 molecules, so photon emission should be proportional to the square of the sulfur content, but it's never an exact match. Also, when there are higher concentrations of organic molecules in the detector, those molecules may occlude some of the emitted photons and stop them from reaching the photomultiplier. These

Table 10.3 Key Features of the Flame Photometric Detector.

Upside	**Advantages**
Detects rate of arrival:	Responds to selected atoms, regardless of molecular species.
Fully selective:	When tuned to sulfur, ignores other atoms.
Responds to sulfur in different molecules:	Measures hydrogen sulfide, carbonyl sulfide, carbon disulfide, sulfur oxides, mercaptans, and disulfides.
Very sensitive:	Can measure ppb concentrations of sulfur gases in the presence of hydrocarbons.
Choice of carrier gas:	Hydrogen, helium, nitrogen, or argon.
Can measure phosphorus:	If tuned for phosphorus fluorescence by choice of optical filter.
Very low internal volume:	Suitable for working with capillary columns.
Low temperature sensitivity:	Doesn't need close temperature control. Can reside in separate detector enclosure.
Burns the sample:	Destroys toxic gases.
Downside	**Disadvantages**
Temperamental:	More difficult to maintain – highly sensitive to contamination or ingress of ambient light.
Sensitive to flow changes:	Needs constant flows of carrier gas, hydrogen fuel, and pure combustion air.
Suffers some hydrocarbon response:	Sulfur response is 10^4 times hydrocarbon response, but large hydrocarbon peaks will still interfere.
Nonlinear response:	Needs a unique calibration curve for each application. Electronic compensation can provide a linear range.
Requires support gases:	Expense and inconvenience of supplying sulfur-free gases: carrier, hydrogen, and pure air.
Additional equipment:	May need purifying devices for support gases.
Potential safety risk:	Cannot be intrinsically safe. Precautions needed due to live hydrogen flame.
Burns the sample:	Need to heat the detector and its vent system to prevent the condensation of water vapor. Can sometimes produce corrosive vent gases.
Potential for flameout:	An additional mode of failure but mitigated by alarm notification and automatic reignition.

uncertainties make it necessary to electronically tweak the detector signal to generate a linear response.

Because these drawbacks are all obstacles to reliability, we expect to see the FPD superseded by the high-sensitivity TCD in future applications that don't need the selectivity or extreme sensitivity of the FPD.

Table 10.3 provides a summary of the performance characteristics of a flame photometric detector.

Other detectors

Over the years, the process gas chromatograph has used several other detectors, usually to achieve an analysis sensitivity that was then unavailable with a TCD. Most of these detectors rely on the **ionization** of the analyte molecules but use different methods of achieving that.

You might find these ionization processes difficult to understand at first, but they are all variations on the same theme: expose the column eluent gas to energetic particles strong enough to knock electrons out of the analyte molecules, then collect the electrons and measure them as a minute electric current.

The three original detectors described below are mostly obsolete in new PGCs, but you may come across one of them still operating in a processing plant. The pulsed discharge detector described last has generally superseded these older devices and can imitate their function with improved performance.

Here, we simply mention the most common devices and explain what they do. Consult the appropriate instruction manual if you need to work on any of these.

Electron capture detector

The electron capture detector (ECD) is a concentration-sensing detector used in process analysis for measuring minute traces of analytes that an FID cannot see. A PGC would most likely use the ECD to measure analytes containing chlorine or other halogen atoms, but the detector is also sensitive to several other organic compounds.

Figure 10.10 illustrates the basic principle of the ECD. The PGC uses high-purity nitrogen as carrier gas or as makeup gas, so nitrogen is always present in the detector. The column eluent gas passes between two polarized electrodes, the negative one having a radionuclide coating that emits β-particles (fast electrons). Some of these high-energy electrons collide with nitrogen molecules, converting them to nitrogen ions by ejecting low-energy electrons. The freed electrons then migrate to the anode and generate a constant current in the external circuit. This is the baseline condition for an ECD.

When analyte molecules having a strong affinity for electrons enter the detector, they snatch some of the low-energy electrons, causing a reduction

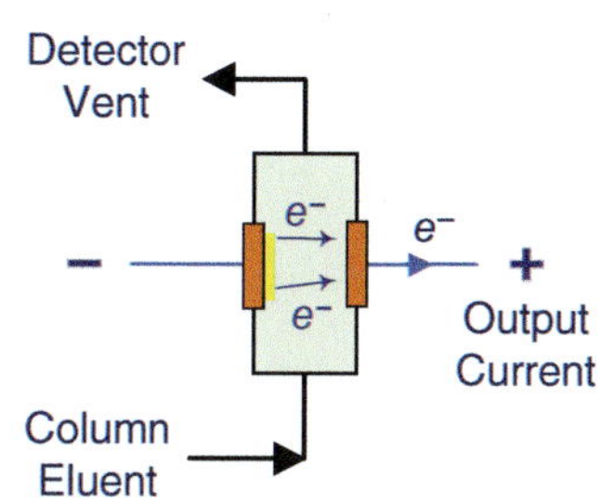

Fast electrons from the radioactive source (yellow) ionize the nitrogen carrier gas, creating a standing current between the polarized electrodes. Certain analytes capture electrons thereby reducing the standing current.

Figure 10.10 Principle of the Electron Capture Detector.

in the output signal. A chromatogram peak is therefore due to a reduction in electron flux.

Most of the negative analyte ions created by electron capture react quickly with the positive nitrogen ions and don't reach the anode.

For process analysis, the ECD is selective to halogen or nitro compounds, and is relatively insensitive to amines, alcohols, and hydrocarbons. Its linear range is good for only about two orders of magnitude when using dc voltage polarization, and the response tends to be unstable due to baseline or calibration drift.

Improved stability and performance accrue from applying microsecond pulses to the collection electrodes instead of a constant dc potential. To get the best performance, an electronic controller moderates the pulse frequency to maintain a constant electrode current. Thus, when the analyte molecules capture electrons, the pulse frequency automatically increases to replace them. The pulse frequency then becomes the chromatogram output signal. This digital approach can yield a linear response over five orders of magnitude (Forsyth 2004).

The ECD is an extremely sensitive detector. As with all such detectors, the gases and their flow paths must be exceptionally pure and clean. Contamination will cause a noisy signal, making analysis inaccurate or impractical.

Helium ionization detector

The helium ionization detector (HID) is a rate-sensitive detector that can measure almost any analyte at extremely low concentration. In practice, though, PGCs use this detector only for the analysis of trace amounts of components that a flame ionization detector cannot detect – the low-level analysis of inorganic gases, for example. The detector responds vigorously to analyte sample concentrations down to 10 ppb and up to about 100 ppm.

The HID uses helium as carrier gas. Inside the detector, fast electrons (β-particles) from a small radioactive source collide with the helium molecules, emitting high-energy photons and forming a cloud of energized helium atoms that can ionize any arriving analyte molecule except neon. Two polarized electrodes collect the ions and electrons formed by the ionization of any analyte molecules. The free electrons migrate to the positive anode creating a current in the external circuit.

The HID is an extremely sensitive and temperamental detector that is difficult to maintain. It reacts to every impurity in the carrier gas, even the minute amount of atmospheric gases that leak through tube fittings (Annino 1993, 100).

The argon ionization detector (AID) is a similar detector that uses argon instead of helium as the ionized gas. The AID is less troublesome than the HID because it responds only to organic compounds, but it still uses a radioactive source, and it can't compete with the simplicity of the FID. The argon ionization detector is now obsolete in process gas chromatographs.

Photoionization detector

The photoionization detector (PID) acts as a concentration detector because the photon radiation destroys very few analyte molecules. In other ways, the PID performs like the flame ionization detector, which it can sometimes replace. Instead of a hydrogen flame, the PID employs ultraviolet photons from a discharge lamp to provide the ionizing energy. The selected ultraviolet lamp irradiates the column eluent gas with enough energy to ionize the desired analyte molecules. The ions and electrons formed then migrate to the charged electrodes, as they do in an FID.

The PID is an extremely sensitive detector with a linearity said to exceed seven orders of magnitude (Guiochon and Guillemin 1988, 471). Depending on the choice of UV lamp, the radiation doesn't have enough energy to ionize certain molecules. For instance, when using a 10.2 eV lamp, the detector gives no response to inorganic gases, light hydrocarbons, and water.

Traces of moisture in the carrier gas will severely dampen the response, so the carrier gas must be thoroughly dry for trace analysis.

Pulsed discharge detector

The pulsed discharge detector (PDD) illustrated in Figure 10.11 is popular in PGCs because it's essentially three detectors in one. In different configurations, it can mimic the functions of the *electron capture detector* (ECD), the *helium ionization detector* (HID), or the *photoionization detector* (PID) – all described above.

Another reason for its popularity is its lack of radioactive material. The absence of a radionuclide eliminates the regulatory procedures and documentation that plague any radioactive device.

Although the PDD is extremely sensitive, less than 0.1 % of analyte molecules ionize so the detector is essentially non-destructive and senses concentration. Therefore, the PDD creates no reaction products in the vent system. Vent heating is unnecessary. On the other hand, any toxic components remain so.

The PDD differs from other ionization detectors by its ionization source. Unlike detectors that use thermal energy or β-radiation, the PDD gets its energy from an electric discharge. The design of the detector ensures that the discharge occurs in a pure gas, either helium or helium doped with another noble gas. The column eluent gas never enters the discharge region, preventing contamination by analyte fragments. The pulsed discharge raises helium atoms to an energetic state while emitting ultraviolet photons. What happens next depends on the detection mode selected (Forsyth 2004).

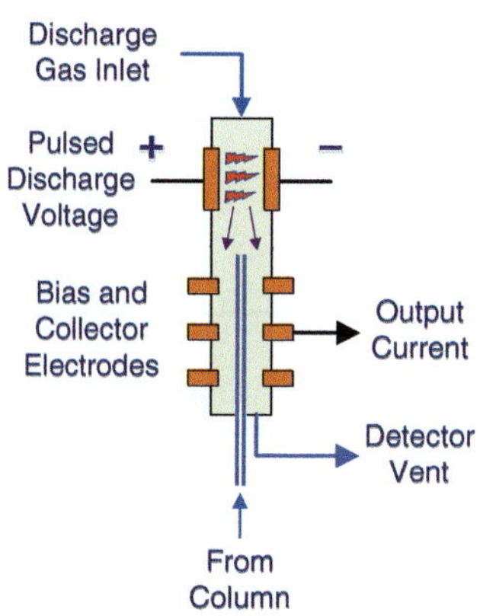

A pulsed electric discharge ionizes the discharge gas which emits ultraviolet photons. The ionized gas then mixes with the column eluent gases and both flow downward across polarized electrodes and out to vent. The ionization of analyte molecules differs according to the selected mode of operation. Each mechanism results in a cloud of free electrons that migrates to the positive collector electrode, providing a sensitive amperometric output signal.

Figure 10.11 Principle of the Pulsed Discharge Detector.

Electron Capture Mode

In the *electron capture mode*, the PDD is a concentration-sensitive detector with extremely high sensitivity to compounds containing chlorine or other halogen atoms, such as freons or pesticides. When the PDD operates in ECD mode, the makeup gas contains an ionizable gas such as methane. High-energy photons from the electric discharge eject free electrons from the methane molecules, thereby converting them to positive ions. A system of bias electrodes focuses the cloud of freed electrons towards the positive anode, where their arrival creates a continuous electron flow in the external circuit. This is the baseline condition of the detector.

When analyte molecules containing electron-deficient atoms like chlorine enter the detector, they are eager to capture electrons and become negative ions. The methane molecules help by colliding with the freed electrons, moderating their energy and facilitating their capture. The capture of electrons reduces the baseline current to produce a downward peak in the chromatogram signal.

Helium Ionization Mode

In the *helium ionization mode*, the PDD is a rate-sensitive detector that performs like an HID. The energy of the pulsed discharge creates metastable helium atoms while emitting high-energy ultraviolet photons. The column eluent gas mixes with the energized helium, which ionizes the analyte molecules and releases free electrons.

In a sense, the detector in HID mode works in an opposite way to the ECD mode. The anode collects the free electrons, generating a positive peak in the detector output current.

Photoionization Mode

In the *photoionization mode*, the PDD is a concentration detector that performs like a PID. The intense ultraviolet radiation from the helium discharge directly ionizes any analyte molecules in the column eluent gas. The bias electrodes focus the released electrons towards the collector anode, where they create an electric current in the external circuit.

In this mode, the detector response to both inorganic and organic compounds is linear over five orders of magnitude. It's possible to tune the detector by adding one of the noble gases as makeup gas, thereby moderating the discharge energy to limit the response to a desired type of analyte, such as all organic compounds or just **aromatics**.

Table 10.4 provides a summary of the performance characteristics of the detectors commonly used in process gas chromatographs.

Table 10.4 Summary of PGC Detectors.

Detector:	Method of Detection:	Measured Variable: [C] = concentration [R] = rate of arrival	Responds to:
Thermal conductivity (TCD)	Heat loss from electrically heated wire or thermistor.	[C] Temperature by electrical resistance.	All components
Flame ionization (FID)	High-temperature ionization of molecules and collection of ions/electrons formed.	[R] Electrode current by electrometer circuit.	Only organic compounds
Flame photometric (FPD)	High-temperature electron excitation and optical selection of 394 nm (S) or 526 nm (P) fluorescent wavelengths.	[R] Photon count by photomultiplier.	All sulfur (S) or all phosphorus (P) atoms
Pulsed discharge (PDD)	High-voltage discharge emits high-energy photons that ionize the reaction gas. Various configurations then mimic the three older technologies listed below.	Reduction or increase in electrode current flow.	Responds per mode of use, as listed below.
Electron capture (ECD)	β-radiation ionizes the carrier gas, and freed electrons create steady current flow. Analytes capture electrons and reduce current flow.	[C] Reduction in electrode current flow.	Oxygen, halogenated hydrocarbons, etc.
Helium ionization (HID)	β-radiation ionizes the helium reaction gas, which in turn ionizes the analyte molecules. Polarized electrodes attract analyte ions and slow electrons, creating a proportionate current flow.	[R] Electrode current by electrometer circuit.	All components.
Photoionization (PID)	Ultraviolet photon ionization of molecules and the collection of ions and electrons formed.	[C] Electrode current by electrometer circuit.	Selected by UV-lamp energy.

Knowledge Gained

Concentration detectors

- *All PGC detectors respond to the difference between pure carrier gas and carrier gas plus analyte.*
- *One kind of PGC detector senses the concentration of analyte molecules in the eluent gas.*
- *The sensitivity of a concentration detector is its signal in volts per analyte concentration in mol/m^3.*
- *A concentration detector doesn't destroy the analyte, so it requires no special vent precautions.*
- *The peak height given by a concentration detector is unaffected by the carrier flow rate.*
- *But if the real peak height changes with carrier flow rate, the detector would respond to that.*
- *The peak area given by a concentration detector is inversely proportional to the carrier flow rate.*
- *Adding a make-up gas to a concentration detector would dilute the analyte and reduce the response.*

Rate-of-arrival detectors

- *Another kind of PGC detector senses the rate of arrival of molecules at the detector.*
- *The technical literature often calls a rate-of-arrival detector a mass-flow detector.*
- *The sensitivity of a rate detector is its signal in amperes per analyte arrival rate in mol/s.*
- *The peak height measured by a rate-sensitive detector is proportional to the eluent gas flow rate.*
- *The peak area measured by a rate-sensitive detector is unaffected by the eluent gas flow rate.*
- *Adding a make-up gas to a rate-of-arrival detector has only a small effect on its response.*
- *The ability to add makeup gas is an advantage for capillary columns working at low carrier flow rate.*
- *A rate detector allows the use of any carrier gas, with fuel or makeup gas added before detection.*
- *The use of multiple detectors helps to simplify the chromatography, improving reliability.*
- *Signal preamplifiers or digitizers are now much closer to the detectors, reducing signal noise.*

Theory

- *To improve the sensitivity of a detector, one must increase the signal-to-noise ratio.*
- *For reliable detection, a peak should be greater than twice the noise level.*

- *In process analysis, the effect of noise on integration accuracy is more important than detectability limit.*
- *Filters can remove high frequency noise, and autozero can correct for baseline drift.*
- *Apply noise filters with caution as they may dampen the response to a fast peak.*
- *Any baseline disturbances that remain must be well-separated from the analyte peaks.*
- *A slow-acting detector will distort the sharp peaks obtained from capillary columns.*
- *The time constant of a detector is its time to output 63 % of a step change of input concentration.*
- *Detector time constant improves with less volume, more flow, and constant sensor temperature.*
- *The response time of a detector is an undefined term that may cause some confusion.*
- *The FID has very low internal volume and no unpurged cavities, so it has a fast response.*
- *Older TCDs have larger volumes, diffusion cavities, and slower thermal change, giving slow response.*

Thermal conductivity detector

- *The thermal conductivity detector (TCD) is an example of a concentration-sensing detector.*
- *The TCD is a universal detector that will run on any carrier gas and measure any analyte.*
- *The TCD doesn't harm the sample, so it's useful as an intercolumn detector to facilitate maintenance.*
- *A TCD typically resides in the column oven to benefit from the close temperature control.*
- *The TCD responds mainly to the differing thermal conductivity of analyte gas and carrier gas.*
- *The TCD has a lesser response to differing specific heats, so the actual response is difficult to predict.*
- *A TCD often uses hydrogen or helium as carrier gas because of their high thermal conductivities.*
- *All TCDs use a reference element to compensate for minute changes in body temperature or gas purity.*
- *Newer TCDs have multiple measuring or reference elements and much less internal volume.*
- *The sensors in a TCD may be thermistors (limited to about 120 °C) or hot wires (up to about 180 °C).*
- *Three modes of electrical operation are constant voltage, constant current, or constant resistance.*
- *The constant resistance mode keeps the sensor temperature constant for faster response.*
- *From many design enhancements, TCD sensitivity and speed of response are now much improved.*
- *Some TCDs for process use have intrinsically safe certification for use in hazardous areas.*

Flame ionization detector

- *The flame ionization detector (FID) is a rate-sensing detector that responds only to organic molecules.*
- *The FID burns a mix of hydrogen and column eluent gas in a small oxygen-rich hydrogen flame.*
- *In the flame, a few of the organic analyte molecules ionize to become positive ions, releasing electrons.*
- *A negative cathode captures positive ions and drives electrons to the burner jet creating a signal.*
- *Purified combustion air flows into the detector at about ten times the hydrogen flow rate.*
- *The FID is a popular detector, being extremely sensitive and linear, and reliable in service.*
- *The FID's low internal volume and tolerance of makeup gas suit it to low-flow capillary columns.*
- *The FID responds linearly to the number of carbon atoms presently in the flame.*
- *Pure hydrogen carrier gas may fuel the FID burner, but a hydrogen-nitrogen mix gives better response.*
- *Adjusting the hydrogen-nitrogen ratio tunes the flame temperature for maximum sensitivity.*
- *The burner creates water vapor that must not condense in the detector or vent piping.*
- *Some sample components create acid vent gases that may need special vent materials.*
- *Never compromise the FID flameout alarm, safety housing, or flame arrester.*
- *A methanator can convert low ppm amounts of carbon oxides to methane, measured by an FID.*

Flame photometric detector

- *The flame photometric detector (FPD) measures low ppm amounts of sulfur compounds, nothing else.*
- *The FPD measures the light emitted when sulfur compounds burn in an oxygen-deficient flame.*
- *An optical filter allows the sulfur wavelengths centered at 394 nm to fall upon a photomultiplier.*
- *The electronic circuits shape the photomultiplier response to make it linear with sulfur content.*
- *The FPD response to hydrocarbons is four orders of magnitudes less than its response to sulfur.*
- *For highly-sensitive work, one may introduce a sulfur gas dopant to swamp baseline noise.*
- *All surfaces in contact with the analyte molecules must be passivated metal and totally clean.*
- *In some applications, the FPD may be replaced by ultra-sensitive thermal conductivity detectors.*

Other ionization detectors

- *The electron capture detector (ECD) is extremely sensitive to analytes containing halogen atoms.*
- *The classic ECD contains only a radioactive cathode and a positive anode.*
- *The β-rays ionize nitrogen, forming nitrogen ions and free electrons that head for the anode.*
- *Upon entering the ECD, analytes containing halogen atoms grab some electrons, forming negative ions.*
- *Then, less electrons flow to the anode, reducing the standing current in the external circuit.*
- *The ECD is now obsolete in PGCs, replaced by the PDD working in ECD mode.*
- *The helium ionization detector (HID) is another highly-sensitive detector using a radionuclide.*
- *In the HID the β-rays energize helium to a metastable state, emitting ultraviolet photons.*
- *The energized helium mixes with the column eluent gas and ionizes every known analyte except neon.*
- *Ionization of analyte molecules releases electrons that migrate to an anode, creating a signal.*
- *The HID is extremely sensitive to all analytes except neon, and is a difficult detector to work with.*
- *The HID is now obsolete in PGCs, replaced by the PDD working in HID mode.*
- *The photoionization detector (PID) is another ionization detector replaced by the PDD.*
- *The PID uses an ultraviolet lamp as the ionizing source but otherwise functions like an FID.*
- *The pulse discharge detector (PDD) has replaced all other ionization detectors in PGCs except the FID.*
- *Instead of a radionuclide, the PDD uses an electric discharge as a source of ionizing radiation.*
- *The absence of a radionuclide simplifies maintenance and eliminates tedious procedures.*
- *The PPD can emulate the ECD, HID, or PID, and tends to perform better due to using short pulses.*

Did you get it?

Self-assessment quiz: SAQ 10

Q1. In what ways does the performance of a concentration-sensing detector differ from the performance of a rate-sensing detector?

Q2. What are the relative merits of thermistors or hot wires as thermal elements in a thermal conductivity detector?

Q3. What advantages and disadvantages are associated with the three different methods of powering hot-wire elements in a thermal conductivity detector?

Q4. What is the purpose of the ~300 V dc voltage in a flame ionization detector?

Q5. Why is it acceptable to add a makeup gas at the inlet of a flame ionization detector, but not at the inlet of a thermal conductivity detector?

Q6. How does a flame photometric detector measure sulfur but not other elements that might be present in the column eluent gas?

Q7. When detecting a peak, what is the output signal of an electron capture detector, and where does it come from?

Check your SAQ answers with those given at the end of the book.

Student evaluation test: SET 10

S1. Why does the detector in a PGC have a simpler task than a detector in a single-purpose analyzer?
Select the one correct answer:

A. A PGC always measures a simple mixture containing only a few components, so it's easy for the detector to measure a single analyte.

B. A PGC separates each analyte from the other components in the sample, so during the measurement of a peak the detector doesn't have to deal with interfering molecules.

C. A PGC detector only measures one kind of molecule, so the presence of other molecules doesn't affect the measurement.

D. A PGC measures each analyte peak by a dedicated detector that never sees any other peaks from the original sample.

S2. According to the text, what improvements in technology have reduced the noise content of PGC detector signals?
Select the two best answers:

A. Detectors are smaller now and much more sensitive than they used to be.

B. Due to improvements in electronic devices, the sensitivity of detectors has improved, increasing the signal-to-noise ratio.

C. Detector amplifiers are now closer to the detector, reducing the length of their sensitive input cables.

D. PGCs now immediately digitize the detector signal and transmit it to the chromatogram processor as serial data.

S3. The classic electron capture detector (ECD) is a simple device that relies on the ionization of nitrogen by β-radiation from a radionuclide and the collection of electrons by a pair of polarized electrodes. Which two of the statements below, when taken together, give an accurate description of how an ECD detects the analytes?
Select the two complementary statements:

A. The ECD ionizes a small fraction of the analyte molecules entering the detector and collects the released electrons to generate a signal representing analyte concentration.

B. The ECD ionizes a small fraction of the nitrogen molecules and in the absence of analyte molecules collects the released electrons to generate a constant "standing current."

C. The electrons released during the ionization of the analyte molecules migrate to the anode creating a steady "standing current."

D. Certain analytes entering the ECD can capture an electron, thus becoming a negative ion. The loss of those electrons reduces the standing current, which becomes a retrograde signal representing analyte concentration.

S4. According to the chapter text, what are the two main categories of detector for process gas chromatographs?
Select the two correct answers:

A. Detectors that respond to the rate of arrival of analyte molecules.
B. Detectors that respond to the ionization of the analyte molecules.
C. Detectors that respond to the concentration of analyte molecules in the sample gas.
D. Detectors that respond to the concentration of analyte molecules in the column eluent gas.

S5. When sensing a peak, the TCD sensor elements can operate in three different controlled modes. Which one of the modes listed below is never employed?
Select the one unemployed mode:

A. Maintaining a constant voltage across the element.
B. Maintaining a constant current through the element.
C. Maintaining a constant power dissipation in the element.
D. Maintaining a constant temperature of the element.

S6. When an FID operates with a potential of −300 V dc on the collector electrode relative to the burner jet, which of the statements listed below are true?
Select all the true answers and none of the false statements:

A. The high voltage inside the detector ionizes any hydrocarbon molecules that arrive in the detector.
B. When the FID detects a peak, electrons flow from the collector electrode to the burner jet via the external electronic circuit.
C. Carbon build-up on the burner jet is less likely than it would be if the jet were negative relative to the collector electrode.
D. Positive ions formed in the flame migrate to the collector, where electrons neutralize them.

S7. Which of the items listed below is a feature of the thermal conductivity detector (TCD)?

Note: In practice there may be some complex interactions when the flow of carrier gas changes. When answering, consider only the effect of the analyte molecules on the detector, ignoring the possible effects of flow rate on detector performance.
Select all the correct answers and none of the incorrect answers:

A. The detector doesn't destroy the analyte molecules.
B. TCD response is linear over about four orders of magnitude.
C. For a constant concentration of analyte in the carrier gas, the detector output signal would be directly proportional to the carrier flow rate.
D. For a chromatogram peak, the measured peak area is inversely proportional to the carrier flow rate.

S8. Which of the items listed below is a feature of the flame ionization detector (FID)?

Note: In practice there may be some complex interactions when the flow of carrier gas changes. When answering, consider only the effect of the analyte molecules on the detector, ignoring the possible effects of flow rate on detector performance.
Select all the correct answers and none of the incorrect answers:

A. The detector doesn't destroy the analyte molecules.
B. FID response is linear over about six orders of magnitude.
C. For a constant concentration of analyte in the carrier gas, the detector output signal would be directly proportional to the carrier flow rate.
D. For a chromatogram peak, the measured peak area is inversely proportional to the carrier flow rate.

S9. Which of the items listed below is a feature of the flame photometric detector (FPD) when measuring sulfur content?

Note: In practice there may be some complex interactions when the flow of carrier gas changes. When answering, consider only the effect of the analyte molecules on the detector, ignoring the possible effects of flow rate on detector performance.

Select all the correct answers and none of the incorrect answers:

A. The detector doesn't destroy the analyte molecules.
B. FPD response is linear over about five orders of magnitude.
C. For a constant concentration of analyte in the carrier gas, the detector output signal would be directly proportional to the carrier flow rate.
D. For a chromatogram peak, the measured peak area is inversely proportional to the carrier flow rate.

References

Cited

Annino, R. (1993). Chromatographs – Gas. In: *Analytical Instrumentation* (ed. B.G. Lipták), 87–108. Radnor, PA: Chilton Book Company.

Annino, R. and Villalobos, R. (1992). *Process Gas Chromatography*. Research Triangle Park, NC: Instrument Society of America.

Clemons, J. (2016). Chromatography in process analysis. In: *Encyclopedia of Analytical Chemistry* (ed. R.A. Meyers). Chichester, UK: Wiley. doi.org/10.1002/9780470027318.a2107.pub3

Conlan, D.A. and Szonntagh, E.L. (1968). The effects of hydrogen carrier gas on thermistor detectors in process gas chromatography. *Journal of Chromatographic Science* 6 (9, September): 485–487. https://doi.org/10.1093/chromsci/6.9.485

Forsyth, D.S. (2004). Pulsed discharge detector: theory and applications. *Journal of Chromatography A* 1050 (1, September 24): 63–68.

Guiochon, G. and Guillemin, C.L. (1988). *Quantitative Gas Chromatography for Laboratory Analyses and On-Line Process Control*. Amsterdam, Netherlands: Elsevier Science Publishers B.V.

Hill, D.W. (1969). Detectors for gas chromatography. In: *Gas Chromatography in Biology and Medicine* (ed. R. Porter). London, UK: J. & A. Churchill, Ltd.

Hinshaw, J.V. (2015). Flame ionization: GC's workhorse detector. *LCGC North America* 33 (7, July): 470–477.

Kania, A. (2011). Achieving improved sensitivity of a gas chromatograph's thermal conductivity detector through optimization of thermal design characteristics. In: *Proceedings of AD 2011: ISA 56th Analysis Division Symposium*, League City, Texas (April 10–14, 2011). Research Triangle Park, NC: International Society of Automation.

Maeda, M. and Kawamura, S. (1981). Analysis of hot wire thermal conductivity detectors for gas chromatography. In: *Proceedings of the 31st Japan Analytical Chemistry Society*, Tokyo (in Japanese).

Miller, J.M. (2005). *Chromatography: Concepts and Contrasts*. Hoboken, NJ: Wiley.

Schupp, E.S. III (1968). *Gas Chromatography*. New York, NY: Interscience division of John Wiley & Sons.

Scott, R.P.W. (2003). *Gas Chromatography Detectors*. Chrom-Ed Book Series, www.chromatography-online.org (accessed 2019-07-11).

Tables

10.1 Key Features of the Thermal Conductivity Detector
10.2 Key Features of the Flame Ionization Detector
10.3 Key Features of the Flame Photometric Detector
10.4 Summary of PGC Detectors

Figures

10.1 Peak Distortion by Slow Detector Response
10.2 Principle of the Thermal Conductivity Detector
10.3 Intrinsically-Safe Thermal Conductivity Detector
10.4 Typical Wheatstone Bridge Circuit
10.5 High-Sensitivity Thermal Conductivity Detector
10.6 PGC Flame Ionization Detector
10.7 Principle of the Flame Ionization Detector
10.8 Typical PGC Flame Photometric Detector
10.9 Principle of the Flame Photometric Detector
10.10 Principle of the Electron Capture Detector
10.11 Principle of the Pulsed Discharge Detector

Equations

Relating to a concentration-sensing detector:

10.1	$S_c = \frac{E}{c}$	Detector sensitivity.
10.2	$E = S_c \cdot c$	Response to an analyte concentration.
10.3	$A_c = S_c \cdot \bar{c} \cdot t_i$	Peak area in volt-seconds.
10.4	$\bar{c} = \frac{M}{V}$	Average concentration during peak integration.
10.5	$V = \dot{V} \cdot t_i$	Volume of carrier gas flow during peak integration.
10.6	$A_c = S_c \cdot \frac{M}{\dot{V}}$	Peak area (inversely dependent on flow rate).

Relating to a rate-sensing detector:

10.7	$S_r = \frac{I}{\dot{M}}$	Detector sensitivity.
10.8	$I = S_r \cdot \dot{M}$	Response to analyte rate of arrival.
10.9	$\bar{i} = S_r \cdot \bar{n}$	Average response during peak integration.
10.10	$A_r = S_r \cdot \bar{n} \cdot t_i$	Peak area in ampere-seconds.
10.11	$A_r = S_r \cdot M$	Peak area (independent of flow rate).

Chemical reactions in a methanator:

Formaldehyde: $CH_2O + 2H_2 = CH_4 + H_2O$

Carbon monoxide: $CO + 3H_2 = CH_4 + H_2O$

Carbon dioxide: $CO_2 + 4H_2 = CH_4 + 2H_2O$

Symbols

Sym.	Variable	Units
A_c	Peak area from a concentration detector	$V \cdot s$
A_r	Peak area from rate-sensitive detector	$A \cdot s$
c	Concentration of analyte in carrier gas	mol/m^3
$\bar{c}$	Average concentration during integration	mol/m^3
E	Response of a concentration detector	V
I	Response of a rate-sensitive detector	A
$\bar{i}$	Average response of rate-sensitive detector	A
M	Quantity of injected analyte molecules	mol
$\dot{M}$	Molar flow of analyte molecules into detector	mol/*s*
$\bar{n}$	Average arrival rate of analyte during integration	mol/s
S_c	Sensitivity of a concentration detector	$V \cdot m^3/mol$
S_r	Sensitivity of a rate-sensitive detector	$A \cdot s/mol$
t_i	Time period of integration	s
T	Temperature	°C
ΔT	Temperature difference	K
τ	Detector time constant	s
V	Volume of carrier gas during integration	m^3
$\dot{V}$	Volume flow rate of carrier gas	m^3/s

New technical terms

When first introduced, these new words and phrases were in bold type. Refer to the Glossary at the end of the book for further information.

aromatics
cation
concentration detector
flame arrestor
flame photometry
flameproof enclosure
fluoresce
hazardous area
hot-wire detector
inorganic compound
intercolumn detector
intrinsically safe
ion
ionization
katharometer
mass-flow detector

methanator
noble gas
optical filter
order of magnitude
organic compound
photomultiplier
rate-of-arrival detector
response time
semi-diffusion
sensitivity
signal noise
specific heat
thermal conductivity
thermistor
time constant
Wheatstone bridge

For information on the names of chemical compounds, refer to the Glossary.

11

Temperature control

"The position of a peak on a chromatogram is very sensitive to a change of column temperature. And the chromatogram baseline may drift up or down if the detector temperature changes. It's therefore important to maintain very precise temperature control".

Need for stability

Temperature is an important variable in gas chromatography, as it affects sample volume, peak retention time, and detector stability.

Sample volume

The effect of temperature starts with the amount of sample injected. The sample valve dispenses a fixed physical volume of the sample fluid, so any change in the sample temperature or pressure may change the number of molecules injected, thereby introducing a measurement error.

Earlier, we saw that a temperature change of ± 0.33 K causes ± 0.1 % change to a gas sample volume, and this contributes ± 0.1 % to the overall uncertainty of the measurement. But the performance of a modern PGC oven is at least ten times better than that, reducing the error to about ± 0.01 % of measurement value.

Retention times

The second and more worrying aspect of temperature variation is how it affects the position of peaks on the chromatogram. Peak retention times may change as much as 3 % per kelvin.

As yet, PGC detectors cannot identify the components they are measuring. They rely on a repeatable retention time to identify each peak and to initiate other timed events on the chromatogram. While it's possible for a

Process Gas Chromatographs: Fundamentals, Design and Implementation, First Edition. Tony Waters.

computer-controlled chromatograph to track the movement of a peak, it's simpler to minimize the variation of retention time by maintaining a constant column temperature.

A constant temperature may also improve the regulation of carrier gas pressure, thereby maintaining a more constant carrier flow rate.

Because of these dependencies, good temperature control is vital to a process gas chromatograph. Having a stable oven temperature enhances the accuracy of peak detection and quantification, and it helps to ensure that chromatographic valves switch at the correct times.

Detector baseline

The third effect of temperature variation is its effect on the detector baseline, which depends on the detection technique employed. A thermal conductivity detector (TCD), for instance, is much more sensitive to temperature change than a flame ionization detector (FID).

When a detector is sensitive to temperature, any variations will show as noise or as regular oscillations on the chromatogram baseline. Ultimately, this signal noise may limit the ability of the PGC to measure analytes at low ppm concentrations. Therefore, precise temperature control of such detectors improves the precision of measurement.

To minimize detector noise, PGCs often have a secondary way to control detector temperature and thereby improve their analytical performance. A temperature-sensitive detector may incorporate a large metal block that, when mounted in the oven, reduces the temperature variation of the detector to about one-tenth of the oven variation. Alternatively, the detector may have its own dedicated temperature controller.

In addition to the direct effect of temperature change on the detector, variations in column temperature affect the rate that a stationary phase will evaporate from a liquid-phase column and "bleed" into the detector. For packed columns, we choose liquid phases that have extremely low vapor pressure, but it's never zero. A small concentration of the vapor is always present in the carrier gas leaving a liquid-phase column – we call it **column bleed**.

A sensitive flame ionization detector may respond to this bleed, offsetting the chromatogram baseline. Random changes in the bleed rate caused by minute changes in column temperature can then cause signal noise and possibly a cyclic baseline variation synchronized with the heater cycle.

Later models of PGC use proportional temperature control, rather than the on-off thermostat of the older models. In addition to stabilizing peak retention times, this improved control of column temperature minimizes the variation in column bleed. Also, the latest capillary columns with bonded stationary phases are somewhat immune from this malady.

Heating technique

Except for the 1976 Foxboro 91-PCT that used steam heating (Annino et al. 1976), PGCs have always used electric heating. Having an electric heater in an industrial plant demands careful design and maintenance because a hot surface can ignite a flammable atmosphere. The choice of heating technique dictates the necessary safety features, which in turn tends to influence the overall instrument design.

PGCs use three different methods of heating:

- An **air-bath oven** uses hot-air heating, which provides a large open space for columns and valves. The air flow also acts as a safety purge to dilute any flammable gases to below their lower flammability limits.
- An **airless oven** incorporates a large metal mass that has high heat capacity and therefore resists temperature change. In the early PGCs using this technique, the small oven space led to predesignated locations for columns and valves, limiting application flexibility.
- A **directly-heated column** has a resistive heating element in direct contact with the column or uses the electrical resistance of the column itself to achieve rapid temperature change. Without a conventional oven, the analyzers can be smaller, lighter, and modular.

As you might expect, each technique has advantages and disadvantages.

The air-bath oven

Heating with air purging

The air-bath oven has always been popular in PGCs as it's reliable, effective, and economical. However, it's not the best way to minimize the overall cost of ownership, due to the high consumption of instrument air. Yet, since the very first instruments back in 1955, most instruments sold have used this heating technique, and it's found in the majority of industrial PGCs working onsite today.

The basic idea of an air-bath design is simple; just circulate a flow of heated air in an insulated enclosure to keep the columns and detectors warm. Due to its similarity to a laboratory oven, everyone calls the heated enclosure an **oven**.

Laboratory gas chromatographs use a resistive wire heater and a fan to rapidly disperse heated air throughout the column oven, but open-wire heaters and fan motors are difficult to use in a hazardous process area. An electric fan motor suitable for use in the potentially explosive atmosphere of an industrial plant would be bulky and expensive.

The solution adopted by Union Carbide in one of the world's first "home-made" gas chromatographs ever to work online (*circa* 1955) is still the most common technique used in PGCs today. Carbide engineers used heated instrument air for the dual roles of protecting the heater and distributing the heat.

In the Union Carbide original design, heated air drove a rotary fan that mixed the incoming air with cooler oven air and distributed it throughout

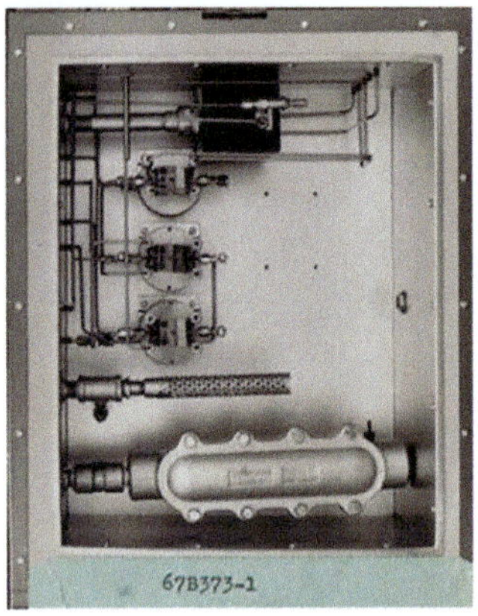

An early air-bath oven, clearly showing the baffled outlet for heated air (columns not yet installed).

Figure 11.1 1967 Beckman Series D Air-Bath Oven. *Source: Beckman Historical Collection, Box 84. Science History Institute. Philadelphia. Reproduced with permission of Rosemount, Inc.*

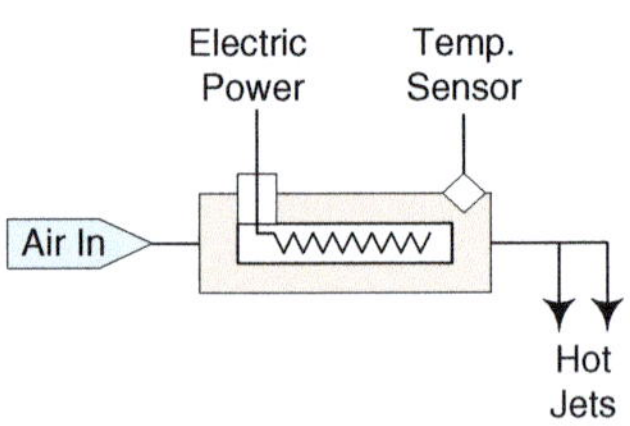

Instrument air cools the heater shell for safety as it becomes heated,and then jets into the column oven to heat and circulate the oven air.

Figure 11.2 Typical Air-Bath Heater.

the enclosure. Apparently, they became concerned that a mechanical fan would not be reliable in continuous service, so they replaced it with a hot-air venturi nozzle having no mechanical parts (Spracklen 1957).

About forty years later, Yokogawa revisited the original Union Carbide concept and reengineered it to perform with high reliability. The newly-designed air-driven heater fan was a key feature of the 1993 Model GC1000 and subsequent Yokogawa products (Matsuura 1994).

Carbide specified their fanless air-bath heating technique for the first commercial PGC ever produced. They entrusted the production of this instrument to the Watts Manufacturing Company in West Virginia. After Beckman acquired Watts in 1956, they retained the air-bath oven for the first Beckman PGC, the Model 120A, and for all subsequent Beckman PGCs. Figure 11.1 shows a hot-air heater installed in a 1967 Beckman Series D PGC Oven.

Meanwhile, some ex-Watts folks who declined to join Beckman in California started a new company in Ronceverte, West Virginia, to manufacture process gas chromatographs: Greenbrier Instruments Inc. This innovative company would greatly influence the future direction of PGC design, yet they continued to incorporate an air-bath heating system in all but one of their models; the 1959 Chroma-Matic 555 used an airless oven heater. The corporate heirs of Greenbrier continued to use air baths in their PGCs, culminating in the main line of PGCs now produced by ABB Process Analytics. Applied Automation and Hokushin quickly followed suit.

Figure 11.2 illustrates the design of a typical air-bath heater. Instrument air flows in the annular space between the heater element and the outer tube, completely isolating the electric heater from the outside air. The flow of air is sufficient to ensure that the outer tube never exceeds the T-rating of the local environment, typically T3 (max. 200 °C).

The heated air enters the oven through a narrow jet that produces a high velocity stream. Due to momentum exchange, a reduced pressure zone develops behind the jet that draws in a larger volume of colder oven air, and then mixes with it to maintain precise temperature control.

To ensure safety, PGC heating systems always incorporate sensors to guard against overheating. Typically, these sensors monitor the air supply pressure and the heater surface temperature. If the air supply pressure fails or the outer metal skin exceeds a safe working temperature, logic circuits will shut off the heater power supply.

Air-bath ovens are economical to manufacture. Using microprocessor control, they achieve excellent temperature stability, typically ±0.02 °C. They heat up and cool down quickly, making them easy to maintain; and they tend to be spacious, with more room for columns, valves, and detectors than the thermal mass type. This flexibility gives the air bath a wide range of application that can include temperature programming when necessary. As pictured in Figure 11.3, some PGCs even feature a temperature-programmed oven within an isothermal oven (Yokogawa 2001, 5).

When compared with an "airless" thermal mass oven, the main disadvantage of an air-bath oven is the large amount of instrument air it consumes and the cost of heating that air to oven temperature. These are ongoing costs of operation. Also of concern are the temperature differences observed at various locations within these ovens, especially near the hot air jets. Although the temperature control is very precise, significant gradients can exist from place to place within the oven. Moving a column within the oven may change the retention time of some peaks.

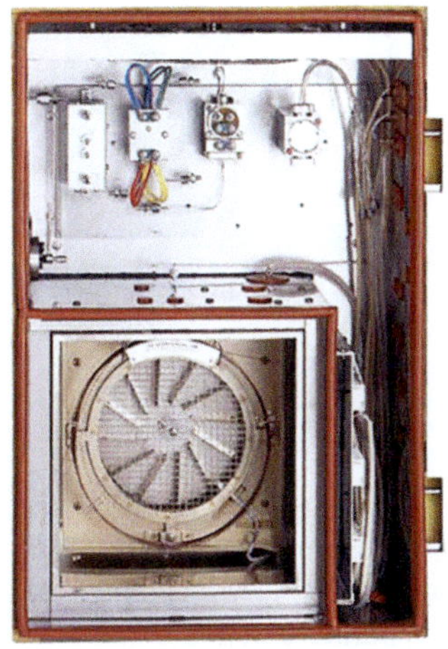

The smaller oven of the 2001 Yokogawa GC1000 Mark II may be temperature programmed.

Figure 11.3 Dual Oven in the Yokogawa GC1000 Mark II. *Source: Yokogawa Electric Corporation. Reproduced with permission.*

The airless oven

Heating without air

The thermal mass approach to temperature control is quite different from the air-bath technique. The basic idea is that a metal block has a high heat capacity, so it must absorb or release a large amount of energy to cause a significant temperature change. It follows that the temperature of a large mass of metal can't change quickly and the simple on-off thermostatic control of oven heater power may provide adequate temperature control.

The simplicity of on-off temperature control was an attractive feature in the early days of process chromatography, as electronic controls were not well-developed at that time. In addition, the need for a large metal block provided a perfect way to house the electrical devices in an explosion-proof enclosure, separate from the non-hazardous valves and columns.

Its inherent reliability and simplicity of fabrication led several manufacturers to choose thermal mass heating systems. Examples are the 1971 GEC Elliot Mark 6 (Figure 11.4) and the 1987 Yokogawa GC8 (Figure 11.5).

GEC Elliot used the thermal-mass approach to temperature control.

Figure 11.4 The 1971 GEC Elliot Chromatograph Mark 6. *Source: Author's Collection.*

At first, thermal mass heating proved itself superior to air bath design. The oven required no air supply, negating a significant cost of an air bath. The encased electric circuits imbued confidence in their safety. The heating elements and controls almost never failed. The temperature control was excellent, with minimal gradients, and the layout gave immediate access to columns and valves.

On the negative side, the thermal mass took a long time to heat up or cool down, increasing the time needed for maintenance. Although oven temperature could rarely exceed 150 °C, this type of heating was adequate for most applications during the first 25 years of PGC development. Then, in the 1980s, improvements in column and valve technology demanded higher oven temperatures and even temperature programming. The thermal mass approach couldn't satisfy those requirements.

A second factor in the decline of the thermal mass oven was the ever-increasing complexity of PGC applications. The second 25 years of PGC development was a time of intense competition between PGC vendors for large new plant projects. To reduce their bid, applications engineers strove to get more measurements out of each PGC, and this led to more

The Yokogawa GC8 also used the thermal mass heating approach. A proportional temperature controller held the oven to within ±0.03 °C of set temperature.

Figure 11.5 The 1987 Yokogawa GC8 PGC Oven. *Source: Yokogawa Electric Corporation. Reproduced with permission.*

columns, more valves, and more detectors in each analyzer. At that time, the thermal mass instruments had a rigid arrangement of valves and detectors mounted on a thermal block and couldn't easily accommodate additional devices. The open and roomy air baths had no such constraint and could easily satisfy the demand for more heated space.

The new millennium brought a spurt of PGC development centered around oven capacity. The new models that evolved had a practically unlimited capacity for valves, columns, and detectors, and all had air-bath ovens – often several of them! Typical examples of this trend include the 1998 AAI-Siemens Maxum, the 2008 ABB PGC5000, and the 2011 Yokogawa GC8000.

Yet the thermal mass design was not dead. It enjoyed a renaissance in the advent of special-purpose PGCs designed for the natural gas industry, such as the 1978 Daniel Danalyzer and the 1998 Rosemount Model GCX. The developers of later transmitter-style chromatographs followed the success of those early pioneers. By focusing on the advantages of the airless oven, several companies developed reliable low-cost instruments that need no air supply, require no shelter, and are ideal for simple applications. Instead of instrument air, these small PGCs use carrier gas or electric solenoids to actuate their injector and column valves.

Small, transmitter-style PGCs that calculate the **heating value** and other flow properties of a gas are perfect for the natural gas transmission and processing industries. The price of natural gas is determined by heating value, so both buyers and sellers of natural gas are highly motivated to get a timely and accurate analysis. That creates a large market for these special-purpose PGCs. In total, it's likely that PGC vendors now sell more instruments with airless heaters than with air baths.

By 2000, the PGC suppliers had split into two warring camps, one preaching the high application flexibility of the air-bath heater and the other championing the merits of the airless oven. Both were successful in their own market. In time, however, it was inevitable that both camps would covet the other's market. So, after long decrying their opponents' technique, they started to adopt it: those of the air-bath persuasion introduced a radical new style of airless PGC, while their well-established competitors came out with much more flexible airless designs – some as large as air baths!

The truth, of course, is that each of the heating techniques has strong advantages in different applications. In summary, here are the main advantages of the airless oven:

- Requires no instrument air – lowering cost of operation.
- Enables a small transmitter-style PGC for dedicated applications.
- Requires no shelter – reducing installation cost.
- Installed near process line – reducing sample system complexity.
- Easy maintenance – swapping a compact chromatographic module.

Airless ovens have now gone full circle and are now available from a majority of companies that manufacture PGCs today. Some examples follow.

The 2008 ABB PGC1000

The ABB Totalflow PGC1000 shown in Figure 11.6 is a small explosionproof PGC designed mainly for the at-line analysis of natural gas. Like other small PGCs of its genre, a PGC1000 can calculate the **density** and heating value of natural gas.

The PGC1000 has 27 standard "column trains" for measuring different analytes, thus eliminating the need for individual application engineering. Each analyzer can accommodate up to four of these trains (ABB 2012).

By the way, an ABB **column train** has the same meaning as a Siemens **applet**. Both terms refer to a valve-column-detector combo dedicated to achieving a limited analysis. If you are interested in the wide range of possible column systems, the ABB brochure referenced above is an excellent source of knowledge, as it displays many column system diagrams and their resulting chromatograms.

This instrument takes modularization to the limit and houses the entire chromatographic system in one replaceable GC Module that is not serviceable on site. The GC Module is a miniature airless column oven that contains two ten-port diaphragm valves and the necessary pressure regulators, conventional columns, thermal conductivity detectors, and local electronics.

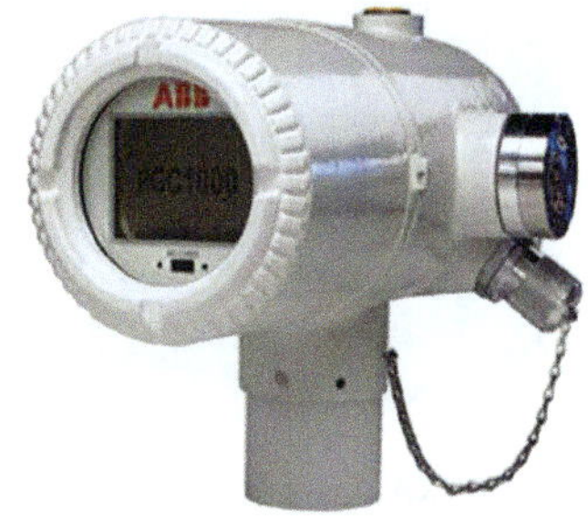

A transmitter-style PGC that can measure several selected analytes in light hydrocarbon streams particularly in the natural gas industry.

Figure 11.6 The 2008 ABB PGC1000 Transmitter. *Source: ABB Process Analytics. Reproduced with permission.*

The 2014 Rosemount Danalyzer 370XA PGC

The 370XA in Figure 11.7 is directly comparable to the ABB PGC1000, and also designed for natural gas analysis. Again, all the chromatographic devices are in a removable module, but this one is a "maintainable module" for repair on site or return to the factory for service. The module incorporates an airless isothermal column oven preset to 80 °C, three six-port diaphragm valves, conventional micropacked columns, and a thermal conductivity detector.

The 370XA is usually set up to measure C_1–C_5 paraffins and C_6+ as a composite peak, all in four minutes. It has advanced software for calculating heating value and other properties of natural gas.

An analytical transmitter for natural gas analysis. The entire chromatographic assembly is a maintainable module for onsite or offsite repair.

Figure 11.7 The 2014 Rosemount 370XA PGC. *Source: Rosemount Inc. Reproduced with permission.*

The 2009 Rosemount 700XA PGC

The 700XA illustrates a recent trend for transmitter-style PGCs to become more flexible in application, allowing them to branch out from their original specialty – natural gas transmission pipelines – and compete with traditional suppliers in the light hydrocarbons market.

The 700XA is ATEX-certified for operation in Zone 1 hazardous areas and doesn't need a shelter. There's a picture of it in Figure 11.8. The airless analytical oven in a 700XA can run up to 120 °C and pack up to six

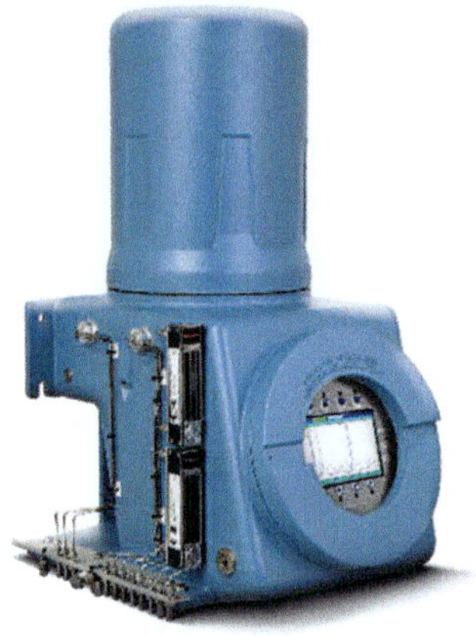

The 700XA has an airless oven with more flexibility for optional valves, columns, and detectors than most transmitter-style PGCs.

Figure 11.8 The 2009 Rosemount 700XA PGC. *Source: Rosemount Inc. Reproduced with permission.*

The 2002 Maxum Edition II has an optional airless oven that is large enough for flexible column and valve arrangements.

Figure 11.9 An Airless Oven and Separate Detector Compartment. *Source: Siemens Analytical Products and Solutions. Reproduced with permission.*

diaphragm valves, several micro-packed columns, and three detectors. This complement of analytical hardware and as many as four programmed methods gives a 700XA the capability to satisfy most applications in the natural gas, petrochemical, and oil refining industries.

The 2002 Maxum Edition II

Around the turn of the millennium, the cost of instrument air became critical, opening a window for small PGCs with airless ovens to penetrate the petroleum markets. Recognizing the threat, all major PGC manufacturers quickly launched an option for airless ovens in their flagship models. These new ovens don't cling to the old principle of an explosion-proof thermal mass because thermal inertia is less important with digital temperature control. The electric heat comes from a small cartridge heater.

As an example of this trend, Figure 11.9 shows an airless oven for the Siemens Maxum Edition II. This PGC can have single or dual independent airless ovens. You can see that the new airless column ovens are light and spacious. Observe the separate enclosure just above the oven. For easy maintenance, this detector bay houses all the detectors except the TCD. The temperature-sensitive thermal conductivity detector remains in the column oven.

Direct column heating

Resistive heating

A third approach to temperature control is direct heating of the column by an electric current. This technique goes by the name of **resistive heating**, although that moniker would seem to fit all forms of electric heating. Here, we are talking about applying heat directly to the column tubing. For electric heating, there are three ways to do that:

- Run a heating wire in direct contact with the column tubing.
- Fit the column into a compact heated cartridge.
- Pass an electric current through the column tubing itself.

All these methods have been employed in laboratory chromatographs, but they pose issues of safety when used in a hazardous plant environment.

Direct column heating is an old idea recently trumpeted as a new development. It is not. The first commercial instrument to offer direct column heating was the 1956 Burrell Kromo-Tog K-2, a laboratory gas chromatograph. It had a nichrome heater wire wound around the column (Guild 1956).

The ABB approach

ABB was the first to use direct column heating in a process gas chromatograph, at first experimentally (Clemons and Thomas 2000), and then as standard in their 2001 PGC2009 and 2017 PGC5009 fast process chromatographs. These instruments pass the heating current through the metal walls of a short micropacked column and employ fast computer control to regulate the column temperature. The column tube is inside a wider tube that allows a flow of vortex-cooled air to rapidly cool the column at the end of each analysis.

The initial application for the Fast PGC was the simulated distillation of wide-boiling hydrocarbon mixtures. This standard method is sometimes known as a "**SimDis**" analysis. It employs a temperature programmed non-polar column that retains components in linear proportion to their boiling points, thus allowing the time axis of the chromatogram to be correlated with **boiling point** temperature (ASTM D2887 or ASTM D7096).

Calibration of the boiling point axis and the detector response comes from the analysis of a known mixture of paraffins. Figure 11.10 is an example of the rapid separation of paraffins that is possible with this instrument. To achieve such an impressive speed, the PGC uses both temperature and pressure programming.

A typical gasoline sample generates over 400 overlapping peaks, none of which have enough resolution for individual measurement. But a distillation analyzer doesn't need to measure the components; it estimates the percentage of the sample that would vaporize at a given temperature. By correlating component retention times to their boiling point temperatures, the chromatogram becomes a plot showing the relative amounts and boiling points of all 400+ components. The PGC estimates **percent-off** at any temperature simply by comparing the total peak area eluting below that boiling temperature with the overall area of all the peaks.

Percent-off is the percentage of the liquid sample that would become a vapor at a specified temperature. When required, an inverse calculation can derive the temperature at which a specified percentage of the sample would boil off. This value is sometimes useful for the automatic control of distillation processes.

The ABB PGC2009 pictured in Figure 11.11 can perform a complete boiling point analysis of gasoline every four minutes, which is about three times faster than a PGC using conventional temperature programming.

Direct column heating eliminates the conventional oven and is a major step in the reduction of instrument size, weight, and cost. In addition, fast heating and cooling make it possible to analyze wide-boiling mixtures. Yet, although direct heating is a useful tool for difficult applications, it's rarely used for the simpler applications common in a petrochemical plant. It's easier and more reliable to use a less complex isothermal column system for those measurements.

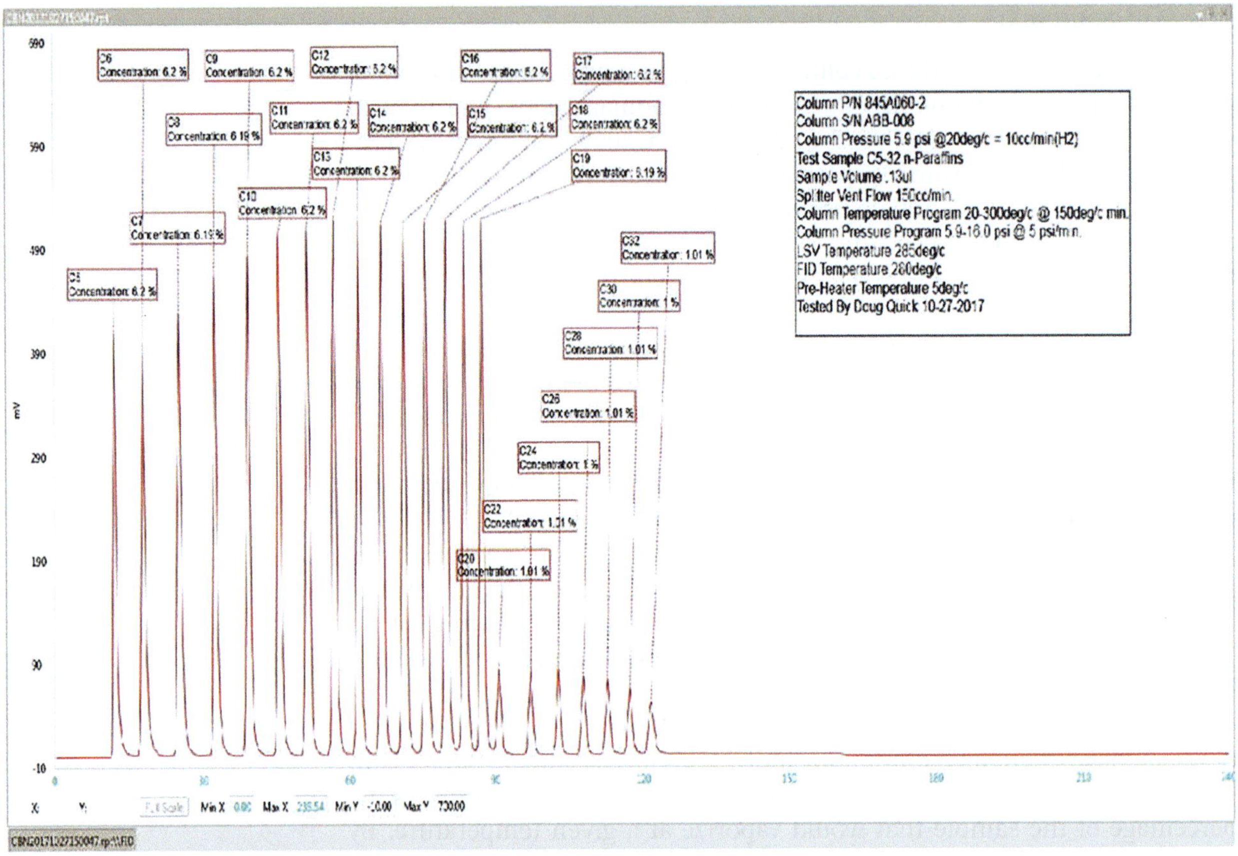

Calibration run on the Fast PGC showing the separation of C5–C32 paraffins in about two minutes. To achieve this fast analysis, ABB employs a linear increase of both temperature and pressure (i.e., flow).

Figure 11.10 Fast PGC Separation of Paraffins.
Source: ABB Process Analytics. Reproduced with permission.

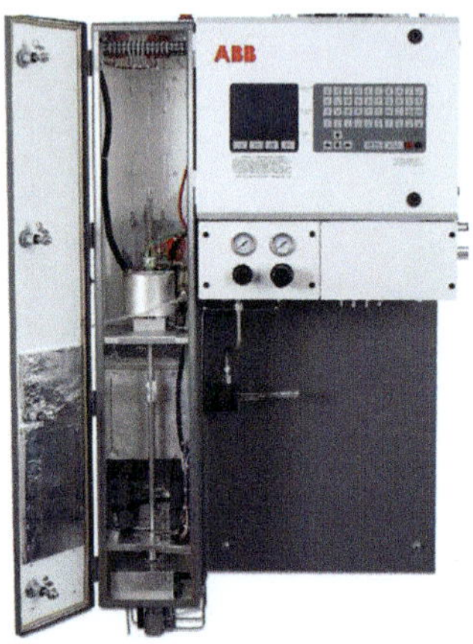

The directly-heated column in the ABB PGC2009 is the short vertical tube visible at lower left.

Figure 11.11 A PGC With Directly-Heated Column.
Source: ABB Process Analytics. Reproduced with permission.

The Teledyne Falcon approach

Figure 11.12 shows the 2007 Falcon *Calidus* from Teledyne Analytical, a portable chromatograph for operation in non-hazardous areas in the laboratory or field. The analyzer has a choice of column modules and detector modules, allowing it to satisfy a wide range of applications. Each module incorporates a digital electronic controller.

This novel product also uses rapid temperature ramping to achieve extremely fast analysis. The column modules contain a resistively-heated steel capillary column with integrated digital temperature control that can achieve the analysis of dozens of analytes under isothermal or programmed temperature conditions. Figure 11.13 is a *Calidus* column module that is capable of ramp rates from 0.1 to 10 K/s up to 350 °C (623 K). When the analyzer uses two column modules, it can also have a column valve for column switching (Crandall et al. 2014).

The *Calidus* uses peak-tracking software to ensure the correct identification of peaks, even when their retention time varies (Rohrback and Ramos 2017). While clearly a benefit, this feature might be a necessity with dozens of peaks on a fast temperature ramp. When peaks elute in seconds, a slight change in the rate of temperature rise may have a significant effect on their retention time.

The *Calidus* analyzer is a general-purpose instrument, not designed for operation in a hazardous environment. When installed onsite it needs to reside in a purged enclosure. This added protection is an adequate short-term fix, but retroactive measures are rarely successful in the long term. We hope the company will soon produce a true process version.

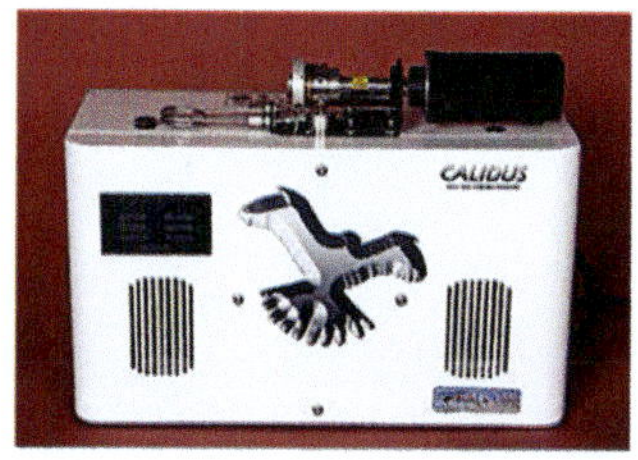

An ultra-fast gas chromatograph using direct column heating.

Figure 11.12 The 2007 Teledyne *Calidus* Portable Analyzer.
Source: Teledyne Analytical Instruments, a business unit of Teledyne Instruments, Inc. Reproduced with permission.

A few cautions

Direct column heating is a powerful technique for the analysis of complex sample mixtures beyond the capabilities of a conventional process chromatograph. The chromatograms are astounding. But the routine and reliable analysis of dozens of components is inherently difficult, and those analyzers may not always work as well as expected.

A major issue with the capillary columns used in most direct heating applications is their limited dynamic range. We often expect our process gas chromatographs to measure components on several different ranges, but fast capillary columns overload very quickly and a sample size suitable for measuring one component on a range of 0–0.1 % may not work well for measuring another component on a range of 0–100 %. The application engineer may find it necessary to carefully tune the injected sample volume to yield an adequate separation of the analyte peaks.

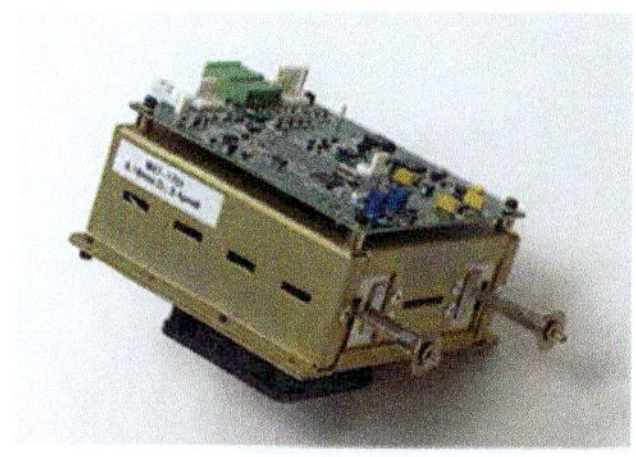

This compact design incorporates 16 m of steel capillary column with direct resistive heating and electronic temperature control.

Figure 11.13 A Direct-Heated Column Module.
Source: Teledyne Analytical Instruments, a business unit of Teledyne Instruments, Inc. Reproduced with permission.

These powerful machines tend to get the difficult applications that no other PGC can do. Yet direct heating can perform some simple analyses in a few seconds and is likely to do so with high reliability. For process applications, that might be a better way to go.

Before buying an ultra-fast process gas chromatograph, evaluate your sampling system: there's no point in having a fast PGC if your sample extraction, transport, and conditioning systems are abysmally slow. Don't forget that process liquids may spend an hour or more in process vessels before they arrive at the sampling tap. No benefit will accrue from connecting a fast PGC to a slow sampling system.

Summary of heating methods

Table 11.1 summarizes the advantages and disadvantages of the various heating methods used in PGC column ovens.

PGC standardization

Realities of the market

The users of PGCs often express their desire for a small analytical device that needs no analyzer shelter, can reside near the sampling tap, and is easy

Table 11.1 Summary of Oven Heating Methods.

Feature		Air-Bath Oven	Airless Oven		Direct Column Heating
			Casting	Enclosure	
Isothermal oven	<150 °C	✓	✓	✓	✓
	>150 °C	✓	✗	✓	✓
Temperature control	Precision	✓✓	✓	✓	✓
	Uniformity	✓	✓✓	✓	✓
Temperature programming	Usage	✓	✗	✗	✓✓✓
	Ramp rate	✓✓	✗	✗	✓✓✓✓
High instrument air usage	For heating	✓✓✓	✗	✗	✗
	For cooling	✓✓	✗	✗	✓
High electric power usage		✓✓✓	✓	✓	✓✓
Fast warm up and cool down		✓✓✓	✓	✓✓	✓✓✓✓
Application flexibility		✓✓✓	✓	✓✓	✓✓
Ease of maintenance		✓	✓✓✓	✓✓	✓✓✓
Analyzer shelter recommended		✓✓✓	✗	✓✓	✓✓✓
Field mounting near tap		✗	✓✓✓	✗	✓
Small transmitter-style PGC		✗	✓✓✓	✗	✗
Safety for hazardous areas		✓	✓	✓	✓
Safety with toxic samples		✓✓	✓	✓	✓

Note: More green checkmarks indicate more of that feature. Data is illustrative of common usage but is not exclusive.

to maintain. We have seen the success of this approach with the repetitive applications found in natural gas processing and transmission. But the history of the PGC doesn't support that philosophy for less repetitive applications. Experience has shown us many times that small limited PGCs have small limited markets.

Vendors sell most of their PGCs analyzers in batches of 10–30 units for new plant construction or process upgrades. For a particular contract, if the vendor cannot accomplish just one of the applications, the entire sale goes to another vendor who can. Neither the end user nor the construction company want the additional cost of handing multiple brands of equipment on site. This loss of market share was a factor in the demise of many transmitter-style PGCs. Examples include the 1973 Foxboro 91-300, the 1971 GCE Elliot Mark 6, the 1976 Foxboro PCT, and the 2002 Siemens MicroSAM.

The applications engineering conundrum

Another way to standardize the product is to eliminate the custom application engineering. We saw how ABB applied this philosophy to the PGC1000, offering many standard column trains. Yet users will always want to measure more and different analytes at lower and lower concentrations, and to do that takes application flexibility in the configuration of columns and the choice of detectors. Which, of course, is exactly what the larger process chromatographs can do.

PGC companies would love to avoid application engineering. It's always difficult for a manufacturing company to make a product that requires customization. Yet the enormous success of the original process chromatographs was entirely due to their application flexibility. Unfortunately, it remains true that any attempt to standardize a PGC will reduce its market size. The only large market for a standard PGC is the natural gas industry.

MEMS technology

For many years now, we have anticipated the development of a chromatograph-on-a-chip produced by microelectromechanical systems (**MEMS**) technology. The basic technique has proven feasible. A PGC with micro-machined injectors and detectors can provide the excellent performance shown in Figure 11.14.

In our dreams, a miniature chromatographic module using an airless oven and MEMS technology could be the rugged and easily-maintained PGC that the end users are looking for. We continue to hope. As yet, experience with miniature PGCs has not been good. The Siemens MicroSAM offers an instructive case study.

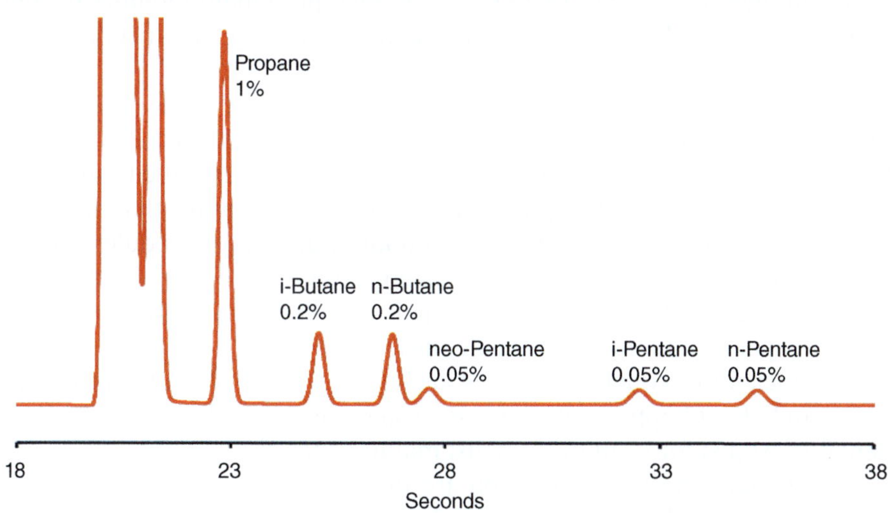

Example of the performance of a MEMS-based injector and detector with foreflush or backflush capabilities and integrated columns in a compact heated housing.

Figure 11.14 Fast Isothermal Analysis of Hydrocarbons. *Source: Qmicro B.V. Reproduced with permission.*

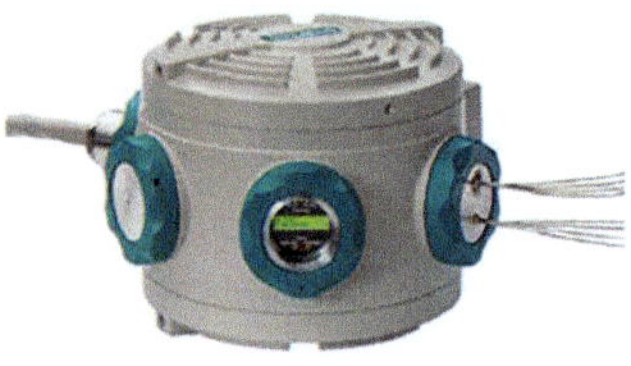

A unique PGC using MEMS technology for chromatographic valves and detectors.

Figure 11.15 The 2002 Siemens MicroSAM PGC. *Source: Siemens Analytical Products and Services. Reproduced with permission.*

The 2002 siemens MicroSAM

Figure 11.15 is a photograph of the Siemens *MicroSAM (Micro Single Analyzer Module)*, a hard-core process instrument featuring an airless column oven in an explosion-proof case for installation in the potentially hazardous process plant environment.

The MicroSAM is the epitome of advanced PGC technology. It uses small-bore fused silica columns and up to eight microelectromechanical (MEMS) thermal conductivity detectors, each with an internal volume of only 0.02 μL. To minimize internal volumes, the analyzer employs a MEMS sample injector and valveless column switching (backflush or heartcut) using the **live-switching** technique mentioned in Chapter 8. There's also a back-up option of using a diaphragm valve for column switching. The MicroSAM can use one of six standard column systems or can have a custom column design.

Siemens discontinued the MicroSAM in 2018, without a successor. Sadly, no highly miniaturized PGC has yet been successful in the industrial process market. MEMS now has a damaged reputation in process analyzer circles, so much so that ABB found it necessary to extoll the virtue of *not* using MEMS-based devices in the PGC1000, but of using conventional chromatographic hardware instead.

A closing thought

This chapter reveals a trend in the PGC industry towards modular maintenance. The idea driving this trend is to build analyzers from selected modules, and then to maintain them by swapping replacement modules having "no user-serviceable parts inside." It's sad that some PGC product literature no longer even mentions the word "column."

The dumbing-down of troubleshooting and maintenance skills is an understandable response to the perennial shortage of skilled PGC practitioners. But it's based on the mistaken premise that unskilled labor can troubleshoot a PGC by swapping modules. For simple faults, that might be true – even if expensive – but the random swapping of modules always deprives technicians of a learning opportunity. If you take away the thinking needed to diagnose and successfully repair the simple faults, how will they ever be capable of fixing the difficult ones?

The inexorable conclusion is that each PGC will die when the last module swap fails. It happens. We have seen so many PGCs shut down with all benefits lost just for want of specialist maintenance care.

It's encouraging to me that you are reading this book. By becoming a skilled PGC specialist, you will always have a well-paid job and a fascinating career.

Knowledge Gained

- *Temperature variations may affect sample volume, peak retention time, and detector stability.*
- *Increased column temperature reduces retention time about 3 % per kelvin, and vice versa.*
- *Most PGCs rely on repeatable retention times to correctly identify the analyte peaks.*
- *Some PGCs have computer programs that track the analyte peaks to avoid misidentification.*
- *Temperature variation may cause drift or noise on the detector baseline, particularly if using a TCD.*
- *The metal block of a TCD is a thermal mass that tends to smooth the oven temperature variation.*
- *The baseline might cycle with temperature variation due to the changing bleed from a packed column.*
- *PGCs use three column heating techniques: air bath oven, airless oven, and direct column heating.*
- *An air bath uses a jet of hot air to heat and circulate the oven air; the airflow also acts as a safety purge.*
- *Air baths are cost-effective and spacious, with room for many valves and columns.*
- *Air baths heat and cool quickly and achieve exceptionally good temperature control.*
- *Air baths may have temperature gradients at different zones within the oven.*
- *Air baths consume a lot of instrument air and heat energy – a significant operating cost.*
- *The airless oven depends on a large thermal mass to hold the temperature constant.*
- *Older thermal mass ovens were also an explosion-proof enclosure to house the electrical components.*
- *The thermal mass design is quite reliable; heating elements and control thermostats rarely fail.*
- *Airless ovens achieve good temperature stability with simple on-off heat control.*
- *The thermal mass approach to column heating is not applicable to temperature programming.*
- *Older airless ovens had less flexibility in the number of valves and columns they could accommodate.*
- *Airless ovens take more time to heat or cool, thereby increasing maintenance and downtime.*
- *Some PGCs use a straight column directly heated by an electric current through the metal tube.*
- *Others house their tightly-wound columns in a self-contained heated cartridge assembly.*
- *Each method of direct heating raises some safety concerns.*
- *Direct column heating eliminates the weight and size of a conventional column oven.*
- *When applied to a capillary column, direct heating gives a fast and repeatable temperature ramp.*

- *Directly heated capillary columns can achieve some extremely fast multicomponent analyses.*
- *Direct heating with digital control may enable the increased use of capillary columns in PGCs.*
- *Miniature gas chromatographs are proving that the fast PGC analysis of process streams is feasible.*
- *A fast PGC needs a fast sampling system and minimal delay between process and sampling tap.*
- *Miniature PGCs using MEMS technology have yet to be successful when analyzing process streams.*

Did you get it?

Self-assessment quiz: SAQ 11

Q1. In a PGC, which of the following chromatogram symptoms are likely to be due to temperature variations:
Select all that apply:

A. Cyclic baseline.
B. Drifting baseline.
C. Variable peak height.
D. Retention time shift.

Q2. The chapter identifies three methods of column heating used in process gas chromatographs.
Which **one** of those column heating methods is most likely to:

A. Achieve ultra-fast analysis?
B. Require instrument air?
C. Have reduced application flexibility?

Q3. Which **one** of the three main heating methods, is most often associated with:

A. An explosion-proof enclosure?
B. An air-purged enclosure?
C. A small and light analyzer?

Q4. Which **one** of the three main PGC heating techniques:

A. Can quickly analyze a wide-boiling mixture?
B. Takes the most time for heating and cooling?
C. Has the flexibility for installing multiple valves and columns?

Q5. Which **two** of the three main PGC heating techniques are:

A. Preferred for analyzing natural gas in gas transmission pipelines?
B. Preferred for applications that need temperature programming?
C. Used in most of the presently installed process gas chromatographs?

Q6. What is the main reason that airbath ovens have become less popular in PGCs?

Q7. What are the three methods of directly heating a chromatographic column?

Check your SAQ answers with those given at the end of the book.

Student evaluation test: SET 11

S1. Consider a peak eluting from a column with a retention time of 300 s. From information given in the text, estimate the *change* of peak retention time when the column temperature falls by 1.0 °C (i.e., 1.0 K).
Express your answer as an integer number of seconds.

S2. According to the text, what is the typical temperature control achieved by an air-bath oven under microprocessor control?
Select one option:

A. ±0.01 °C.
B. ±0.02 °C.
C. ±0.033 °C.
D. ±0.05 °C.

S3. From the list below, select the one kind of heating never used to control the temperature of a PGC column?
Select one option:

A. Blowing electrically heated air through the oven.
B. Blowing steam-heated air through the oven.
C. Mounting the column close to a radiant heating element.
D. Attaching the column to an electrically heated metal mass.
E. Having an electric heater element in direct contact with the column.
F. Passing an electric current through a metal column.

S4. When comparing an air-bath oven with an airless oven, which two of the listed features are advantages of the air-bath oven?
Select two options:

A. Lower ongoing cost of operation.
B. Fast warmup and cool down, facilitating maintenance.
C. Suitable for temperature programming.
D. Simple, effective, and reliable temperature control.

S5. When comparing an air-bath oven with an airless oven, which two of the listed features are advantages of the airless oven?
Select two options:

A. Lower ongoing cost of operation.
B. Fast warm up and cool down, facilitating maintenance.
C. Suitable for temperature programming.
D. Simple, effective, and reliable temperature control.

S6. What large market for PGCs has well-defined analytical measurement requirements easily satisfied by a simple modular PGC?
Select one option:

A. The natural gas industry.
B. The oil refining industry.
C. The petrochemical industry.
D. The chemical industry.

S7. According to the text, what safety alarms are typically available to monitor the heater in an air-bath oven?

Select all the correct answers and none of the incorrect answers:

A. Heater surface temperature – high alarm.
B. Heater surface temperature – low alarm.
C. Heater air flow rate – low alarm.
D. Heater air supply pressure – low alarm.

S8. In what ways does a directly-heated column surpass the performance of a similar column in a temperature-programmed air-bath oven?
Select two options:

A. The directly-heated column is capable of separating analytes that traditional temperature programming cannot separate.
B. The directly-heated column is capable of a faster rate of temperature change for heating and cooling.
C. The directly-heated column is capable of resolving lower concentrations of the analytes.
D. The directly-heated column can achieve lower analysis times when separating wide-boiling mixtures.

S9. Why is it essential for the temperature of the analytical devices in a process gas chromatograph to be tightly controlled?
Select all that apply:

A. Because small variations may affect the number of molecules injected onto the column, particularly when injecting a gas sample.
B. Because small variations may affect the retention time of the peaks, and most PGCs rely on constant retention time to identify and measure each peak.
C. Because small variations may upset the chromatogram baseline, particularly with a thermal conductivity detector.
D. Because small variations may increase the equipment failure rate and make the analyzer less reliable.

References

Cited

ABB Totalflow (2012). *PGC100 Analytical offerings*. Product brochure. Houston, TX: ABB Totalflow Products.

Annino, R., Curren, J., Kalinowski, R., Karas, E., Lindquist, R., and Prescott, R. (1976). Totally pneumatic gas chromatographic process stream analyzer. *Journal of Chromatography A* 126, 301–314. https://doi.org/10.1016/S0021-9673(01)84082-6

ASTM D2887-18 (2018). *Standard Test Method for Boiling Range Distribution of Petroleum Fractions by Gas Chromatography*. West Conshohocken, PA: ASTM International. doi: https://doi.org/10.1520/D2887-18

ASTM D7096-16 (2016). *Standard Test Method for Determination of the Boiling Range Distribution of Gasoline by Wide-Bore Capillary Gas Chromatography*. West Conshohocken, PA: ASTM International. doi: https://doi.org/10.1520/D7096-16

Clemons, J.M. and Thomas, T.J. (2000). Fast process gas chromatography. *Proceedings of the 45th Annual ISA Analysis Division Symposium*, Vol. 33, 115–128.

Charleston, WV, April 9–12, 2000. Research Triangle Park, NC: International Society of Automation.

Crandall, J., Roques, N., and Bostic, S. (2014). Rethinking the anatomy of a gas chromatograph. In: *Proceedings of the ISA Analysis Division Symposium*, May 5, 2014, *Baton Rouge, Louisiana*. Research Triangle Park, NC: International Society of Automation.

Ettre, L.S. (2005). *The evolution of gas chromatography at Perkin Elmer*. White Paper 006068C_01, 1–16. Norwalk, CT: Perkin Elmer, Inc.

Guild, L.V. (1956). Some practical applications of gas chromatography. *Presented at the 7th Pittsburgh Conference,* Pittsburgh, PA, February 27–March 2, 1956.

Matsuura, T. (1994). *Pressurized Oven*. Japanese Patent JP,06-186218,A(1994), awarded July 8, 1994.

Rohrback, S. and Ramos, S. (2017). Making chromatography more robust. ISA 62nd Analysis Division Symposium, Pasadena, CA, 2017, *Session 09-2, 1–14*. Research Triangle Park, NC: International Society of Automation.

Spracklen, S.B. (1957). The development of gas chromatographs, *ISA Journal* (November), 514–517.

Yokogawa (2001). *Process Gas Chromatograph Mark II*. Product Brochure. Bulletin 11B03A03-01E, Tokyo, Japan: Yokogawa Electric Corporation.

Table

Figures

New technical terms

When first introduced, these new words and phrases were in bold type. Further information is in the Glossary at the end of the book.

air-bath oven
airless oven
applet
boiling point
column bleed
column train
density
directly-heated column
heating value
live switching
MEMS
oven
percent-off
resistive heating
SimDis

Consult the Glossary for information about chemical names.

Part Three

PGC control

Figure 12.1 The 1979 Beckman Model 6750 Chromatograph Data Processor.
Source: Science History Institute. Philadelphia. https://digital.sciencehistory.org/works/0p096733n. Reproduced with permission of Rosemount, Inc.

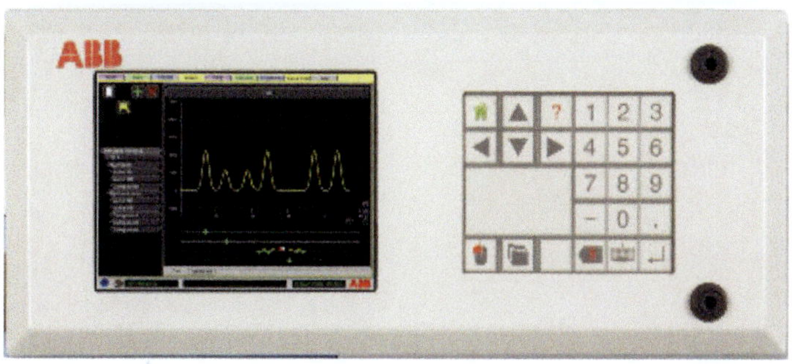

Figure 12.2 The 2008 ABB Model PGC5000A Master Controller.
Source: ABB Process Analytics Reproduced with permission.

"It would appear that we have reached the limits of what it is possible to achieve with computer technology, although one should be careful with such statements, as they tend to sound pretty silly in 5 years."

John von Neumann
circa 1949
(Edwards 2012)

Process Gas Chromatographs: Fundamentals, Design and Implementation, First Edition. Tony Waters.

Why study this?

Part Three of the book introduces the Control Unit of the PGC, the device we originally called a Programmer. Since process analyzer specialists may need to maintain or troubleshoot older equipment in the field, these four chapters review the full range of technologies used over the years to control the operation of a process gas chromatograph. This study will complete your basic knowledge of PGCs and allow you to progress to higher-level courses on PGC maintenance, application engineering, or troubleshooting.

12

Event scheduling

"A chromatographic analysis is not instantaneous or continuous, it gradually unfolds over time, and a series of timed events control its progress. It all starts with a valve action – sample injection – and it may require other valve actions to occur at exactly the right time after that. PGCs also use elapsed time to identify the peaks and thence to quantify the desired analytes. So, timing is the essence of process gas chromatography; you must get it right."

A sequence of actions

Program timing

For a PGC to operate automatically, a repetitive sequence of timed events must occur during each analysis. Therefore, every PGC needs a timing device to control these events. Older PGCs had a simple **programmer** like the early example seen in Figure 12.3.

Modern analyzers are quite different: they have a microprocessor to memorize and run the timing program. We shall see that the microprocessor can perform many more functions than the simple timing of events, so it's now known as a **control unit**. Figure 12.4 shows one of the first microprocessor control units.

Since older analyzers are still working in the field and may also need some care, you may need to work with modern or traditional PGC control devices. Therefore, we will review most types of timing devices found in installed instruments.

In a process gas chromatograph, scheduled functions typically include the following time-based functions, but not necessarily in this order:

- Set the detector baseline to desired zero point.
- Measure baseline noise.
- Initiate atmospheric referencing for gas samples.

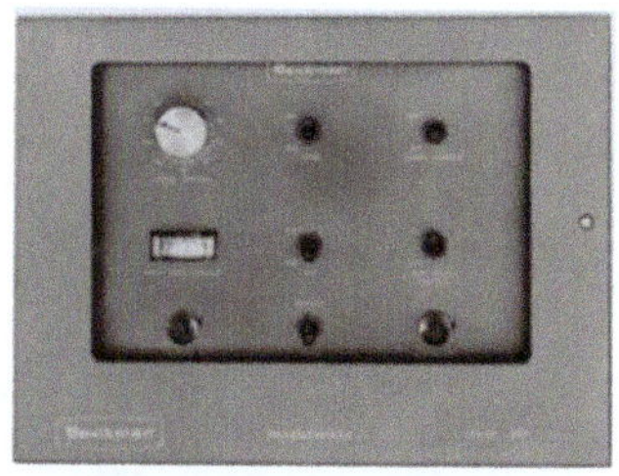

How far we have come!

Figure 12.3 Circa 1964 Beckman Model 620 Programmer.
Source: Beckman Historical Collection, Box 77. Science History Institute, Philadelphia. Reproduced with permission of Rosemount, Inc.

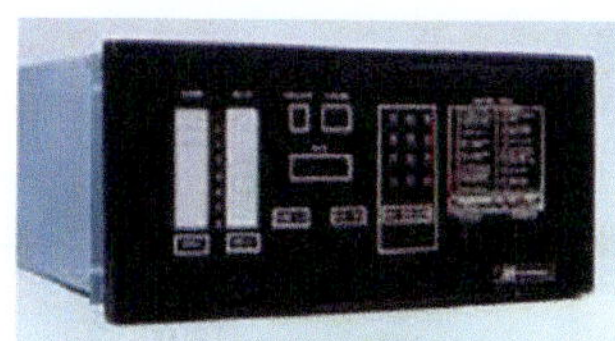

An early microprocessor PGC control Unit.

Figure 12.4 The 1978 AAI Optichrom 2100 Control Unit.
Source: Siemens Analytical Products and Solutions. Reproduced with permission.

Process Gas Chromatographs: Fundamentals, Design and Implementation, First Edition. Tony Waters.

- Sample injection (start of analysis).
- Terminate atmospheric referencing.
- Switch to the next-listed stream in preparation for its analysis.
- Change a carrier gas pressure by electronic pressure regulator.
- Reset the sample injector valve to the sample fill position.
- Switch column valve on and off at the correct times.
- Turn Peak *#1* gate on and gate off at correct times.
- Turn Peak *#n* gate on and gate off at correct times.
- Initiate calculation programs (e.g., barometric pressure correction).
- Display and transmit data.
- Issue alarm notifications.
- End the analysis and initiate a repeat of the cycle.

As you can see, the control unit is a busy device. In the following paragraphs we examine these functions in more detail.

In a modern control unit, there is no need to enter the scheduled functions in serial order. When specifying an individual action, you simply enter the time that you want it to occur. For example, the entries in Figure 12.5

	Enable	Name	ID	Ret. Time (sec)	Window (sec)	Unit	Quantitate	STD Mult.	RT Update	Ref. ID#	STD ID#	Level 1	Level 2	Level 3	Level 4	Level 5	Level 6	Level 7	Level 8	Level 9
3	☑	METHANE	3	125.1	2.372	PPMV	Area	1	Calib	0	0	0	99.99	0						
4	☑	C2's	13	138.7	3.834	PPMV	Area	1	Calib	0	0	0	199.98	0						
5	☑	PROPANE	4	164.9	5.075	PPMV	Area	1	Calib	0	0	0	99.98	0						
6	☑	PROPYLENE	5	180.2	4.079	PPMV	Area	1	Calib	0	0	0	99.98	0						

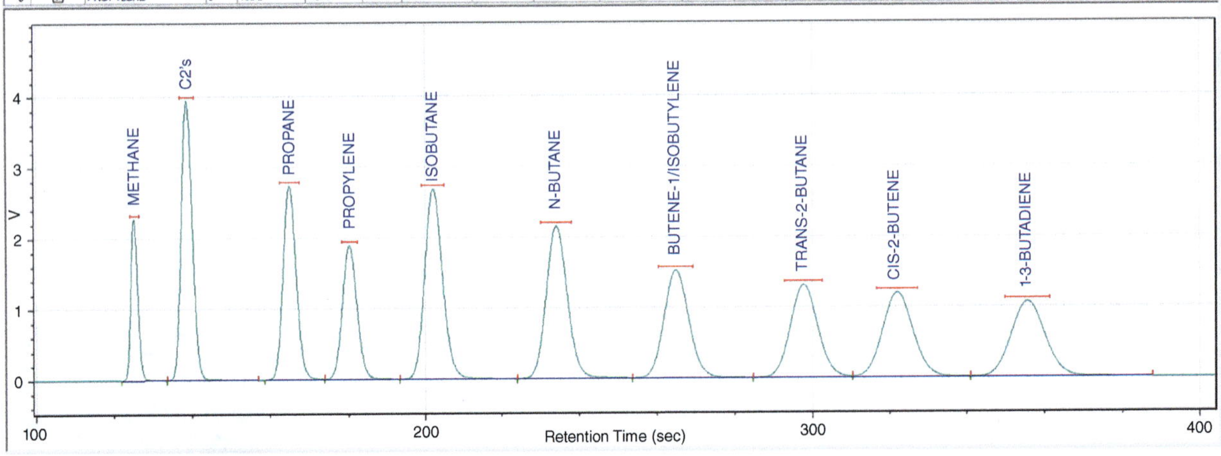

As an example, this screenshot of a partial chromatogram illustrates some typical program settings for the Maxum Edition II PGC. You would enter the settings in the table above or use a graphical method of adjusting them. On the chromatogram, the red horizontal bars indicate the acceptable retention time window for each peak, while the short vertical downward and upward tick marks show the automatically detected start and end times for the integration of each peak. The "blue" line across the base of a peak shows the baseline used for quantitation.

Figure 12.5 Typical Method of Setting Peak Parameters.
Source: Siemens Analytical Products and Solutions. Reproduced with permission.

include the analyte names, the expected peak retention times, and the window of retention time variation considered acceptable for each peak.

Autozero

The **autozero** function resets the chromatogram baseline to the desired position on the chart. In older electromechanical systems the autozero may need a start time and a stop time, whereas digital systems need just a single instruction. The next chapter discusses autozero in more detail.

Atmospheric referencing

It is now common to vent gas samples into the flare header or another collection manifold for disposal. If so, the pressure in the gas sample loop will be about the same as the pressure in the disposal system, which is unlikely to be constant. In such cases, you can improve analysis repeatability by briefly venting the sample loop to the outside air before injection, thereby letting the gas in the sample loop come to atmospheric pressure.

To vent the sample, PGCs use atmospheric referencing valves (ARV), sometimes known as atmospheric balancing valves. Look in Chapter 7 to see some typical ARV valve arrangements.

Please note the need to vent the ARV to the outside air, *not* to the flare header! We see this mistake often in installed systems on site.

It's important for the sample loop to remain connected to the outside air until sample injection has occurred. The ARV may then return to its normal state, with sample gas flowing through the sample valve. The sample gas will not flow through the sample loop until the injector valve resets to its sample-fill position.

Usually, the control unit must energize the ARV to connect the sample loop to the outside air, but in some fail-safe systems, the ARV will work in the opposite direction and must be deenergized for atmospheric referencing.

The complete ARV function is not necessary when the sample gas vents directly to atmosphere. Yet even then, some systems will block the sample flow before injection, just to be sure there is no build-up of pressure in the sample loop. There are also some applications that don't need the ARV because they control the sample loop pressure with a downstream backpressure regulator, or with an upstream pressure regulator and a downstream needle valve.

In summary, an ARV function requires two timed events: one occurring before sample injection and one after.

Sample injection

The instant of sample injection corresponds to time zero on the chromatogram, but a particular control unit may not set it to zero on the system clock. This is because other functions such as automatic zero or ARV actuation precede the injection of a sample. Another control unit might set sample injection to zero time and allow negative time settings for the pre-injection functions.

It's also possible for a second sample injection to occur within the analysis cycle. However, the two injections may need to be simultaneous if they produce complementary data for the analysis of a single process stream, as in natural gas analysis, for instance (ISO 6974).

The control unit actuates the sample injector valve (SIV) to inject a sample into the column system. The SIV must then remain actuated until the carrier gas has completely transferred the sample onto the column. That said, it can sometimes be helpful to deactivate the SIV as quickly as possible to minimize peak tailing caused by sample molecules adhering to the walls of the sample volume.

If the SIV has no other function, you can reset it as soon as sample injection is complete. A gas sample injection is a fast process, and the SIV can reset within a few seconds. A liquid sample may need a little more time to vaporize; if you reset the SIV too soon, it may reduce the size of some peaks and produce inaccurate measurements.

When the injector valve has an additional function, though, it's not possible to minimize the valve actuation time. There are many injector valve configurations that also switch columns, mainly to backflush the first column. In those systems, the valve must remain energized until all the analyte peaks have eluted from the first column. Mostly, however, such systems are for gas samples or volatile liquids that are easy to inject, so the longer actuation time is not an issue. The more difficult liquid sample injections always use a dedicated injector valve.

In summary, the sample injector valve requires two timed events; valve on and valve off. The first of these is the true start of the chromatogram, and the second depends on the valve configuration and the intent of the application.

Step stream

When a PGC is set up to analyze more than one stream, there are optional ways to sequence those streams. The default program is a simple numerical sequence with each stream analyzed in turn, but it's usually possible to analyze the streams in any desired order and frequency. In older chromatographs, the sequence may be hard-wired and difficult to change, while modern control units will let you enter any preferred sequence for analysis.

The **step stream** command closes the active stream valve and opens the next one in the programmed sequence, thereby allowing a different process sample to flow through the injector valve. To ensure a representative sample of the next process stream, the step stream command occurs soon after sample injection, which allows the new process stream to purge the sample lines and the injector volume while the PGC is analyzing the previously injected sample.

As a safety measure, though, you will sometimes find that all the stream valves remain off and no sample flows for the first half the analysis cycle.

Then, at about half-way through the current analysis, the appropriate stream valve opens and the next sample stream purges the old sample out of the lines, hopefully in time for the injection of a representative sample.

Shutting off the sample supply may not be good practice for all samples. One can understand the need for caution when analyzing toxic or flammable samples, but analytical measurements are usually more reliable when the sample fluid flows continuously.

As noted in Chapter 9, some multistream PGCs are set up to analyze several streams concurrently using separate applets, each with its own sample injector, columns, and detector: a technique sometimes dubbed parallel chromatography. Effectively, such systems are equivalent to two or three single-stream chromatographs running continuously, so they don't need stream selection valves or a step stream command. The technique works well for two or three streams but becomes impractical for a larger number of process streams.

Column switching

In systems that have one or more independent column valves, the control unit needs to repeatably activate and deactivate each of those valves at exactly the right time. Each column valve needs a valve-on and valve-off command. When the analyzer uses the heartcut technique, the analysis method may call for the same valve to switch on and off several times.

Look back at Chapter 9 for a detailed discussion of column switching techniques.

Peak gating

Two scheduled events bracket the time that an analyte peak is due to appear on the chromatogram: a start time and a stop time. Entering these two times in the analysis method defines a **peak gate**. A peak gate is an instruction to the processor to measure a peak appearing between the gate start time and the gate stop time. In many systems there are two ways to do this, and you must select the one you want:

- A **forced integration** gate measures the entire area of the chromatogram above baseline level. Area integration continues for the full width of the gate and may encompass one peak or several. If measuring peak height, it will measure the highest peak present.
- A **slope-detect** gate allows the processor to detect the beginning and ending of each peak. It primes the system to look for the actual start, crest, and end of each peak on the chromatogram. The processor then measures the first peak whose detected crest time is within the specified gate.

For the moment, we are focused on the scheduled events, gate-on and gate-off. The next chapter looks at the peak measurement techniques in more detail.

PGC processors will optionally mark the peak gates on the chromatogram. Have a look at the chromatogram in Figure 12.5. A small red downtick indicates the gate start time, and a small green uptick indicates the gate stop time.

When setting up a peak gate, there are many decisions to make relating to the peak detection criteria, the method of measurement, and the calibration of that peak. The following chapters of the book review all this in more detail.

Initiate a calculation

This optional instruction initiates the calculation of a desired stream property from the measured sample composition, using a preprogrammed algorithm. Two common examples are normalization and weight percent concentration.

Chapter 15 discusses calibration methods in more detail.

Data transmission

To request a **host computer** to read new measurements, an older programmer may need to issue a **priority interrupt** or **come read** signal. Modern control units just send their data at will via a serial-data communications network.

Alarm notifications

A controller should issue an alarm notification when any measurement result or physical condition is outside preassigned tolerances.

End-of-cycle

As the preset analysis time ends, the programmer will immediately start a new analysis. If you don't want that, most control units can wait a preset time before restarting the analysis.

Event markers

Most PGCs mark the timed or found events as a small blip on the chromatogram record. The extent of this chromatogram marking was a fixed function of the early programmers, but in modern systems it's adjustable via software control, so you can set the desired occurrence and height of the markers.

Calendar events

In addition to the repetitive events occurring in each analysis cycle, the program may schedule certain functions to occur at a time of day or calendar date. In a computer-controlled PGC, these functions can include alarm processing, report writing, and calibration routines. The time-of-day calendar clock doesn't need the high resolution used for the analysis timebase: ±1 s is typically sufficient.

Timing mechanisms

Mechanical programmers

In the first PGC systems, a programmer unit provided event timing by means of an electromechanical device. Most PGC users installed these programmers inside the plant control room.

Mechanical timers were available until the late 1980s, and some are still in use. They use a rotary timing device driven by a synchronous electric motor. Cams on the rotor depress and release microswitches at appropriate times, and the speed of the motor determines the overall analysis cycle time.

Some programmers have a separate cam having adjustable on and off positions for each function. In others, sequential cams operate a single microswitch to advance a stepping switch. Then, each position of the stepper switch performs an assigned function, sometimes with the help of a simple relay logic circuit.

In subsequent models, optical sensors replaced the mechanical switches, but the drive motor remained.

Electronic timers

Eventually, solid-state electronic timers superseded the optomechanical devices. A programmer with a solid-state timer often uses a diode patchboard for setting event times. You can see the patchboard on the front panel of 1985 Servomex Model CP 409 Programmer pictured in Figure 12.6.

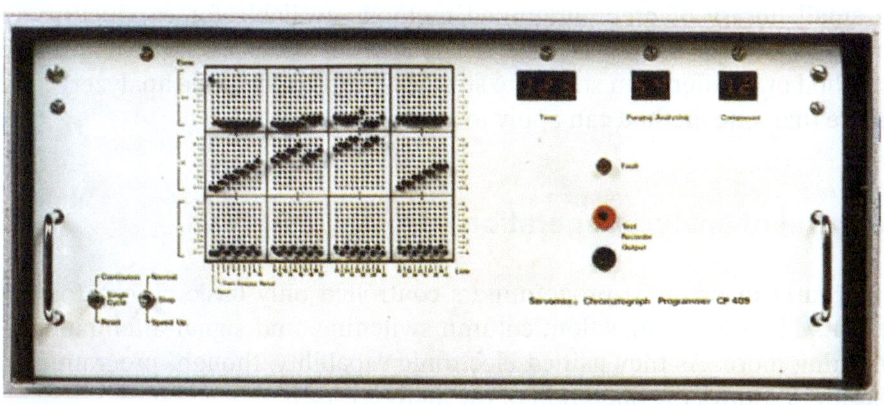

A digital electronic programmer with a diode patchboard for setting program times.

Figure 12.6 The 1985 Servomex Model CP 409 Programmer.
Source: Servomex Ltd. Reproduced with permission.

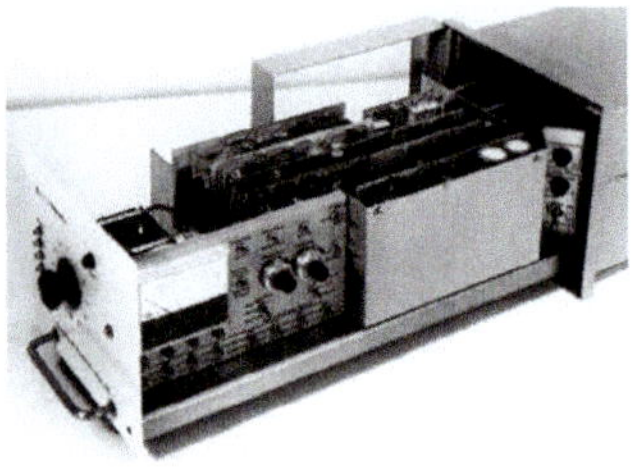

This was one of the last cam timers offered for sale.

Figure 12.7 The 1985 Servomex Model CP 403 Programmer.
Source: Servomex Ltd. Reproduced with permission.

During the late 1970s, manufacturers started to offer digital electronic timers alongside the traditional cam timers. Jobsite installations dating from the 1980s may have digital or mechanical timers, or even both. For instance, there are three versions of the 1985 Servomex Series 400 Programmer. The Model CP 409 and the simpler Model CP 406 for single-stream applications are both solid-state devices with digital timing. Yet the lower-cost Model CP 403 pictured in Figure 12.7 runs with a cam timer, reflecting the still-high cost of digital electronics at that time (Servomex 1985).

Microprocessor control

By 1980 most PGC vendors had a microprocessor-based programmer. Examples include the 1978 Foxboro-Arcas 935, the 1978 Bendix 9000, the 1979 Hokushin V601, the 1979 Yokogawa GC6P (Figure 12.8), and the 1980 Amscor Mark IV,

The model dates cited above evidence the rapid adoption of microprocessors. In a microprocessor-based control unit, the analysis cycle clock may be set for any desired timebase in minutes or seconds, sometimes to a resolution of 0.01 s.

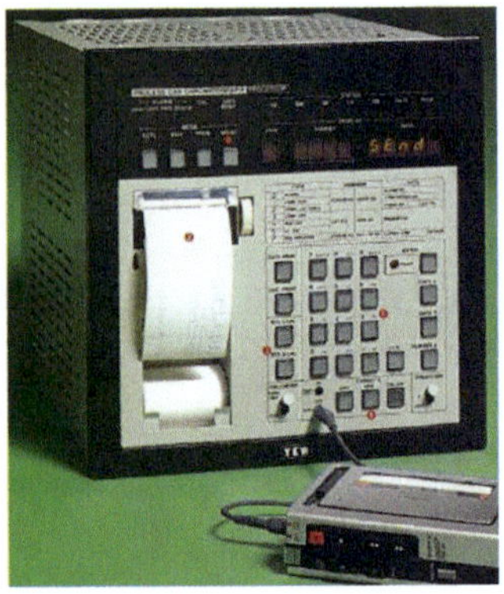

One of the first microprocessor-based PGC control units.

Figure 12.8 The 1979 Yokogawa Model GC6P Control Unit.
Source: Yokogawa Electric Corporation. Reproduced with permission.

The program or method

In an early PGC, the **program** of timed events is simply the mechanical position of cams on a motor-driven shaft, disk, or belt. Even when the PGC had a digital timer, the analysis programs were relatively simple.

In a microcomputer-based PGC, you can view the program of scheduled events on a display screen. To set or change a program, simply enter the times that you want events to occur. In addition to start and stop times, the system may also need times for processing results or transmitting data to a host computer system.

A computer-based PGC can perform many other functions during an analysis. For each function you'll choose between options and specify the timing. The PGC control unit then saves your listing of choices in nonvolatile memory as a unique analytical **method**. An analyzer may have a small library of preprogrammed methods available for conducting an analysis, and it follows the one designated for each process stream. The method may differ from stream to stream – even on the same analyzer – and more than one method can apply to a single stream.

Control of analyzer operation

The early mechanical programmers controlled only basic chromatogram events like sample injection, column switching, and signal attenuation – nothing more. As they gained electronic capability, though, programmers began to control the operation of the analyzer itself.

Temperature control

You may find that older analyzers at the jobsite have a separate thermostat for oven temperature control. The newer microprocessor control units can actively control the temperature for as many as four separate zones. These zones might include one or more column ovens, detector enclosures, or sample conditioning devices.

Isothermal temperature controllers don't need an event schedule because they are always on. Temperature-programmed column ovens need to be set up by entering several timed events into the analysis method. Each pair of events will typically define a duration with a unique pattern of temperature control. Most programs include an initial period of constant temperature, one or more temperature ramp rates, an optional hold period at the highest temperature, and a final cooldown period.

Pressure control

Many PGCs still use mechanical regulators for controlling carrier gas pressures, finding them sufficiently stable when the inlet gas pressure and the regulator body temperature are both held constant. They cost less than electronic pressure regulators and contribute to safety, having no electronic parts.

Electronic pressure controllers (EPC) sometimes provide a more constant carrier gas pressure, and their set pressure is conveniently under software control. This is easier than using flow restrictors to drop the carrier pressure – as is necessary in backflush and heartcut column systems, for instance. If the analysis needs a carrier pressure to change at some point during the analysis, there will be a timed event in the analysis method to make that change.

Systems employing the live-switching technique instead of valves use differential pressure to switch the direction of carrier gas flow. Then, the analytical method must contain the timed events and the commands to reset those pressures. From a timing perspective, this is much the same as setting valve switching times.

It's even possible for a PGC application to employ a gradually increasing carrier gas pressure. The increasing flow rate thus caused is analogous to a temperature ramp. As yet, process gas chromatographs rarely use this technique.

Peak identification

The correct identification of analyte peaks is of fundamental importance. Clearly, it's possible to set up a PGC to measure a wrong peak. That would be a grave mistake, but it wouldn't be the first time that it happened!

Process chromatographs currently lack the ability to identify the molecules in a detector, so a chromatogram peak is just an indication that something other than the carrier gas just passed through the detector. To infer the identity of that something, PGCs always use the retention time of the peak.

> *Be sure to understand that the PGC detector cannot identify the peaks for you. It's your responsibility to ensure that you are measuring the correct peaks.*

With a constant flow rate of carrier gas, a repetitive temperature profile, and an enduring column performance, the retention time of each component should be constant and therefore a reliable indicator of peak identity. In practice, retention times may slowly change due to the aging or contamination of a column, or from minor variations in column operating conditions. And when column switching is employed, other causes of retention time shift might occur, such as variations in valve timing.

Most PGCs control their operating conditions well enough to ensure predicable retention times, and then use those retention times to identify the peaks. A few PGCs are more creative and attempt to predict the elution times of each analyte peak.

Fixed-time gating

The traditional way to identify a peak is to bracket a time period on the chromatogram within which you expect the peak to elute. This technique is known as peak gating and requires two timed events in the program, one for the peak start time and one for the peak end time. For instance, if you enter a peak gate for propane between 100 s and 110 s after sample injection, the control unit will assume that any peak eluting in that time span is propane.

Since the identity of a peak is based solely on its retention time, there's clearly a potential for misidentification. It's the responsibility of the application engineer to ensure that no other peak is likely to elute in the same gate as an analyte peak.

The proactive application engineering of a PGC highlights a significant difference between process and laboratory methods. The PGC applications engineer knows the nominal composition of the process fluid and can identify or predict potential interferants. This prescience increases our confidence of identifying peaks by their retention times. The composition of a laboratory sample is less predictable, so the laboratory instrument is subject to a wider range of potential interferants.

The first PGC applications were simple, with only a few peaks on the chromatogram and not much chance of misidentification. But as users started to apply PGCs to analyze more complex process streams, the column systems also became complex, increasing the risk of selecting the wrong peak.

What's more, the chromatogram of a complex mixture may exhibit a flurry of peaks, again increasing the chance of misidentification. Consider, for example, an ultra-fast temperature-programmed chromatograph separating a dozen peaks per minute. With so many peaks, a small variation in their retention times might be enough to move a chromatogram peak to a place where another peak used to be.

Misidentification of a peak is *the most serious error* that a PGC can make. An inherent advantage of chromatography over spectrometric methods is that chromatography separates the molecules in the sample, allowing the independent measurement of each selected type of molecule. But a chromatograph cannot identify the molecules in its detector; it doesn't know what it's measuring.

> *A PGC detector can't identify the molecules it's measuring, so it totally depends on each peak eluting in its designated gate. If another peak appears in the propane gate, the PGC will measure it and incorrectly call it propane.*

Retention time tracking

To compensate for minor variations in retention time, some PGCs can track the movement of one or two reference peaks and adjust the analyte peak gates in proportion to any detected retention time drift.

Recall from Chapter 1 that the *adjusted retention time* of a peak is its retention time as measured from an air peak. For any two peaks on a given column, the ratio of their adjusted retention times is constant. Therefore, one way to ensure correct identification is to move the peak gate times in proportion to the adjusted retention time of a reference peak.

With the 1972 Model 91-300 Compact Process Chromatograph, Foxboro may have been the first to introduce relative peak timing to a process chromatograph. When this instrument operates in ratio-time mode, it computes peak gating times relative to the observed retention time of a non-retained peak (such as air) and a trigger peak. These two reference peaks may be substances endemic in the process sample or simultaneously injected by a second sample loop on the injector valve (Jones 1976, 254–262).

The 1976 Foxboro Model 91-PCT Pneumatic Composition Transmitter also uses ratio peak timing, uniquely accomplished by pneumatic logic circuits (Annino et al. 1976).

The idea of using reference peaks to predict analyte peaks was an interesting innovation, yet few manufacturers adopted the technique. While it worked well for a simple single- or dual-component analysis, it was hard to find a convenient reference peak in a complex analysis. And anyway, PGC hardware performance was rapidly improving, obviating the need for relative peak timing.

PGC analyzers installed in the 21st century have retention time repeatability that is good enough to nix the fear of peak misidentification. Mechanical thermostats and volatile liquid phases are gone. Instead, digital controllers precisely regulate column temperatures, and bonded stationary phases greatly enhance column stability.

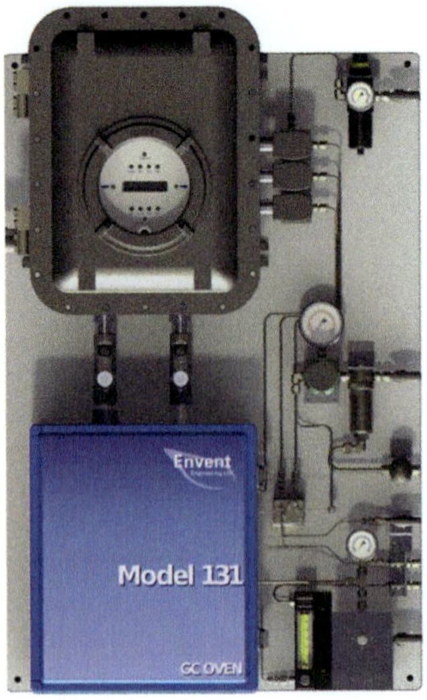

This small PGC makes limited adjustments to the expected retention time of each peak based on previous experience.

Figure 12.9 The 2010 Envent Model 131 Natural Gas PGC. *Source: Courtesy of Envent Engineering, Ltd. Reproduced with pemission.*

The new PGCs are stable enough to avoid peak misidentification in *traditional* process chromatography. But GC technology continues to advance, and the more recent technique of ultra-fast temperature programming, which can separate scores of peaks in just a few minutes, is again susceptible to the possibility of peak misidentification. Yet it's now easy for a microcomputer to compensate for shifting elution times because chromatographic software systems always record the retention time of the peaks, and they can use that information to adjust the gate times accordingly.

For example, the 2010 Envent Model 131 PGC shown in Figure 12.9 uses a simple procedure to track a slowly drifting peak retention time. During the initial setup procedure, the operator enters the expected retention time of each analyte peak and the amount of deviation considered acceptable. In operation, the processor measures the actual retention time of each peak and compares it to the expected time. If the retention time is within the range of acceptable values, the processor accepts the peak and updates the expected time with the newly found one. Of course, there are preset limits to the amount of adjustment allowed. By updating the expected retention time during each analysis cycle, the processor continues to track all movements of the peak that are within allowable limits.

It's a nice approach, but even this technique might not suffice for some analyses. In a fast analysis with variable sample composition, a more complex tracking technique may be necessary, one that engages every peak on the chromatogram.

One such advanced technique is a feature of the 2007 *Calidus*, a portable gas chromatograph from Teledyne Analytical. This analyzer can produce a very rapid analysis, thereby incurring some risk of peak misidentification. To mitigate that risk, the microprocessor applies a multivariate correlation algorithm to all the observed peak retention times. This procedure aligns all the peaks on successive chromatograms and can reprocess multiple stored chromatograms to ensure the correct identity of each analyte peak (Rohrback and Ramos 2017).

Knowledge Gained

- *A chromatographic analysis is not instantaneous or continuous; it gradually unfolds over time.*
- *A repetitive sequence of accurately timed events must occur to successfully perform an analysis.*
- *Some functions occur at a desired frequency, or by time of day or date.*
- *Scheduled events may perform housekeeping functions like atmospheric referencing or autozero.*
- *Atmospheric referencing ensures that injected gas samples are at relatively consistent pressure.*
- *Autozero sets the chromatogram baseline to a level appropriate for the application.*

- *The analysis starts with sample injection and ends when the PGC is ready to receive another sample.*
- *The sample injector valve must stay energized long enough for all of the sample to enter the column.*
- *A multistream application may select streams for analysis in any predefined sequence.*
- *After injection, the program closes the current stream valve and opens the next one in sequence.*
- *A PGC may use parallel columns to analyze several streams at once, without need of stream switching.*
- *The timer must energize and deenergize the column valves at the exact times needed by the separation.*
- *A peak gate is set up for each analyte peak with start and stop times that bracket its elution time.*
- *In some systems, data calculation and transmission events occur before the end-of-cycle time.*
- *Older PGC programmers have a mechanical timer device driven by a synchronous motor.*
- *Mechanical timing devices include dedicated cams, optical sensors, rotating discs, and moving belts.*
- *Digital timers using solid-state electronics eventually superseded motor-driven cam timers.*
- *Microprocessor-based PGC control units appeared in the late 1970s and soon became standard.*
- *The microprocessor also improved the control of column temperature and carrier gas pressure.*
- *An analytical method contains all the information necessary to automatically analyze a sample.*
- *The PGC detector doesn't know what it's measuring, so it identifies the peaks by their retention times.*
- *Most PGCs use a fixed-time gate to identify each analyte peak on the chromatogram.*
- *Each peak gate belongs to a named analyte peak, and no other peak should elute within it.*
- *Instead of timed gates, microcomputer PGCs may identify peaks by detecting their elution times.*
- *Some PGCs adjust the peak gate timing based on the position of known reference peaks.*
- *Another way to track a peak is to update the stored retention time with each newly found value.*
- *Be aware that a peak appearing at the expected retention time may not be the desired analyte!*
- *The worst mistake you can make with a process chromatograph is to measure the wrong peak.*
- *Applications using a complex column system are more likely to select the wrong peak.*
- *Newer PGCs achieve more constant retention times by better oven control and more durable columns.*
- *Applications with many peaks on the chromatogram are more likely to suffer peak misidentification.*
- *Ultra-fast temperature ramping may require peak alignment software to ensure correct identification.*

Did you get it?

Self-assessment quiz: SAQ 12

Q1. Over the years, process gas chromatograph programmers and control units have used several different timing devices. What are these different devices?

Q2. PGCs commonly use a ten-port valve configuration for gas sample injection and Column 1 backflush. For this valve arrangement, how many timed events are necessary in the analysis method?

Q3. How can a PGC measure two process streams concurrently, without using stream selector valves?

Q4. What is a fixed-time peak gate?

Q5. What is an analysis method?

Q6. How does a PGC identify the peaks on the chromatogram?

Q7. What is the purpose of retention time tracking?

Check your SAQ answers with those given at the end of the book.

Student evaluation test: SET 12

S1. Which one of the PGC functions listed below is never a scheduled event?
Select one option:

A. Atmospheric referencing
B. Autozero
C. Temperature setting for an isothermal column oven
D. Temperature settings for a temperature-ramped oven

S2. Which one of the functions listed below is the least likely to be a timed event?
Select one option:

A. Sample injector valve actuation and deactivation times
B. Column valve actuation and deactivation times
C. Analyte peak apex time
D. Analyte peak gate-on and gate-off times

S3. Which one of the devices listed below does not measure the elapsed time of analysis in any process gas chromatograph?
Select one option:

A. Synchronous motor
B. Diode patchboard
C. Solid-state counter
D. Microprocessor

S4. Consider a PGC analyzing a gas sample that vents into a flare header that has a variable pressure. If the PGC has an atmospheric referencing valve (ARV), which **one** of the statements listed below is incorrect?
Select the one incorrect statement:

A. When actuated, the ARV isolates the gas sample vent from the flare header.
B. When actuated, the ARV prevents the gas sample from flowing into the sample loop.
C. When actuated, the ARV connects the gas sample loop to the flare header.
D. When actuated, the ARV connects the gas sample loop to the outside air.

S5. It may be advantageous to reset the sample injector valve to the sample fill position a few seconds after injection, but this is not always possible. Which of the two sample injector valve configurations described below would prevent a quick reset?

> *Note: This question requires you to employ knowledge gained from both this chapter and previous chapters.*

Select one option:

A. The sample injector is a six-port valve with an external sample loop and only one port connected to a column.

B. The sample injector is a ten-port valve with an external sample loop and two or more ports connected to one or more columns.

S6. In a process gas chromatograph, there may be several peaks on the chromatogram. How can we be certain that the PGC correctly identifies the chemical compound responsible for each peak?

To receive a score, you must select all the correct answers and none of the incorrect answers:

A. A process gas chromatograph cannot identify the peaks on its chromatogram with absolute certainty.

B. Given stable operation, a peak that appears at the expected retention time of an analyte must be that analyte.

C. Given stable operation, a peak that appears at the expected retention time of an analyte is probably that analyte.

D. The microcomputer always prints the correct name of the component peak on the chromatogram display.

S7. Microcomputer control units are very versatile and can be set up in many different ways. But which of the following items would never be part of a PGC timed-event schedule?

To receive a score, you must select all the correct answers and none of the incorrect answers:

A. A gradual increase in column temperature.

B. A gradual decrease in column temperature.

C. A gradual increase in carrier gas pressure.

D. A gradual decrease in carrier gas pressure.

S8. What is the purpose of a fixed-time peak gate?

To receive a score, you must select all the correct answers and none of the incorrect answers:

A. To ensure that an analyte peak passes through the detector during an assigned time window so the processor can measure its height or its area.

B. To identify an analyte peak on the chromatogram, assuming that its retention time will be adequately constant.

C. To delineate a fixed peak integration period when the method demands forced (fixed-time) integration.

D. To instruct the processor to look for the start, crest, and end of a peak when the method demands automatic peak detection.

S9. What chromatographic principles have been employed in peak retention-time tracking systems?

To receive a score, you must select all the correct answers and none of the incorrect answers:

A. For any two peaks on a given column, the ratio of their adjusted retention times is constant.

B. The resolution between two peaks is their separation divided by their average base width.

C. The appropriate gate times for a given peak are proportional to small observed changes in the retention time of that peak.

D. A multivariate correlation algorithm can automatically align all the peaks on successive chromatograms.

References

Cited

Annino, R., Curren, J. Jr., Kalinowski, R., Karas, E., Lindquist, R., and Prescott, R. (1976). Totally pneumatic gas chromatographic process stream analyzer. *Journal of Chromatography A,* 126, 301–314. https://doi.org/10.1016/S0021-9673(01)84082-6

Edwards, J.R. (2012). *A History of Early Computing at Princeton*. Turing Centennial Celebration (May 10–12, 2012). Princeton, NJ: Princeton University. Accessed on March 2, 2019 at: https://webrecorder.io/mudd/turing/20180328150956/https://www.princeton.edu/turing/alan/history-of-computing-at-p/

ISO 6974 (2014). *Natural Gas – Determination of Composition and Associated Uncertainty by Gas Chromatography* (in six parts). Geneva, Switzerland: International Organization for Standardization.

Jones, E.B. (1976). *Instrument Technology*, vol. 2. London, UK: Butterworths.

Rohrback, S., and Ramos, S. (2017). Making chromatography more robust. ISA 62nd Analysis Division Symposium, Pasadena CA, 2017. Session 09-2, 1–'4. Research Triangle Park, NC: International Society of Automation.

Servomex (1985). *Servomex Process Gas Chromatographs*, Product Brochure #7981-3147, (June): 1–4. Crowborough, UK: Servomex Ltd., Sybron Analytical Products Division.

Figures

12.1	The 1979 Beckman Model 6750 Chromatograph Data Processor
12.2	The 2008 ABB Model PGC5000A Master Controller
12.3	Circa 1964 Beckman Model 620 Programmer
12.4	The 1978 AAI Optichrom 2100 Control Unit
12.5	Typical Method of Setting Peak Parameters
12.6	The 1985 Servomex Model CP 409 Programmer
12.7	The 1985 Servomex Model CP 403 Programmer
12.8	The 1979 Yokogawa Model GC6P Control Unit
12.9	The 2010 Envent Model 131 Natural Gas PGC

New technical terms

When first introduced, these new words and phrases were in bold type. For further information, refer to the Glossary at the end of the book.

autozero	**peak gate**
come read	**priority interrupt**
control unit	**program**
forced integration	**programmer**
host computer	**slope detect**
method	**step stream**

Consult the Glossary for information about chemical names.

13

Data display techniques

"There are four ways to present analysis data to the process operators or report measurements to the process control system. At first, only the chromatogram and bargraph readouts were available. As technology improved, the trend record became the most useful output because each analyte then had its own measurement signal, just like other process variables. Then, when microcomputers came along, the numerical analyte concentrations could be printed on a teletypewriter and eventually sent directly to the process control computer".

The chromatogram display

Detector signal

The raw signal from a chromatographic detector is usually electronic, although one detector mentioned in Chapter 10 has a pneumatic output. Most detectors give a voltage output signal, either inherently or after the first stage of amplification. To produce a chromatogram, the first gas chromatographs simply connected that live voltage signal to a potentiometric recorder running with a chart speed of about 25 mm/min.

Figure 13.1a illustrates a typical chromatogram display. For any chromatograph, the chromatogram is an invaluable maintenance tool, essential for evaluating the column performance or for diagnosing an instrument malfunction. Yet for the early process chromatographs it was so much more than that; the chromatogram was the *only* measurement signal available – and it was needed in the control room! So all the original PGCs had to convert the chromatogram to an analog output signal compatible with control panel instrumentation; originally 3–15 psig or 10–50 mA, then standardized as 4–20 mA.

Process Gas Chromatographs: Fundamentals, Design and Implementation, First Edition. Tony Waters.

Digitized chromatograms

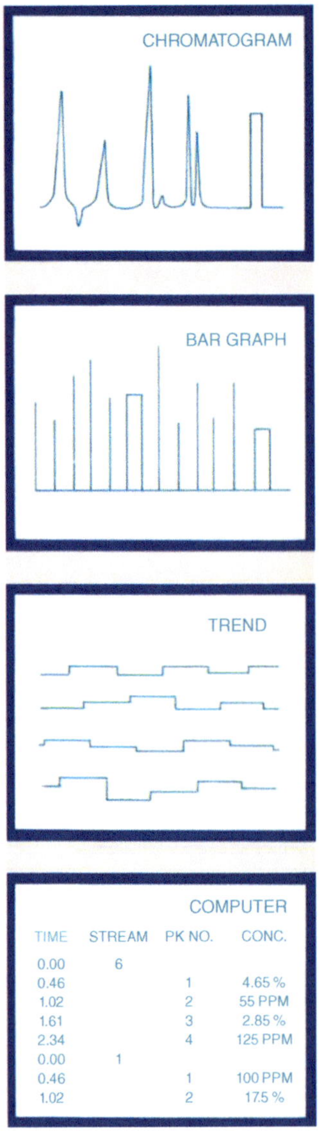

(a) Chromatogram
(b) Bargraph Display
(c) Trend Record
(d) Computer Printout

Figure 13.1 Historical PGC Data Displays.
Source: From Servomex (1985) Reproduced with permission.

When PGCs adopted microprocessors, all the necessary data processing could be performed locally in the programmer. Concurrent advances in data transmission technology allowed the numerical analysis results to be transmitted to the control room and printed on a teletypewriter, as shown in Figure 13.1d. With the elimination of analog measurement signals, it then seemed pointless to transmit an analog chromatogram signal to the control room.

This was a time when we feared that the chromatogram might be discarded as a worthless antique: *Who needs an analog display when all processing is digital?* Luckily, good sense prevailed, and every PGC can display a chromatogram.

The chromatogram is an essential tool for evaluating the resolution of analyte peaks and for checking the correct setting of the peak gates. The chromatogram is also necessary for troubleshooting, as all chromatographic faults show visible symptoms there. We really need that chromatogram:

> *The chromatogram is a fundamental troubleshooting tool without which it would be nearly impossible to diagnose measurement errors or instrument malfunctions.*

In the age of microprocessors, the entire chromatogram is digitized and stored in memory. The microprocessor can process, reprocess, or transmit a memorized chromatogram to another device. The chromatogram can be printed, displayed on a screen, expanded in either direction, or overlaid onto a previously-saved reference chromatogram.

These new tools are a huge aid to troubleshooting, but it's worth remembering that some older data processing systems reconstruct the chromatogram from digital data after noise suppression and digital sampling. If so, it may not show all the tell-tale nuances of a live analog chromatogram (Annino 2003).

More recent systems save original chromatograms as raw data files and apply noise suppression when processing that data for peak measurement. The raw data chromatogram is preferred during troubleshooting to ensure that noise filtering can't mask irregularities.

The development of digital communication networks has given maintenance staff the ability to monitor or control dozens of PGCs from a single convenient location, so plants often install one or more maintenance consoles in an office environment remote from the analyzers. To effectively maintain the PGCs, the maintenance personnel still need a chromatogram, so each analyzer in the field transmits a digital chromatogram to the maintenance console. The PGC specialist can then do routine oversight and perform troubleshooting tasks without visiting the process area.

In addition, diagnostic software is available to check the status of PGC variables and track any chromatogram variation over time. These useful tools can monitor PGC health and predict when maintenance will be needed, and may even remind the technicians to order replacement parts.

Chromatogram autozero

In Chapter 1, we saw that the chromatogram signal exhibits a series of peaks on a more-or-less constant baseline signal. To display the chromatogram, we need to offset that signal so we can see any negative deviations in the baseline.

To generate the zero offset, the early laboratory instruments simply added an opposing millivoltage from a manually adjusted potentiometer. Typically, it was set to about 5 % of the recorder scale.

But a process instrument operates continuously, and some long-term **baseline drift** is inevitable due to detector aging, temperature variation, or **column bleed**—the small number of liquid-phase molecules that continually escape from the column as vapor in the carrier gas. Improvements in instrument performance and column stability have reduced the severity of baseline drift but have not eliminated it.

Therefore, to compensate for slow baseline drift all process gas chromatographs need an *automatic* zero correction. This **autozero** function is scheduled to occur at a time when no peaks or baseline disturbances are expected, often just before or just after sample injection.

During each analysis cycle, autozero resets the chromatogram baseline to the desired zero point. At first, the autozero adjustment was mechanical, a motor driving a variable resistor. Later, a solid-state circuit did it more reliably.

In an older PGC using analog data processing, the function of the autozero circuit depends upon the detector in use. For instance, with a thermal conductivity detector the autozero might directly balance the detector bridge circuit, whereas with a flame ionization detector it simply adds a voltage to offset the baseline signal from the detector.

All modern control units rely on their microprocessor to compensate for any zero offset. The exact mechanism varies by manufacturer. To avoid an overcorrection that might obscure a real fault condition, the processor should limit the extent of zero adjustment and issue warning alarms when the adjustment approaches that limit.

Peak height calibration

Another feature of a process GC is the individual attenuation of peak heights. When a chromatogram has a constant sensitivity throughout, some peaks are hardly visible, and some are offscale. Adjusting the detector sensitivity for each peak allows all the desired analyte peaks to display on a constant recorder scale; such as 0–10 mV, for instance. If desired, the individual peak heights can then be calibrated to indicate the concentration of each analyte.

A process GC must do this automatically, so early programmers have a calibration attenuator (a potentiometer) dedicated to each analyte peak. During each peak gate, the programmer switches the appropriate attenuator into the detector output circuit. You can see a line of potentiometers at top left in the 1960 Greenbrier Attenuator-Timer Unit pictured in Figure 13.2.

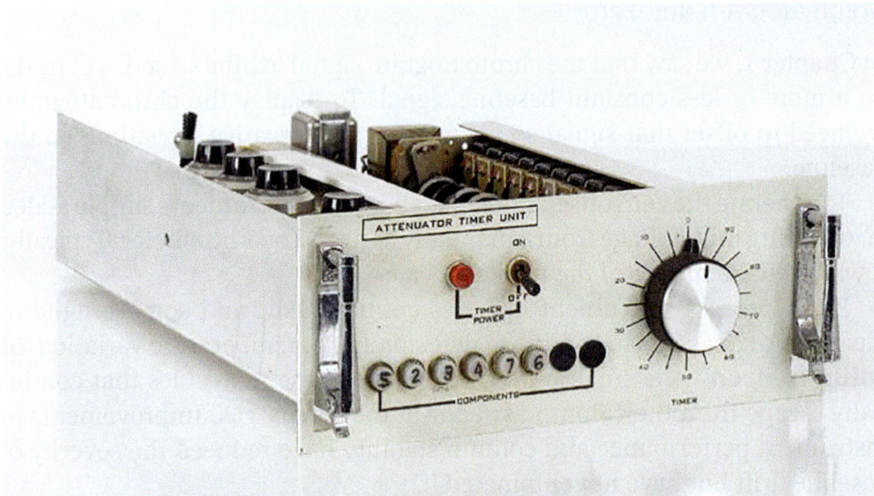

Figure 13.2 The 1960 Greenbrier Chroma-Matic III-A Programmer. *Source: ABB Process Analytics. Reproduced with permission.*

This was one of the first PGC programmers, an electromechanical device. The timer cams are on the same shaft as the TIMER knob and are just visible above the word UNIT on the front panel. At top left you can see several black-and-silver locking knobs for setting the chromatogram signal attenuation. These potentiometers are set to give the correct peak height during calibration.

Contrary to popular opinion, **peak height** calibration is an acceptable way to quantify analyte concentrations. There are even situations, when the columns can't achieve a full resolution, that peak height is a superior way to measure one or more of the overlapping peaks. However, the linearity of height measurement depends on the shape of the peak. Perfectly symmetrical peaks would be linear, but no peaks are perfect. The more asymmetric a peak is, the less likely its peak height will be linear with analyte concentration.

Peak area calibration

Every part of a peak corresponds to the certain number of analyte molecules in the detector at that instant of time. Given a linear detector, it follows that the total area under the peak is proportional to the number of those molecules in the original sample. This is true even for an asymmetric peak. So, with a constant sample injection volume, **peak area** is an excellent measure of concentration for each analyte in the sample.

If the peak width changes between the time that the standards are run and the time that the unknown samples are run, then peak area measurement will be more accurate and reliable than peak height measurement. On a linear detector, peak area is always proportional to the quantity of analyte, even if the peak width changes.

Most PGC detectors provide a linear response to low analyte concentrations in the carrier gas, but they may become somewhat non-linear when the analyte concentration is higher. If that happens, a peak area measurement will be more accurate than peak height because most of the area is detected at low concentrations, whereas the peak height occurs at the instant the peak reaches its highest concentration and suffers its greatest deviation from linearity.

It's harder to measure peak area than peak height because of the difficulty of accurately detecting the baseline at peak start and stop times, especially when the analyte peak is not fully resolved. In the laboratory, we used to cut the peak out of the paper chart and weigh it! The first automatic device to measure peak area was quite unsuited to process use, being a mechanical contraption attached to the chromatogram recorder (Ball et al. 1967).

The first PGC to measure the area under a chromatogram peak used a solid-state *op-amp* with capacitor storage. It worked, but due to the inflated effect of baseline variation upon measured peak area, analog peak integration was often less accurate than the peak height measurement it tried to replace. Luckily, the digital techniques described in Chapter 14 quickly superseded those analog devices and made peak area measurement both accurate and reliable.

Most PGCs now measure peak area, though it's good to remember that peak height might occasionally give superior results.

The bargraph display

The paper saver

Of course, it's immediately obvious that process operators don't want or need a chromatogram; they just need to know the concentration of selected analytes. It follows that the programmer should select those peaks of interest, ignoring any others that might be present. Once selected, the peaks should be calibrated to accurately represent the concentration of the analytes. Having done all this, one problem remains: even on slow speed, a continuous chromatogram consumes a lot of chart paper.

The **bargraph** display illustrated in Figure 13.1b was devised to conserve paper and give process operators a visible record of plant variations over time. When in bargraph mode, autozero sets the baseline to zero on the recorder chart.

To form a bargraph, the programmer turns off the chart motor while waiting for a peak. After each peak passes and the recorder pen returns to zero, the chart motor runs just long enough to advance the chart a few millimeters. Each peak then appears as a bar on the chart, with the bar height indicating the analyte concentration. To maximize the clarity of display, only the peaks of interest are shown on the bargraph readout; all others are attenuated to zero.

To show the analysis start time, bargraph systems sometimes mark a square bar on the chart. The programmer adds a small voltage to the chromatogram signal, advances the chart, and then removes the impressed voltage. When a bargraph record is for multiple streams, the programmer may set the height of the square bar to indicate the stream number of the following bars. You can see these stream indications in Figure 13.1b.

Bargraph displays are obsolete for modern chromatographs, but there might still be a few old ones out there.

The trend record

Analog peak processing

The bargraph was soon replaced by the **trend record**, a better way to display the measurements to process operators and essential for process control. A trend record is an individual measurement signal for each analyte that can be separately recorded. At first, each trend signal simply retransmitted the height of a selected analyte peak and remained constant until updated by a new analysis result. Later, digital processing techniques allowed the trend signal to represent the integrated peak area. Figure 13.1c shows a typical trend record; the regular updates of the measurement value appear as steps in the recorder trace.

To generate a trend output from a chromatogram signal, the early PGCs needed two new devices; a **peak picker** to measure the height of each peak and a **memory unit** for each analyte to hold the output signal constant until updated with new input data.

An analog peak picker acts like a diode. Figure 13.3 illustrates its function. When a chromatogram signal is applied to its input, its output voltage follows the input voltage until it reaches the peak apex. It then holds that value until reset. Once the output signal from the peak picker becomes constant, it's transferred to a memory unit dedicated to that peak, overwriting any previously memorized value. Then, the memory unit outputs a continuous trend signal representing the most recent analyte measurement.

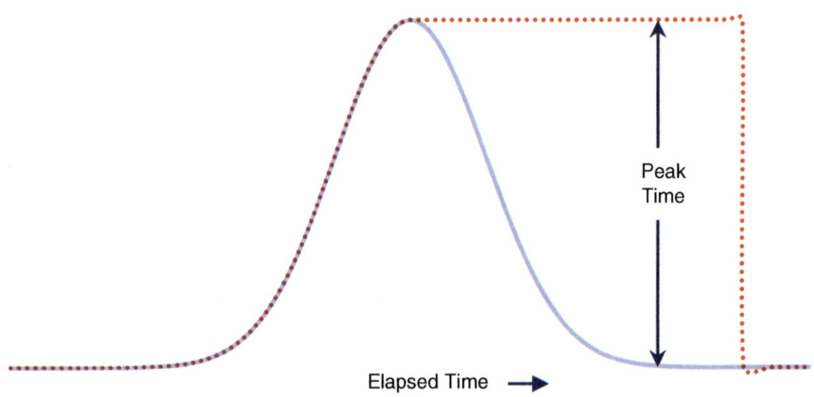

The peak picker output (red dotted line) follows the peak value (blue solid line) as it rises to a maximum height but doesn't fall when the peak declines. The detected peak height is then transferred to a dedicated memory device that outputs a continuous measurement signal.

Figure 13.3 Principle of an Analog Peak Picker.

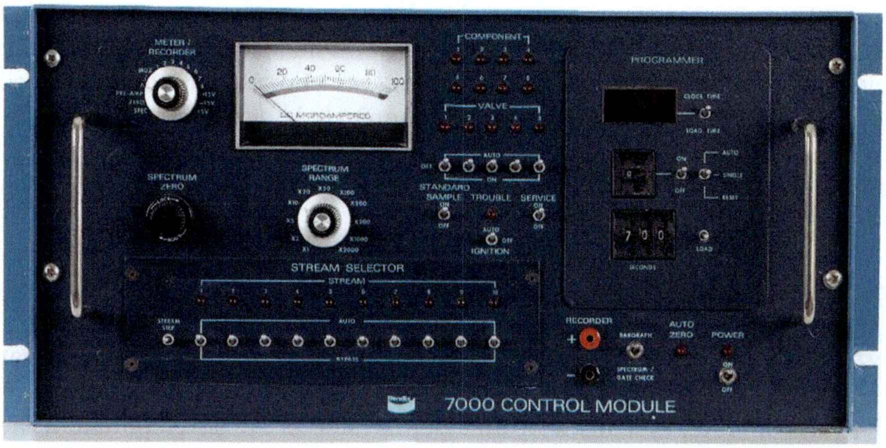

Typical of its genre, the Bendix 7000 features a digital clock, digital signal conditioner, and digital peak memory. D/A converters provide analog outputs (Bendix 1975).

Figure 13.4 The 1975 Bendix 7000 Digital Programmer. *Source: AAB Process Analytics. Reproduced with permission.*

The first peak pickers and memory units were pneumatic (Fuller 1956), but solid-state electronics was developing fast, and the new *op amp* was ideal for the peak picking and memory functions. Thus, you will find that analog signal processors from that era feature solid-state amplifiers, peak pickers, memory units, and power supplies. To perform each of these functions, the early models had a separate electronic unit mounted with other electronic devices on an indoor rack.

Later models using smaller electronic devices include those functions inside the programmer itself. A good example is the Bendix 7000 Digital Programmer seen in Figure 13.4. This instrument can process a detector signal, provide calibrated trend outputs, and control the schedule of events. Today, we would call it a PGC **control unit**.

Digital signal processing

Process gas chromatographs were the first process field instruments to employ digital signal processing and serial-data transmission. That was an unusual role reversal, as most process analyzers adopt proven laboratory techniques. After a while, the laboratory chromatographs caught up and eventually surpassed the chromatogram-processing ability of the process instruments. Ironically, the two approaches are now rapidly merging, and many of today's PGCs use software originally developed for laboratory instruments.

The minicomputer story

The minicomputer control of multiple PGCs is a sidebar in the story of PGC data processing. As the story begins, the microprocessor doesn't yet exist; but by the time the story ends, the microprocessor's rapidly increasing power and diminishing size has eclipsed the minicomputer entirely.

In 1966, Bendix Greenbrier exhibited the first minicomputer-controlled PGC at the ISA exhibition in Chicago. Conventional programmers and peak pickers were gone. Seven Bendix process gas chromatographs interfaced with a DEC minicomputer that communicated with a Honeywell distributed control system. A teletypewriter printed out the analysis of stream composition when requested, and when any out-of-limit alarm condition occurred (Dietz 1987).

The idea of a central computer to process all the chromatographic data persisted for about 20 years. For instance, in 1973 Applied Automation came out with the OPTICHOM 2C minicomputer system controlling up to 32 analyzers, followed in 1984 by Combustion Engineering's Central Data Manager that could handle as many as 40 PGCs.

Minicomputer systems were an essential step in the development of the modern process chromatograph. They established the feasibility of PGC computer control and a methodology for doing it. All subsequent electronic developments used computer technology to improve PGC application flexibility and performance.

Yet there's a good lesson to learn here. Over the years, we have found that the time-sharing of PGC equipment is rarely a successful strategy because any downtime of the shared resource impacts the measurements from several analyzers. One failure in a shared controller might suddenly eliminate a hundred vital indicators of process quality.

> *It's an oft-told tale in process analysis that any device that is common to multiple measurements is a potential disaster. This principle applies to simple things like filters and stream selector valves, and to complex things like minicomputers.*

When a computer is the sole control device and the sole data path from multiple PGCs to the process control system, its occasional downtime shuts down all the analyzers and blinds process operators to any variation in the composition of their process fluids. The instantaneous loss of information may cause issues with product quality or process throughput and might even compromise plant safety.

The ubiquitous microprocessor

In the technical literature, the first use of a microprocessor in a PGC programmer was noted by Villalobos (1975). The notion was further developed by Thomas et al. (1977), who defined the desirable functions of a microprocessor control unit and correctly predicted that:

> *"The marriage of the process GC and microcomputer will no doubt be considered one of the more significant milestones of the decade."*

The authors were right, but they probably could not have predicted the frenetic pace of development over the next few years. For the record, Table 13.1 lists some of the microcomputer control units offered for sale by 1980.

Table 13.1 The First Microprocessor-Based PGC Control Units.

Released	Company	Model	Citation
1978	Applied Automation	Optichrom 2100	McCoy (1978)
1978	Bendix	9000	Clemons (2017)
1978	Foxboro Arcas	Arcas PMC 2000	Foxboro (1978)
1979	Yokogawa	GC6P	Yokogawa (1981)
1979	Hokushin	V610	Hokushin (1979)
1979	Amscor	Mark IV	Amscor (1979)
1980	Siemens	P101	Ball (1980)

The microprocessor-based control unit gave some relief from relying on a single minicomputer to control dozens of chromatographs. Unfortunately, though, the first microprocessor chips were expensive, so product engineers again resorted to time sharing – but not so many this time. Most of the new control units could control no more than six PGCs.

Since digital processors naturally generate digital composition data, these control units could transmit numerical values to the process control system via a serial data link. But process engineers saw digital transmission of process variables as an untried concept and were often concerned about reliability and security. As insurance, they required key analysis data to be also transmitted by the trusted 4–20 mA analog trend signals, even though the wiring for those individual signals added a considerable expense to their installations.

To output an analog trend signal, a digital control unit needs a dedicated digital-to-analog (D/A) device for each stored measurement, again increasing the cost. Ironically, once those 4–20 mA trend signals reached the control room, they were re-digitized for input to the process control computer.

Over time, plant data networks became more secure, and the usage of digital data transmission rapidly increased, so the need for analog trend signals plummeted. When needed, a computer-generated graphical trend is visible on the process operator's console or on the display screen of a maintenance workstation.

Reliable digital transmission has eliminated most analog trend signals. In a newer PGC, the chromatogram is the only analog display that remains, and even that may come from reconstituted digital data.

A regression perhaps?

As the cost of electronic devices came down, it soon became common for each PGC to sport its own microcomputer. The 1977 Beckman 6750 was perhaps the first PGC system to have a central data processor unit controlling up to six PGCs, each with its own autonomous microprocessor (see Chapter 12, Figure 12.1).

The following twenty years may turn out to be the Golden Age of the industrial process chromatograph. There is little doubt that a dedicated microprocessor for each PGC is the optimum configuration for long-term reliability.

But, sadly, by their very nature, all *Golden Ages* end.

In the closing years of the second millennium, each of the three major producers of process gas chromatographs introduced new PGC models. These bright new analyzers feature some advanced concepts; but, in defiance of history, they all have a single microprocessor controlling up to four column ovens potentially housing lots of valves and detectors.

In these PGCs, the control unit sequences dozens of timed events on multiple chromatograms, controls the temperature of several ovens, and processes a multitude of detector signals to get the desired measurements. It also monitors the health and safety of the instrument itself and takes appropriate action when any parameter falls outside preselected limits. In addition, these new PGCs may transmit measurements to and accept commands from the process control computer, and even communicate with other PGCs and personal computers.

It all sounds very impressive until you realize that these gargantuan machines, like the shared microcomputers before them, depend totally on one processor.

Once again, we are sacrificing long-term reliability to the omnipresent deity of reduced capital cost. Purchasing a PGC with three ovens that contain perhaps six parallel applications is not a route to high reliability – quite the opposite, in fact.

Let's say it again:

> *Any device that is common to multiple measurements is a potential disaster. This principle applies to simple things like filters and stream selector valves, and to complex things like microcomputers.*

The PGC manufacturers are keenly aware of this principle, but struggle to offer the best long-term performance while negotiating with buyers who see only a one-time capital investment.

At the time of writing, one PGC manufacturer (ABB) has moved to ameliorate the common device problem. With the latest electronic components, their computer board has become small enough for each column oven to have its own dedicated controller, a flashback to the Beckman 6750 configuration we lauded earlier. Each controller-oven pair operate as an autonomous PGC, but lack a human interface, so a separate user interface unit serves several analyzers. This commonality doesn't reduce reliability because the user interface is not active during normal operation, it's used only occasionally for maintenance checking or troubleshooting. ABB offers a choice of three interface devices; a hard-wired control unit used only as a user interface, a local or remote personal computer connected into the PGC ethernet network, or a WiFi-enabled portable tablet rated for operation in the hazardous area.

The rapid rate of change of electronic systems continues to provide opportunities and headaches to the end-users. The opportunity to improve functionality and reliability is welcome, but they must weigh those advantages against more rapid obsolescence, increased cost of spare parts, and the enduring need for maintenance training.

Central maintenance station

The ability to monitor the performance of several PGCs from a central control room or office location was an attractive feature of minicomputer systems. Thus, when PGCs became microcomputer controlled, the popular concept of a central maintenance station survived.

To enable a central maintenance system, each PGC must be capable of accepting digital commands and transmitting digital data. An Ethernet or proprietary data link connects all the PGCs located in one plant area to the central maintenance station and to each other, forming a two-way communications channel.

In modern parlance, the central station is known as an **analyzer management system** and typically runs on a master PGC control unit or a personal computer with the necessary software to transmit commands to, and receive data from, any of the remote PGCs. This allows the maintenance staff to check and update any selected analyzer within the safety and comfort of the control house. They conduct routine activities such as stream selection, trend recording, validation, statistical analysis of performance, and method development without going into the process unit.

Maintenance personnel can initiate and complete many troubleshooting tasks at the central station, sometimes assisted by proprietary diagnostic software. To facilitate such tasks, the ability to display and print a diagnostic chromatogram from any detector in any analyzer on the PGC network is a prime design requirement of a central maintenance station.

The networking of analyzers also allows an authorized person to access any PGC from a personal computer connected physically or wirelessly to the network. In advanced systems, it's even possible to access and control any PGC from the front panel of any other PGC on the same network.

Continuous analyzer controllers

In times past, continuous process analyzers and sensors had only analog output signals and needed a way to transmit their measurement data to the process computer. To accomplish this task, the PGC network would include a separate **continuous analyzer controller** to digitize those analog signals and transmit the data. Now, it's less common to interface external sensors via the PGC network because most field instruments can communicate digitally.

PGC function alarms

Indicators

These days, all PGCs have a way to indicate their overall health. Some PGCs have externally visible light-emitting diode (LED) indicators, often in multiple colors; others post notifications on their integral display screens.

To assist in troubleshooting, all PGCs have colored LED indicators on individual circuit boards. Usually, a green light indicates nominal operation and nothing to worry about, a yellow light warns of a non-fatal anomaly, and a red light means trouble. LEDs may be lit or pulsed to show different conditions, and combinations of color can indicate more specific maladies. For instance, each circuit board in the Siemens Maxum Edition II has three LEDs that in various combinations of color indicate six operational states. The ABB Model 5000 PGC goes one further and even winks out a numerical error code.

Since each PGC has a distinct set of status indicators, you'll need to consult the manufacturer's instruction manual to discover their meaning in your instrument.

Alarm notifications

Unlike the earlier electromechanical PGCs, a microcomputer-controlled PGC can monitor its own operating condition and performance. The processor can issue warning or malfunction alarms and, in more advanced instruments, might even take corrective action to adapt to a change.

Since the digital detection and transmission of alarm notifications is essentially free, the number and scope of the alarms has increased at a vigorous rate. Many PGCs now feature hundreds of alarms that hopefully provide more clarity than confusion to the hapless troubleshooter. One PGC instruction manual we consulted had a table of alarms that occupied 48 pages.

To reduce human confusion, the alarm software may cluster these multifarious alarms under a few broad categories. Typical categories might include:

- Safety system fail
- Oven temperature control error
- Electronic pressure control error
- Detector signal quality
- Component peak error
- Calibration or validation error
- Scheduling error or notification
- Software error

- Digital communications error
- Sampling error

For those alarms that have settable parameters, the setup procedure is similar to that used for the analysis method. Usually, it comes down to making entries in a table displayed on the screen of the control unit or maintenance console. Each PGC model has its own inventory of alarms that cover virtually any change in its operating conditions.

Knowledge Gained

- *The detector signal is usually a voltage, either inherently or after the first stage of amplification.*
- *A chromatogram display is a record of the detector signal during each complete analysis.*
- *The chromatogram is an essential maintenance and troubleshooting tool.*
- *Early PGCs transmitted the analog chromatogram signal to the process control room.*
- *Microprocessor control units can digitize and store the entire chromatogram for later processing.*
- *The digital chromatogram can be printed, displayed, or compared with a saved reference chromatogram.*
- *Digitized chromatograms may not show all the fine structure of the original detector signal.*
- *Baseline drift is inevitable due to detector aging, temperature change, or gradual column bleed.*
- *Autozero nulls baseline drift by resetting the detector zero at least once in each analysis cycle.*
- *Early PGCs had a dedicated attenuator to calibrate the height of each chromatogram peak.*
- *For symmetrical overlapping peaks, calibration by peak height is superior to calibration by peak area.*
- *Peak area calibration is more fundamental and is unaffected by peak shape.*
- *It's harder to measure peak area due to uncertainty in the detection of the peak start and stop times.*
- *The original analog peak integrator used a solid-state "op amp" and capacitor storage.*
- *A bargraph display, now obsolete, provided peak height information without wasting chart paper.*
- *Each bar is a calibrated chromatogram peak drawn without chart movement.*
- *Between bars, the chart moves a short distance while the signal is held at zero.*
- *A square bar indicates the analysis start time and may be calibrated to show the stream number.*
- *Only the analyte peaks are displayed on a bargraph display; all others are zeroed out.*
- *A trend record is an analog signal dedicated to one analyte measurement, like other process variables.*
- *An analog peak picker holds the maximum peak height and transfers it to a memory device.*

- *The memory device outputs a constant analog signal until updated by a new analysis result.*
- *The pneumatic peak pickers and memory units were soon replaced by solid-state electronic devices.*
- *The first digital systems used a single minicomputer to control multiple PGCs and transmit their results.*
- *Unless it's virtually fault-free, a common device reduces the overall reliability of a system.*
- *Digital processors now transmit numerical data, so analog trend signals may be unnecessary.*
- *Early digital systems also had redundant analog trend signals to avoid losing vital measurements.*
- *The numerical data was first printed on a teletype and later directly input to the process computer.*
- *The first microprocessor-based controllers were also time-shared to control several PGCs.*
- *Later PGCs have an integral microprocessor, although it may also control other local PGCs.*
- *The new millennium PGCs have multiple column ovens but only a single, shared computer.*
- *Microprocessors are now so small that each column oven can have a dedicated computer.*
- *Microcomputer PGC systems adopted the concept of a central maintenance station (CMS).*
- *Chromatograms from all the PGCs are available at the central maintenance station.*
- *Many routine maintenance tasks can be performed at the CMS without entering the process unit.*
- *Colored LED indicators show PGC analyzer status and the functional health of their circuit boards.*
- *Microcomputer PGCs produce a swarm of status alarms, often organized into several categories.*
- *Status messages appear on the local PGC display panels and on the central maintenance station.*

Did you get it?

Self-assessment quiz: SAQ13

Q1. Over the years, process gas chromatographs have used several forms of data display. What are the four main types of display?

Q2. Why is the chromatogram an essential display?

Q3. What is the fundamental reason that a measurement of peak area is the best estimate of analyte concentration?

Q4. Under what circumstances would peak height measurement be used, instead of peak area measurement?

Q5. What is an analog memory unit?

Q6. Why is it undesirable for a PGC to contain a single device that is common to multiple measurements?

Q7. What are the advantages of connecting PGCs on a digital data highway?

Check your SAQ answers with those given at the end of the book.

Student evaluation questions: SET-13

S1. Which one of the items listed below has never been used as a data display for an industrial process gas chromatograph?
Select one option:

A. Chromatogram.
B. Bargraph.
C. Trend record.
D. Serial data link.
E. Computer printout.

S2. Which one of the items listed below has never acted as a chromatogram signal to the final readout device?
Select one option:

A. An analog voltage signal.
B. An analog current signal.
C. An analog pneumatic signal.
D. An analog hydraulic signal.
E. A digital electronic signal.

S3. Which one of the statements listed below correctly describes the movement of the recorder chart paper when switched to bargraph mode?
Select one option:

A. The chart doesn't move in bargraph mode.
B. The chart briefly moves after each peak returns to baseline.
C. The chart briefly moves when the peak reaches its maximum.
D. The chart moves continuously, but very slowly.

S4. Many older PGCs used an analog peak picker … for what?
To receive a score, you must select all the correct statements and none of the incorrect statements:

A. Choosing which peaks to measure.
B. Distinguishing peaks from baseline noise.
C. Sensing the height of an analyte peak.
D. Switching analyte peaks into another column.

S5. In a process gas chromatograph, what is an autozero?
Select the one best description:

A. It's an automatic adjustment that always sets the detector baseline signal to zero on the chromatogram display.
B. It's an automatic adjustment that sets the detector baseline signal to a selected value, typically just above zero on the chromatogram display.
C. It's an automatic adjustment that sets the peak memory to zero before it accepts a new analysis value.
D. It's an automatic adjustment to zero-out any peaks that should not be shown on the bargraph readout.

S6. In a modern microprocessor-controlled PGC, why is it important for a chromatogram display to be available to plant personnel?
To receive a score, you must select all the correct statements and none of the incorrect statements:

A. Technicians need the chromatogram display to check the timing of peak gates, often indicated by marker spikes.
B. Technicians need the chromatogram display for troubleshooting, as all chromatographic faults show visible symptoms there.

C. Technicians need the chromatogram display for measuring the concentration of the analyte peaks that are visible there.
D. Plant operators need the chromatogram to show that the analyzer is running and working properly.

S7. In a modern microprocessor-controlled PGC, when might peak height be a better measurement of analyte concentration than peak area?
To receive a score, you must select all the correct statements and none of the incorrect statements:
A. When the analyte peak has an asymmetric shape.
B. When the analyte peak is symmetrical, but incompletely separated from an unmeasured peak.
C. When the analyte is a backflushed peak, such as a "total heavies" peak, comprising several component peaks that don't have perfectly coincident elution times.
D. When the detector signal starts to become nonlinear at higher concentrations of analyte molecules within the detector.

S8. A single microprocessor-based PGC control unit can easily control several local column ovens or several independent gas chromatographs, each analyzing different components or different streams. That being true, why is this multitasking arrangement considered to be poor system design?
To receive a score, you must select all the correct statements and none of the incorrect statements:
A. Because a microcomputer operates too slowly to communicate with PGCs installed at different locations in the plant.
B. Because a single failure in the control unit will cause the loss of many process measurements.
C. Because the whole system is out of service when one of the column ovens fails, causing the loss of many process measurements.
D. Because each hardware device that is a resource shared by several measurements always reduces the reliability of the whole system.

S9. Consider an advanced system comprising many microcomputer-controlled process gas chromatographs networked to a central maintenance station and reporting analytical results to a process control computer. Where in this system would an analyzer specialist be able to look at a chromatogram from one of the analyzers?
To receive a score, you must select all the correct statements and none of the incorrect statements:
A. On the display screen of the local control unit for that analyzer (assuming it has a display screen).
B. On the display screen of a personal computer attached to the network.
C. On a display screen or printout at the central maintenance station.
D. On the display screen of the process operator's console.

References

Cited

Amscor (1979). Introducing the most advanced process chromatography system in the world. In: Product Advertisement, *InTech Magazine, (September): 117*. Pittsburgh, PA: Instrument Society of America.

Annino, R. (2003). Chromatographs: Gas. In: *Instrument Engineers' Handbook: Process Measurement and Analysis,* Fourth Edition, Volume 1, Section 8.12. (ed. B. Liptak). Originally published in 1995. *Boca Raton,* FL: CRC Press.

Ball, H. (1980). *Process Gas Chromatography.* Siemens analytical application note 291, *Bulletin 5961 IF 09863, 63–95.* Karlsruhe: Germany, Siemens Aktiengesellschaft.

Ball, D., Harris, W., and Habgood, H. (1967). Errors in manual integration techniques for chromatographic peaks. *Journal of Chromatographic Science* 5 (12, December 1): 613–620, https://doi.org/10.1093/chromsci/5.12.613

Beckman (1979). *Model 6750 Microprocessor-Based Process Gas Chromatograph.* Product Bulletin 4189. Fullerton, CA: Beckman Process Instruments.

Bendix (1975). *Bendix 7000 Gas Chromatograph Programmer.* Product Brochure B2290SB375, 1–4. Lewisburg, WV: Bendix Process Instruments Division.

Clemons, J.M. (2017). Personal Communication: *Bendix Process Analytics, Model 9000 Microprocessor Controller for Process Gas Chromatograph Analyzer Control, 1978, three temperature control zones, five valves, controls one analyzer.*

Dietz, J.M. (1987). The Forum: Technology review 1987 – applications of gas chromatography. 33rd Analysis Instrumentation Symposium 1987 in Houston, Texas (April 21, 1987). Research Triangle Park, NC: Instrument Society of America.

Foxboro (1978). *Arcas PMC 2000 Process Micro Computer.* Product Bulletin PMC-0578, 1–8. Houston, TX: Foxboro Analytical.

Fuller, D.H. (1956). Gas chromatography in plant streams. *ISA Journal,* 3, 440–444.

Hokushin (1979). *Process Gas Chromatograph V600 Series.* Product Catalog No. E084 (September). Tokyo, Japan: Hokushin Electric Works, Ltd.

McCoy, R.D. (1978). Adding capabilities to process chromatography with microprocessor-based programmers. In: *Analysis Instrumentation,* Vol. *16,* 155–162. Proceedings of the Joint Spring Conference, in Houston, Texas (May 22–26, 1978). *Pittsburgh,* PA: Instrument Society of America.

Servomex (1985). *Process Gas Chromatographs.* Product Brochure 7981-3147, (June): 1–4. Crowborough, UK: Sybron Analytical Products Division.

Thomas, T.J., Martin, W.M., and van Vuuren, P. (1977). Design criteria for a microcomputer based chromatograph. In: *Analysis Instrumentation,* Vol. 15, 17–21. Proceedings of the 23rd Annual ISA Analysis Instrumentation Symposium in Charleston, VW (May 17–19, 1977). Pittsburgh, PA: Instrument Society of America.

Villalobos, R. (1975). Process control by gas chromatography. *Analytical Chemistry* 47 (September): 983A–995A.

Villalobos, R., Porter, R., LeBlanc, R., and Hearn, R. (1978). New concepts in microprocessor-based process chromatographs. In: *Proceedings of 33rd Annual Symposium on Instrumentation for the Process Industries,* at Texas A & M University (January 1978).

Yokogawa (1981). *Model GC6 Process Gas Chromatograph.* Product Bulletin 11B1A2-E (October 1). Tokyo, Japan: Yokogawa Electric Works, Ltd.

Table

13.1 The First Microprocessor-Based PGC Control Units

Figures

New technical terms

When first introduced, these new words and phrases were in bold type. For further information, refer to the Glossary at the end of the book.

analyzer management system
autozero
bargraph
baseline drift
column bleed
continuous analyzer controller
control unit
memory unit
peak area
peak height
peak picker
trend record

Consult the Glossary for information about chemical names.

14

Peak area integration

"The detailed method of programming a computer-controlled PGC varies widely by manufacturer, and even the terminology varies by brand. Yet there are many similarities, and we try to capture this common ground. Our intent here is to introduce typical setup procedures using simple language that is easy to understand. You'll need to adapt these procedures to the unique requirements and terminology of your own model of PGC".

Digital chromatogram processing

Pulse frequency digitization

The need to integrate peak area was a strong incentive to apply digital techniques to the process gas chromatograph. As a result, PGCs were using digital electronics long before other process instruments.

A few of the original digital integrators may still be in service. In these, a solid-state electronic device converts the millivolt detector signal to a proportional pulse frequency. When an analyte peak passes through the detector, the pulse rate rises to a maximum at the top of the peak and then drops back to minimum as the chromatogram signal returns to baseline. To measure the area of the peak, a solid-state counter circuit accumulates the total number of pulses produced by the peak and holds this count in a short-term digital memory.

These older systems have two alternative ways to transfer the calibrated digital result to a process control computer. The programmer might issue a "come read" request after each peak, instructing the process computer to interrupt what it is doing and read the peak data held in the short-term memory; or it may transfer the data to a digital memory dedicated to that peak and allow the process control computer to read the data at its convenience. The former method is cheaper as it requires less memory.

When computer interfacing is unnecessary, the programmer transfers the short-term memory data to a long-term memory dedicated to that analyte

Process Gas Chromatographs: Fundamentals, Design and Implementation, First Edition. Tony Waters.

that outputs a sustained analog signal proportional to the concentration of that analyte in the original sample. This method costs more than digital transmission as it requires a dedicated memory for each analyte in each stream.

Sampled-data digitization

Modern PGCs take a different approach. The speed of the microprocessor allows it to instantaneously measure the detector output voltage at a frequency as high as 1,000 times per second (Hz), although most systems run slower than that. The frequency of taking measurements is the **sampling rate**. Given sufficient memory, the entire chromatogram can be digitized and stored in memory. The computer can then display or reprocess the **raw chromatogram** as if it were a live signal to find the best settings for the application.

During peak area measurement, the processor may average consecutive sets of the **raw data** to improve the precision of each measurement. Thus, the number of data points used for peak integration might be less than the number of raw data points collected for that peak. PGC practitioners consider 40 data points per peak to be the minimum for accurate measurement. To accumulate that number of readings, narrow peaks need rapid data collection but wide peaks do not, so the rate of data collection is an important setup parameter.

In effect, the sampling procedure converts a chromatogram peak into an array of equal-width rectangles, like those in Figure 14.1. The height of each rectangle is stored in memory to form a digital representation of the entire chromatogram.

We shall see that when the processor is measuring a peak, it estimates the level of the baseline and subtracts an appropriate amount from each stored value. Then, since the stored values are equidistant in time, the sum of all the corrected values is proportional to the peak area.

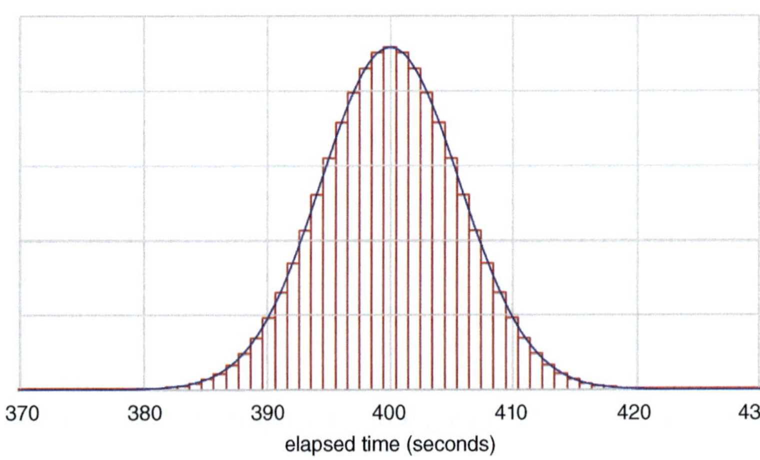

To measure the area under a peak, the processor reads the detector voltage signal (blue curve) at least 40 times. The figure illustrates the approximation that results from converting a symmetrical peak into 40 equal-width rectangles (red histogram).

Figure 14.1 Sampled-Data Method of Digitizing a Peak.

Note that each stored data point represents a *height*, not an area. Therefore, the highest of the stored values is an estimate of the peak height. In practice, most systems use a prediction algorithm to interpolate the true height and position of the peak apex from a few data points on either side of the maximum recorded value. They use a similar process in reverse to estimate the true height and position of a **valley point** between partially-resolved peaks.

Signal noise measurement

The high signal-sampling speed of a PGC microprocessor allows it to measure the detector noise and compensate for its effect on the measurement. Thus, the method table may include a *noise measurement* period at a time when a flat baseline is expected. For a few seconds, the processor samples the detector signal and temporarily stores enough measured values for a statistical analysis of the noise; about 20 baseline readings is adequate, although some processors may take up to 100 samples. From these values, the processor estimates the overall peak-to-peak signal noise and might calculate the **standard deviation** (σ) of the noise data.

Some PGCs use the measured noise level to automatically establish a minimum acceptable peak height; such as twice the noise level, with the noise level defined as $\pm 3\ \sigma$. Alternatively, a PGC may allow manual entry of the minimum peak height during its setup procedure. The processor will ignore any peak that fails to exceed this threshold value. Another way to distinguish peaks from noise is to require real peaks to have a minimum peak width or a minimum peak area.

Signal noise reduction

All microprocessor PGCs use noise reduction procedures, but the exact technique of noise filtering differs by brand. Mostly, it's done by averaging several readings of the raw chromatogram.

To avoid synchronizing with any 50 Hz or 60 Hz noise induced from power fields, a PGC processor reads the detector signal at a different frequency. A sampling rate of 40 readings per second is fast enough to measure a peak that is only one second wide. A faster data sampling rate may be appropriate for very narrow peaks. When measuring slower peaks, though, that high rate of data acquisition would generate far more data than necessary for accurate measurement and would use a lot of memory.

Most processors sample the chromatogram at a higher rate than is necessary to integrate a peak and then memorize the average of a "bunch" of consecutive readings. Averaging tends to cancel random noise and reduces memory usage.

The 2010 Rosemount Model 2350A pictured in Figure 14.2 is an example of a PGC controller that uses a simple averaging technique. This processor samples the detector signal at 40 Hz and converts each reading to a 16-bit digital number stored in a temporary buffer memory. Then, it averages each

This explosion-proof control unit is an option for use with the Rosemount Model 500 PGC.

Figure 14.2 The 2010 Rosemount 2350A Control Unit.
Source: Rosemount Analytical. Reproduced with permission.

N consecutive readings from the buffer and stores those average values for further processing (Rosemount 2010, 1–10).

The factor *N* is the **peak width parameter**, an integer number that you enter into the method table. The entered parameter is not the real peak width; it's just an easy way to set the averaging procedure. Here's how it works: if you estimate that an analyte peak is about five seconds wide, set the *peak width* parameter to "5." This entry instructs the processor to average each five consecutive readings. Since this processor has a fixed sampling rate of 40 Hz, setting the peak width parameter to "5" generates eight averaged data points per second, whereas a peak width setting of "10" produces only four averaged data points per second. Notice that both of these settings provide 40 data points per peak, but for the slower peak each stored value is the average of ten readings.

Averaging *N* results reduces the standard deviation of random noise by $\sqrt{N}$, so a setting of N=4 cuts the noise level in half.

The *peak width* command should occur well ahead of the peak to be measured and usually stays in effect until the processor encounters another *peak width* command. If using a single *peak width* command, set it to the width in seconds of the narrowest peak on the chromatogram. When *peak width* is not set, some instruments will measure the peak width automatically, adopting the observed width at 5 % or 10 % of the peak height.

Another averaging technique is the **moving average**. This method doesn't reduce the number of readings per peak, so it doesn't need a *peak width* command. An ultra-fast chromatogram using the moving average technique might sample 100 raw data points during a one-minute peak, but a lower sampling rate is adequate for the wider peaks of isothermal chromatography.

The processor reads the chromatogram signal at the selected sampling rate and stores each reading in buffer memory. Then, it smooths the data by computing the average of each *M* consecutive readings, where *M* is typically an odd number between 3 and 9. Once the average is safely stored, the processor adds the next data point, drops the oldest one, and repeats the procedure. Thus, there are the same number of data points in the smoothed chromatogram, but each one is the average of several readings.

This PGC can compute the weighted average of nine readings for each chromatogram data point.

Figure 14.3 The 2002 Siemens Maxum™ Edition II PGC.
Source: Siemens Analytical Products and Solutions. Reproduced with permission.

The mathematical procedure used to calculate the moving average depends on the model of PGC. The simplest method takes the arithmetic average of three or five readings and stores the average as the center data point. This simple averaging procedure certainly cancels out random noise, but it slightly distorts the peak shape. More sophisticated procedures apply a weighting factor to each raw data point before computing their average. The weighting factors depend only on the number of data points being averaged and are therefore constant for a given PGC processor.

For instance, the 2002 Siemens Maxum™ Edition II in Figure 14.3 has an option to compute the weighted average of nine readings for each chromatogram data point. The weighting factors used are the well-known values derived by Savitzky and Golay (1964).

The technique of averaging analog data obtained at a constant sampling rate is known technically as **signal smoothing**. Since smoothing computes the average value of the past M readings, each chromatogram point is recorded shortly after the real-time detector signal. When working with reconstructed chromatograms, be aware that the whole chromatogram has been time-shifted by up to a few seconds and may be slightly out of synch with the real events in the columns.

Signal smoothing routines act as an electronic **low-pass filter**; low-frequency peaks pass through with minimal height attenuation while high-frequency noise is irreversibly blocked.

All moving-average techniques reduce the signal noise level without changing the true area under a peak, but they also slow the speed of response. Be careful: applying excessive dampening to the chromatogram signal will tend to flatten the faster peaks and reduce their sensitivity of measurement. Small peaks may then be rejected by the preset peak-height threshold.

Never use data-smoothing techniques to solve a noise problem – once set, it tends to be permanent and desensitizes the PGC forever after.

While all detector signals exhibit some noise, an excessively noisy signal indicates that something is wrong; it might be detector pressure fluctuations, contaminated utility gases, damage to the detector sensing elements, or a fault in the detector electronics. Instead of applying more signal smoothing, investigate the cause of the noise and eliminate it.

On many routine applications, process chromatographers find signal smoothing unnecessary and deactivate the feature entirely.

Quantifying the analyte peaks

A process gas chromatograph can calculate analyte concentration from a measurement of **peak height** or **peak area**. Microprocessor-based PGCs automatically have both measurements available for selection in the method table.

Each brand of PGC has its own distinct way to set up a peak for measurement, so this text illustrates only general principles. Before making any adjustments to a PGC program, you must learn the correct procedures for your instrument. The processor will follow the instructions you enter in the method table, and it's easy to introduce significant errors by specifying the wrong method of integration.

Before making any changes to an analysis method, be sure to save a backup copy of the original. This is a sensible precaution, best practice, and face-saver.

After making changes to the method, be sure to check the settings with samples that contain different concentrations of components. Settings that work well for one mixture of components may work badly when those concentrations change.

Typically, a computer-controlled process gas chromatograph offers two distinct ways to measure the area under a peak: **forced integration** or **automatic peak detection**. The *Method Table* tells the processor which one to use.

Forced integration

If selecting forced integration, you will manually set up a **peak gate** for each analyte peak by specifying an integration start time and an integration stop time.

Ideally, the **start time** will occur before the peak begins to rise, and the **stop time** will occur after the peak has returned to baseline. In practice, however, it's not always possible to achieve a baseline resolution between peaks. When an analyte peak doesn't return to the baseline, the start or stop time may be located at a valley point between overlapping peaks. If so, you must tell the processor what to do.

In the method table, for each analyte peak, enter the time when you want the integration to start and specify the condition that will exist at that time; usually, you can choose *baseline* or *valley*. Then, enter the time you want the integration to stop and the condition that will exist at that time; again, *baseline* or *valley*.

Your choice of *baseline* or *valley* is important because it defines the **zero line** underneath the peak that the processor will use when reading the height of each slice of the peak; i.e., the height of the rectangles in Figure 14.1.

When the method specifies that a *baseline* exists at both start time and stop time, the processor measures the chromatogram signal at the beginning and end of the gate and projects a straight zero line between them. The zero line may be flat, sloping up, or sloping down, depending on the actual baseline levels before and after the peak.

When the method specifies that one of the gates as a *valley*, the processor assumes that the zero line is horizontal and uses the actual baseline value it sees at the other end of the gate.

Some manufacturers call the baseline sensing procedure a **software autozero** (SAZ), but the processor doesn't really set the zero, it just measures and stores the baseline voltage.

Forced integration continues between the specified gate times and accumulates the entire area of the chromatogram above the projected zero line. A measured area can be due to one peak, several peaks, or no peaks at all. Forced integration can therefore be set up to group several peaks together. Note, however, that the grouping of peaks can give an accurate measurement only when their response factors are similar. We discuss the calibration of composite peaks in Chapter 15.

Note that a processor in forced integration mode doesn't detect the valley points automatically. Therefore, if you want to measure to a valley, you must set the gate to the exact time of the valley.

Unfortunately, the actual valley time may shift due to small variations in the peak retention times, or even due to a change in peak heights. This is a disadvantage of using forced integration mode for measuring incompletely resolved peaks; the slope-detection mode might be a better choice.

When using forced integration, be careful to select stable reference points for each baseline measurement. Forced integration won't work well if the baseline reference point is subject to disturbances caused by signal noise, stray peaks, or column switching.

Slope detection

Most computer-controlled PGCs opt for dynamic peak detection. This technique, which is also common in modern laboratory gas chromatographs, locates peaks automatically by detecting changes in the slope of the detector signal.

For each peak detected, the computer stores the observed start time, apex time, and end time. It then infers the identity of the peak from the peak apex time; i.e., its retention time. For each desired analyte, the user specifies the name of the peak and sets a range of retention times considered acceptable.

Slope-detection algorithms may allow a laboratory chromatograph to detect and measure every peak on the chromatogram. Being more conservative, though, PGCs usually opt for gate control of the peak detection process, even when using slope detection. By limiting slope detection to the period between fixed gates, there is less chance that the processor will misidentify the peak, even if the peak happens to be zero.

Therefore, peak gating for **slope detection** has a different function than gating for forced integration. The start gate initiates the search process. Enter a *start time* in the analysis method well before the peak is due and specify *slope detect* to start looking for a peak. Peak area integration will not start until the processor detects a peak.

The processor detects the onset of a peak by monitoring the rate of change of the baseline slope, mathematically the **second derivative** of the detector signal. As soon as this rate of change exceeds a preset **threshold value**, the processor recognizes the start of a peak. Therefore, the setting of the threshold value is important.

Some processors measure the signal noise level and calculate a default value for the slope threshold. Be wary about setting the threshold level manually, thereby overriding the default value:

- If you set the threshold too low, spurious baseline noise will trigger the detection of multiple false peaks that are visible on the chromatogram as a swarm of event marker blips.
- If you set the threshold too high, the integration will start late and finish early, which will cut a horizontal strip off the bottom of the peak and cause a significant measurement error.

Typically, a processor will find the top of the peak by detecting when the slope of the detector signal passes through zero, going from positive to negative. It will employ the same technique in reverse to accurately locate a valley point between two partially resolved peaks.

Errors from baseline disturbances

With slope detection, any chromatogram baseline disturbances occurring within the peak gate may cause measurement errors.

Positive deviations from the baseline may be due to detector noise, stray peaks, or pressure pulses from column switching. Positive deviations are rarely a problem. Any real peaks should be fully separated and properly gated so they don't appear within an analyte peak gate. Processors can help by rejecting spurious peaks that fail to comply with minimum height, width, or area requirements, and these thresholds may be adjustable by the operator.

It's more difficult for a processor to reliably reject negative baseline deviations. Such excursions may come from flow disturbances during column switching or from real peaks for which the detector has a negative response. When the negative excursion is inside a peak gate, the processor detects the slope change at the bottom of the disturbance as the start or end of the analyte peak and uses the baseline at that point to project the integration zero line. Figure 14.4 illustrates the outcome of this mistake: the processor incorrectly includes the area below the true baseline in the accumulated peak area causing a significant measurement error.

To avoid these errors, ensure that unavoidable baseline disturbances are well-separated from the measured peaks and do not occur within an analyte gate. A useful technique is to run a blank chromatogram with only carrier

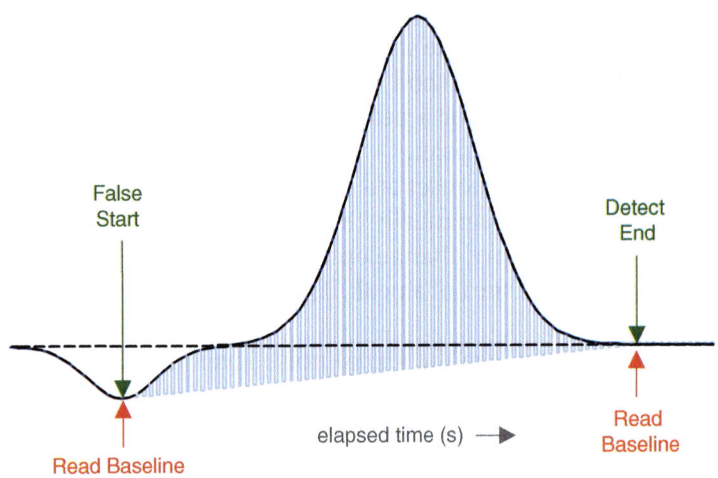

When a negative baseline disturbance occurs within a peak gate, the processor may see the slope change at the bottom of the disturbance as the start or end of the analyte peak. The integrator projects a straight zero line under the peak and measures the area above that line. The total measured area (shaded blue) mistakenly includes a large area lying beneath the true baseline (dotted black), causing a significant measurement error.

Figure 14.4 Error by Negative Peak in Peak Gate.

gas as sample. By overlaying the blank and process chromatograms, it's then possible to determine which baseline upsets are due to column valve switching and which are due to real peaks in the process sample.

Baseline noise may also trigger slope detection. In trace analysis, a small noise spike on the peak tail could trigger early termination of the peak measurement. Careful adjustment of the peak threshold or data smoothing function may be necessary to obtain the best compromise.

Troubleshooting aids

While integrating an analyte peak, a processor can easily monitor the shape of that peak, but not all PGCs offer this additional diagnostic. When equipped, the processor notifies the operator when the shape of an analyte peak has changed beyond preset limits. Even without an alarm, the information to evaluate peak shape may be available in the integration data tabulated by your processor. You may find this information useful during troubleshooting activities.

All digital integrators store the height and area of an analyte peak. Dividing the area of the peak by its height derives a parameter indicative of the **peak width**, which makes a good check on chromatographic performance. If your processor doesn't already monitor this parameter, you can add it as a custom program or just choose to track it manually. A gradually increasing peak width would warn of incipient malfunction: it might indicate slow sample injection, reduced column efficiency, or detector malfunction. Thus, a peak width alarm would be a useful diagnostic aid.

Another useful ratio is found from the recorded start time (t_1), apex time (t_2), and end time (t_3) of a peak. All digital integrators store these values. For a perfectly symmetrical peak, the rise time ($t_2 - t_1$) would be equal to the fall time ($t_3 - t_2$), but most real peaks are slightly asymmetric with the fall time being a little longer than the rise time. Once established, though, the peak shape shouldn't change, so the ratio of fall time to rise time should stay constant. You can manually track this **asymmetry ratio** or calculate it during troubleshooting. Again, an automatic alarm to indicate an increase in asymmetry would be a useful diagnostic aid.

Since the processor already has data on retention times and peak widths, it can even monitor peak resolution. Recall from Chapter 1 that resolution is the ratio of peak separation to average peak width. The same options are open as before; follow the resolution manually or program the processor to do it for you.

Measuring overlapping peaks

Avoiding the problem

Most chromatographic software systems now include procedures for allocating the overlap area between partially-separated peaks. Unfortunately, these powerful processing techniques can lead people to think that the microprocessor can make up for poor chromatography, so why bother to get a better resolution?

Not true:

> *Computer software can never substitute for good chromatography!*

Ideally, the PGC should completely separate all analyte peaks from each other and from other peaks present on the chromatogram; this is the only way to get the highest measurement accuracy. For this reason, end user engineering standards may limit the acceptance of overlapping peaks, sometimes specifying a minimum amount of flat baseline before and after an analyte peak. This is laudable. Yet, those same users often demand a faster analysis, and their dual aspirations may be mutually exclusive:

> *The fastest analysis always produces a marginal resolution of at least one analyte peak: that's the definition of the fastest analysis.*

We may not like overlapping peaks – we don't – but the continual need for faster analysis makes overlapping peaks inevitable in process gas chromatography, so we need a way to measure them.

Tackling the problem

When two peaks overlap, the chromatogram doesn't return to baseline between them, but it must dip to a **valley point**. In the laboratory, a gas chromatograph may have to estimate the concentration of a **shoulder peak**, one that is just a hump on the side of another peak and shows no valley whatsoever. Shoulder peaks are unacceptable in process analysis: get a better resolution.

During a peak overlap, the detector is responding to both components. At each instant, the detector is responding to the total number of molecules it is seeing. It's important to understand that the overlapping chromatogram peaks do not superimpose, they accumulate. Perhaps it's better to call them **fused peaks**.

To become an expert chromatographer, look carefully at the overlap portion of the chromatogram and imagine the line each peak would take if the other peak was not there. We show these lines as dashed black lines in the figures below.

It would be nice if our processors would draw those dashed lines on our chromatograms, yet there is no perfect way to know which part of the overlap area is due to each peak. The chromatogram from a single detector just doesn't contain that information. Yet we still need the analysis, so PGC processors use mathematical techniques to get the best estimate of each peak area. Given a separation that shows a distinct valley between two peaks, PGC software can estimate the concentration of one or both of those components. The estimate will never be perfect: the deeper the valley, the better the estimate.

Here, we examine three typical software techniques for allocating the area between two fused peaks and point out the errors caused by misapplying them. The three techniques are:

- Perpendicular drop method
- Non-vertical drop method
- Tangent skim method

Perpendicular drop method

The **perpendicular drop** technique is a common strategy for allocating the shared area between two overlapping peaks of comparable size. The technique allocates peak areas by a virtual perpendicular line descending from the valley point to the projected zero line, as shown in Figure 14.5.

For symmetrical peaks on a horizontal baseline, the perpendicular drop procedure would be a perfect way to measure two or more peaks having about the same height. Since the peaks are close, their widths are almost identical, and the area allocation error is negligible, as you can see in Figure 14.5.

For overlapping peaks that differ in height, the perpendicular drop is not so good because the smaller peak tends to lose a disproportionate amount of the shared area, as shown in Figure 14.6. Even so, it's still the best method for measuring overlapping peaks when the larger peak is less than ten times the height of the smaller peak (Bicking 2006a; Dolan 2009).

The overall effect of the perpendicular drop method is more complex when the peaks are asymmetric, as is often the case. Natural peaks typically rise to a crest a little quicker than they return to baseline. This topography

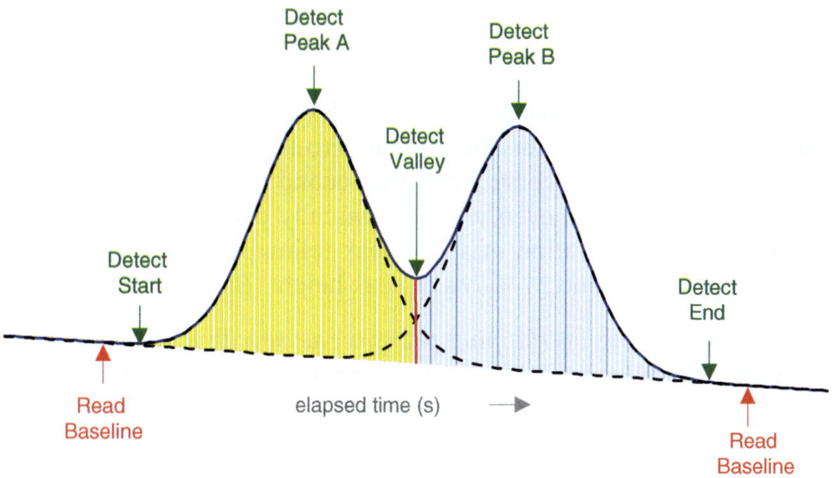

When the chromatogram baseline is straight but gradually drifting up or down, a straight zero line projected across two overlapping peaks gives acceptable accuracy. If the two peaks are symmetrical and similar in size, the area lost from one peak is about the same as the area gained from the other peak.

Figure 14.5 Area Allocation by Perpendicular Drop.

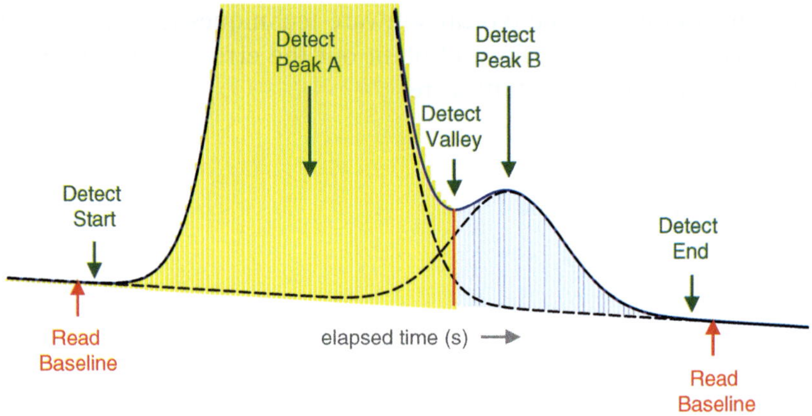

For two peaks with a size ratio of 9:1 or greater, the perpendicular drop method of area allocation leads to an integration error. Note that the measured area of Peak B (blue integral) is now an underestimate of its true area because the valley point has moved closer to the smaller peak. The smaller peak now loses more area than it gains from the larger peak. The error in measuring Peak A (yellow integral) is less significant as it's a smaller percentage of the peak area.

Figure 14.6 Error by Perpendicular Drop for 9:1 Peak Areas.

causes the first peak to lose more area than it gains from the second peak, underestimating the area of the first peak and overestimating the area of the second peak. Since this outcome is the opposite of the size-ratio effect, they tend to compensate for each other. This may explain why perpendicular drop is the most accurate method of integrating real overlapping peaks of comparable size.

Keep in mind that peak-height calibration may outperform peak-area calibration when measuring partially resolved peaks. Peak heights are often unaffected by a partial overlap. Working with different degrees of overlap, Bicking (2006a) found that peak height measurement was consistently more accurate than peak area.

Don't integrate to a valley point

It's usually a serious mistake to specify a valley between overlapping peaks as the reference point for the integration zero line. Figure 14.7 illustrates the problem, and it gets worse when the second peak is smaller than the first peak. In one report, even with a nominally perfect resolution of 1.5, valley-to-baseline integration of a 2 % peak following a 98 % peak caused 32 % loss of peak area and 17 % loss of peak height (Bicking 2006b).

> *When measuring overlapping peaks, never specify a valley point as the beginning or end of an integration zero line. Doing so will cause an enormous measurement error, as illustrated in Figure* 14.7.

It's even worse to specify valley-to-valley integration. Valley reference points are terrible for measuring multiple overlapping peaks. Consider the three peaks in Figure 14.8: valley-to-valley integration of the center peak omits the entire base of the peak, causing about 50 % measurement error.

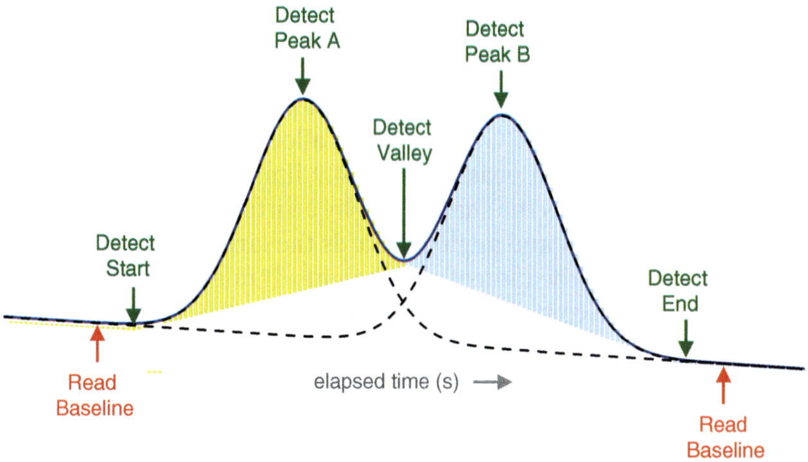

When measuring either one or both overlapping peaks, selecting the valley point for the integration zero line leads to severe errors in measurement. The white area is not counted in the area of either peak. Always use the perpendicular drop method shown in Figure 14.6 to integrate overlapping peaks of comparable size.

Figure 14.7 Error by Integrating to a Valley Point.

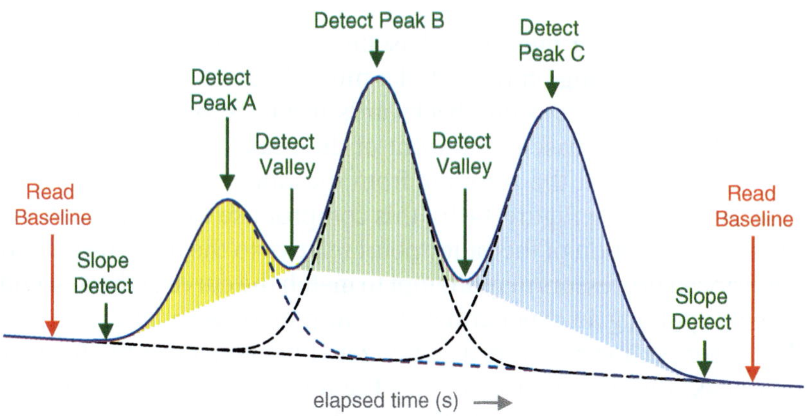

When measuring overlapping peaks, selecting valley points for the integration zero line leads to severe errors in measurement. The white area is not counted in the area of any peak. Use the perpendicular drop method shown in Figure 14.6 to integrate overlapping peaks of comparable size. For measuring multiple peaks, most processors can project a virtual straight baseline between the start and stop gates.

Figure 14.8 Error by Integrating Valley-to-Valley.

As a general rule, you should never set peak integration to use a valley point, but there is one exception: valley point integration is reasonable when a known and reproducible baseline disturbance occurs under the peaks. In Figure 14.8, for example, imagine the white area under the peaks is mostly due to a baseline bump that occurs even without a sample injection. If so, the peaks are not really overlapping, and their valley heights give the best estimate of the true baseline under each peak.

Once again, we see that electronics can't substitute for good chromatography. The real solution is to improve the chromatography and eliminate the baseline disturbance entirely.

Angular drop or allocated area

Above, we noted that the perpendicular drop method of measuring overlapping peaks is prone to error when the peaks are not the same size. To minimize this error, some PGC software can allocate the shared area in proportion to the size of each peak. Figure 14.9 illustrates this technique. In effect, the procedure splits the shared area by a virtual straight line descending from the valley point and angled towards the larger peak, with an angle of drop proportional to the ratio of peak heights. It might even be possible to tune the split angle to compensate for peak asymmetry (Kipiniak 1981).

To improve the area allocation for overlapped peaks, some advanced software attempts to simulate the true shape of each overlapped peak shown as dashed lines in the figures. As yet, we don't have a perfect simulation. When calculating the area allocation, current software assumes that the dashed lines are either **exponential** or **Gaussian** in shape. In practice, it's found that the Gaussian shape (like the bell curve of statistics) provides a better fit than the exponential shape.

Tangent skim method

Use the **tangent skim** method of baseline correction to measure a small **rider peak** that's sitting on the tail of a much larger peak. Ideally, tangent skim would project a zero line backwards from the peak-end baseline to touch the descending baseline in front of the peak. In practice, processors find it easier to project the zero line from the valley point to the peak-end baseline. As shown in Figure 14.10, this compromise causes a small loss of area due to the location of the valley point relative to the baseline curvature. Some advanced processors may attempt to match the curve of the descending baseline by applying an exponential curvature to the zero line.

The tangent skim procedure uses two zero lines concurrently. To measure the first peak, the processor computes the total area of the two peaks and then subtracts the area of the skimmed peak.

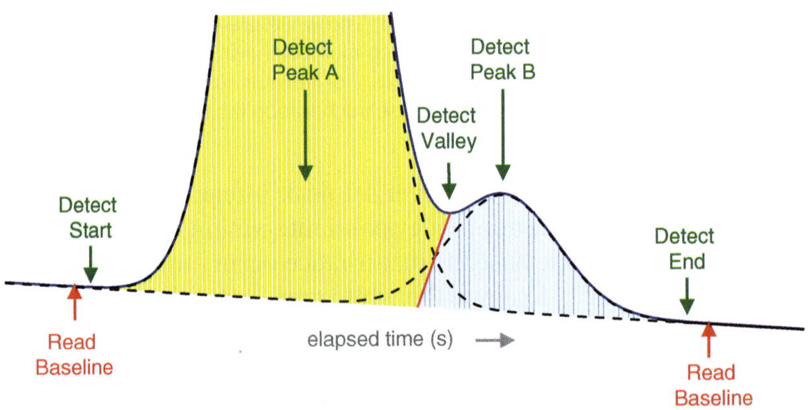

The processor allocates more area to the smaller peak, compensating for the error that would be caused by a vertical drop. Effectively, it drops an angled line from the valley point, where the angle is proportional to the peak height ratio. The split line becomes vertical when the peaks are equal in height.

Figure 14.9 Angular Drop Method for 9:1 Peak Areas.

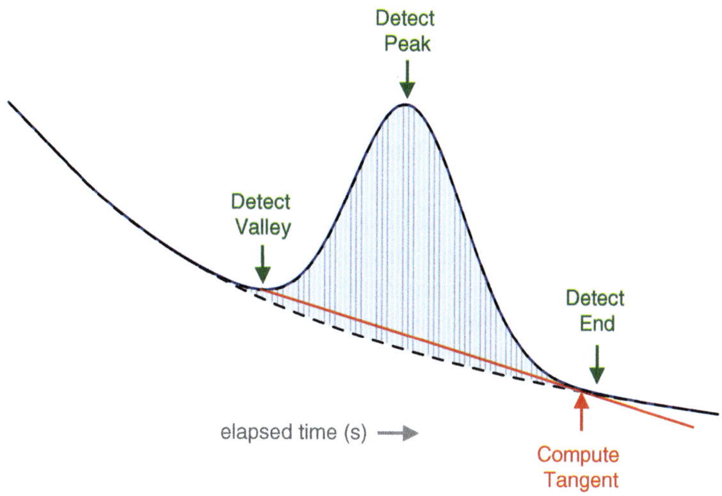

The baseline is falling from a previous large peak. The processor projects a zero line from the valley point to just touch the baseline after the peak. Don't use tangent skim unless the baseline after the analyte peak is still falling.

Figure 14.10 Tangent Skim for a Rider Peak.

Of all the methods of area allocation, tangent skim is the one most often used incorrectly. As a general rule, don't specify tangent skim if:

- You're measuring the first peak of two overlapping peaks.
- The peaks are normal in shape and the first peak is less than 100 times the size of the second peak.
- There's no valley point after the first peak and all you can see is a shoulder peak.
- The baseline after the second peak is horizontal or flat, instead of a natural continuation of the downward curve seen before the peak.

For success with tangent skim, ensure the peak is sitting on a curved baseline that starts before the peak and continues afterwards, as in Figure 14.10. Imagine the peak is not there: does the baseline continue to gradually fall, or does it quickly become horizontal?

Specifying tangent skim for overlapping peaks is a common setup mistake that causes significant measurement error. When set that way, the zero line obtained from the tangent skim is almost the same as the zero line under the second peak in Figure 14.7. The two mistakes are similar; both cause the processor to miss a substantial portion of the peak base and record a low peak area.

Effect of setup mistakes

In the end, the accuracy of your analysis depends on certain parameters entered in the method table. These entries tell your PGC how to integrate each analyte peak, including those that overlap other peaks. Only well-trained professionals should handle such sharp tools. When software tools are applied incorrectly, they can cause enormous measurement errors, sometimes exceeding 50 % of the true value.

Be aware that mistakes made in the setup of the peak integration methods may go unnoticed, even during factory acceptance tests. This can happen because the calibration of an analyte peak tends to compensate for mistakes made in its area measurement.

The PGC processes the calibration chromatogram by the same area-allocation routines as the process chromatogram and is likely to make the same errors. To evaluate the effect of those errors, consider two cases; when the calibration fluid doesn't include the overlapping component, and when it does:

- When the calibration chromatogram doesn't include the overlapping peak, the analyte peak calibration will be accurate over the whole measurement range. This is good, but when analyzing a process sample that contains the overlapping peak, an error in peak area allocation may occur, causing a low analyte measurement.
- When the calibration chromatogram includes the overlapping peak, the calibration of the analyte peak is based on its *measured area*, even though it may not be the true peak area. Using the measured area for calibration tends to compensate for errors in peak area allocation. Unfortunately, the instrument is then accurate only for the exact size of those peaks in the calibration fluid. Any other combination of peak sizes would incur a different integration error and require another calibration; in other words, the complete calibration curve is non-linear, perhaps grossly so.

Always remember that calibration may be masking the effects of setup mistakes and causing unexpected non-linearities in your measurements. Check that each peak is set for the optimum area-allocation procedure.

Knowledge Gained

- *PGCs were the first field instruments to employ digital processing and serial-data transmission.*
- *The first digital peak integrators used solid-state electronic circuits.*
- *Early peak integrators converted the detector signal to a pulse frequency and counted the pulses.*
- *Microprocessors integrate peaks by sampling the detector signal voltage at constant time intervals.*
- *In effect, the sampling routine converts the peak into an array of equal-width rectangles.*
- *At least 40 data points are necessary to faithfully represent the shape of a peak.*

- *Each memorized data point represents the height of a rectangle and is therefore proportional to its area.*
- *The data-sampling speed can be as high as 1000 Hz, but is mostly in the 40–100 Hz range.*
- *The processor can also sample the baseline signal to estimate the standard deviation of random noise.*
- *The microprocessor can average several readings to minimize noise, but this may distort a fast peak.*
- *Some processors use a Peak Width parameter to specify the number of readings averaged.*
- *When the processor uses a moving average, it doesn't reduce the number of data points.*
- *The value stored as the central value of the group is the average of an odd number of data points.*
- *To minimize peak distortion, some systems adjust the values by a weighting factor before averaging.*
- *Signal smoothing may obscure some fine detail and desynchronize the chromatogram from real events.*
- *PGC processors measure peak height and peak area, and you can select either one for calibration.*
- *The ratio of peak area to peak height indicates the width of a peak and is a useful alarm parameter.*
- *Peak height calibration is effective for analyte peaks that overlap other peaks on the chromatogram.*
- *Forced integration measures the area of all peaks occurring between the time gates.*
- *Forced integration can project a straight zero line between gates or a flat zero line from one gate.*
- *Forced integration results in severe error if the zero line runs to a disturbed baseline or valley point.*
- *Processors can detect the start, apex, and end of a peak by changes in slope of the detector signal.*
- *Slope detection might mistake a negative baseline dip to be the initial or final baseline under a peak.*
- *The ratio of fall time to rise time indicates peak asymmetry and may be a useful alarm parameter.*
- *Peaks have predictable retention times, but PGCs often use timed gates to ensure peak identity.*
- *Processors can estimate the areas of overlapping peaks, but it's better to improve the resolution.*
- *When peaks overlap, their chromatogram curves are added, not superimposed.*
- *A shoulder peak exhibiting no valley is unacceptable for measurement; get a better resolution.*
- *The perpendicular drop method splits overlapping peaks at the valley point occurring between them.*
- *The drop method achieves adequate accuracy when the smaller peak height is >10 % of the larger peak.*
- *Select good baseline references on both sides of the overlapping peaks; never use a valley point.*
- *Never run an integration between two valley points unless dealing with a reproducible baseline upset.*
- *Advanced software may allow the split between peaks to be nonvertical, exponential, or Gaussian.*

- *Tangent skim can measure a peak on the tail of a larger peak, but it's often misapplied.*
- *Tangent skim projects a zero line from the preceding valley to touch the chromatogram after the peak.*
- *Don't specify tangent skim unless the baseline continues to fall after the measured peak.*
- *Never specify tangent skim for overlapping peaks of comparable size.*
- *Calibration may compensate for integration errors, but cause linearity errors at other concentrations.*

Did you get it?

Self-assessment quiz: SAQ 14

Q1. If a PGC processor is set up with a peak width parameter of 16, what is the reduction in detector noise signal?

Q2. If a microprocessor samples the chromatogram signal at 40 Hz and the peak width parameter is set to eight seconds, about how many data points will be used to integrate a peak that triggers a slope-detected start at 31 s and a slope-detected end at 41 s?

Q3. If a processor samples the detector signal at 40 Hz and outputs a reconstructed chromatogram signal using the moving average of nine raw data points, to what extent are events on the reconstructed chromatogram shifted from real time?

Q4. How is it possible for slope detection to cause an error in peak integration, leading to a falsely low value of measured peak area.

Q5. What is the correct peak integration setting for:

- **A.** Two slightly overlapping analyte peaks of comparable size.
- **B.** A shoulder peak on the side of a larger peak with no valley point between them.
- **C.** A small analyte peak on the tail of a previous large peak, where the large peak tail continues to fall exponentially after the small peak has eluted.

Q6. Generally, it's not advised to integrate to a valley point, but there are exceptions. Under what circumstances might it be appropriate to specify valley-to-valley integration?

Q7. Generally, it's not advised to integrate to a valley point, but there are exceptions. Under what circumstances is a valley point always used at one end of the integration zero line?

Check your SAQ answers with those given at the end of the book.

Student evaluation test: SET 14

S1. According to the textbook, which one of the following observations indicates the fastest analysis?

Select the one correct answer:

- **A.** When each analyte peak is fully resolved from other peaks on the chromatogram.
- **B.** When at least one analyte peak exhibits marginal resolution.

S2. What is a shoulder peak?
Select the one correct answer:

A. One of a pair of same-sized peaks that appear on either side of a larger peak and look like a person's shoulders.
B. A wedge-shaped peak that falls much faster than it rises.
C. A peak on the side of a larger peak, with only a high valley visible between them.
D. A peak on the side of a larger peak, with no valley visible between them.

S3. Which two of the techniques listed below have been used by a PGC to measure the area of a selected peak on a chromatogram?
Select the two best answers:

A. By measuring the width of a peak and multiplying that width by half the peak height.
B. By converting the detector voltage signal to a pulse frequency signal and counting the pulses.
C. By sampling the digital voltage of the detector signal at constant intervals as a peak passes through the detector.
D. By transferring the data from short-term memory into a long-term memory device dedicated to that analyte.

S4. When a modern microprocessor-controlled PGC digitizes a chromatogram peak, what digital values does it store in memory?
To receive a score, you must select all the correct statements and none of the incorrect statements:

A. The measured value of the detector signal in voltage units at each uniform increment of time.
B. The measured value of the detector signal in voltage-second units at each uniform increment of time.
C. The height of each rectangular slice of the peak.
D. The area of each rectangular slice of the peak.

S5. Which of these statements are true?
To receive a score you must select all the true statements and none of the false statements:

A. Averaging a few readings reduces signal noise.
B. A moving average reduces the sampling frequency.
C. Signal smoothing may flatten the peak height.
D. Signal smoothing may reduce the peak area.

S6. What are the two acceptable methods for measuring two partially overlapping peaks of comparable size?
Select the two correct options:

A. By peak areas, with baseline-to-baseline integration of the first peak and tangent skim integration for the second peak.
B. By peak areas, with baseline-to-baseline integration and area allocation by perpendicular drop from the valley point.
C. By peak heights, with baseline-to-baseline zero-line projection.
D. By peak heights, with baseline-to-valley zero-line projections.

S7. When a peak is set for slope detection and the relevant parameters are not entered correctly, what errors might occur?
To receive a score, you must select all the correct statements and none of the incorrect statements:

A. If a negative baseline disturbance occurs within the peak gate, it may trigger the start of integration, resulting in a measured peak area higher than the true value.
B. If the slope detection threshold is set too low, the integration will start late and finish early, which will cause significant measurement error.
C. If the slope detection threshold is set too high, spurious baseline noise might trigger the detection of multiple false peaks.
D. If the peak height (or area) threshold is set too high, the processor might ignore a genuine small peak, treating it as signal noise.

S8. If computed and stored by the microprocessor, which of the following mathematical ratios might be useful for troubleshooting a PGC?
To receive a score, you must select all the correct statements and none of the incorrect statements:

A. For each analyte peak, the ratio of peak width to baseline noise level.
B. For each analyte peak, the ratio of peak area to peak height.
C. For each analyte peak, the ratio of peak fall-time to peak rise-time.
D. For two adjacent peaks, the ratio of peak separation to average peak width.

S9. For peak area integration of an overlapping peak, an incorrect setting of the start and stop reference points will generate an inappropriate zero-line, and thus cause a large error in the measured peak area. Yet, even when set that way, a maintenance technician could carefully complete a single-point calibration of the PGC without noticing anything amiss.
How is this possible?
Select the one best answer:

A. Because the maintenance technician will allow for the missing peak area when doing the calibration.
B. Because the calibration procedure always calibrates the measured peak area (even if it's not the correct area), so the analyte peak calibration is always accurate at that time.
C. Because the calibration procedure compensates for the error in peak area and will be accurate from that time onwards.
D. Because the same fraction of the peak area will be missing, whatever the peak size, so the calibration is correct for any analyte concentration.

References

Cited

Bicking, M.K.L. (2006a). Integration errors in chromatographic analysis; Part I: peaks of approximately equal size. *LCGC North America* 24 (4, April 1): 402–414.

Bicking, M.K.L. (2006b). Integration errors in chromatographic analysis; Part II: large peak size ratios. *LCGC North America* 24 (6, June 1): 604–616.

Dolan, J.W. (2009). Integration problems. *LCGC North America* 27 (10, October 10): 892–899.

Kipiniak, W. (1981). A basic problem – the measurement of height and area. *Journal of Chromatographic Science* 19, 332–337.

Rosemount (2010). *Model 500 Gas Chromatograph: Hardware Reference Manual, Revision K (July)*. Houston, TX: Rosemount Division of Emerson Electric Company.

Savitzky, A. and Golay, M.J.E. (1964). Smoothing and differentiation of data by simplified least squares procedures. *Analytical Chemistry* 36 (8): 1627–1639. https://doi:10.1021/ac60214a047.

Figures

New technical terms

When first introduced, these new words and phrases were in bold type. For further information, refer to the Glossary at the end of the book.

asymmetry ratio
automatic peak detection
exponential
forced integration
fused peaks
Gaussian
low-pass filter
moving average
peak area
peak gate
peak height
peak width parameter
perpendicular drop
raw chromatogram
raw data
rider peak
sampling rate
second derivative
shoulder peak
signal smoothing
slope detection
software autozero
standard deviation
start time
stop time
tangent skim
threshold value
valley point
zero line

Consult the Glossary for information on chemical names.

15

Calibration

"Most process analyzers are not absolute measuring instruments; they are comparative devices. A process gas chromatograph doesn't even know what it's measuring or how much of that substance is present in the sample, it simply compares the size of the analyte peaks on the current process chromatogram with their size on the latest calibration chromatogram. Then it makes deductions".

Measurement principles

Terminology

Like other technologies, the science of measurement has its own terminology to enable communication without ambiguity.

A measurement is always a *quantity* of something that can vary. The measured **variable** is known as the **measurand** and must be a **continuous variable**, one that has an infinite number of possible values. In process analysis, each measurand is the concentration of an analyte in the process fluid. Concentration is certainly a continuous variable.

Not all variables are continuous; for example, money doesn't have an infinite number of possible values. We can't measure discontinuous variables like that, so we count them. When done properly, a **count** is exact.

A measurement can never be exact; to exactly represent an infinite number of values would require an infinite number of digits. Since we don't have an infinite number of digits, all measurements are *estimates* and somewhat uncertain; they are subject to error. The two kinds of error that commonly affect an analytical measurement are **random error** and **systematic error**.

An archery target provides a good illustration of these errors. Random error causes the spread of hits on the target and a narrow spread indicates very little random error. *We would call that* precise *shooting*. However, the narrow cluster of hits might not be at the center of the target, it could be off center horizontally or vertically, or both. A constant offset from the bull's eye

Process Gas Chromatographs: Fundamentals, Design and Implementation, First Edition. Tony Waters.

is a systematic error or bias. An archer can correct for an observed bias by calibrating their bow sights, thereby moving the cluster of hits to the center of the target. *We would call that* accurate *shooting*.

Note that accuracy depends upon first achieving precision, and then calibrating the outcome to minimize bias. That's true for all measuring instruments, including process gas chromatographs.

Random error

Although we try to hold conditions constant during measurement, some variation is inevitable. In practice, a host of random variations occur in dynamic fluid flows and electronic circuits. A myriad of tiny variations combine to create errors of measurement that are truly random in occurrence and effect.

Random error makes our measurements less repeatable, thereby reducing the **precision** of measurement. After decades of product engineering and development, a modern PGC is a very stable instrument and may return almost the same value for multiple measurements of the same sample. But that doesn't make it accurate, it makes it precise. A precisely measured value may be wrong.

Systematic error

Systematic error consistently biases the measured value from the true value of the measurand. Usually the bias is due to a constant offset (zero error), an incorrect response factor (span error), or a non-linearity in the instrument response.

In principle, we can calibrate a precise instrument to minimize systematic bias, but the calibration procedure itself suffers measurement error, and that can bias the calibration. One common source of such systematic error is the uncertain concentration of analyte in the calibration sample.

The proximity of a measurement value to the true value of the measurand is the **trueness** of that measurement. Accuracy is the sum of trueness and precision (ISO 5725-1 1994).

Uncertainty

Because of random and systematic error, we can never know the true value of the analyte concentration, but we can stipulate a range of values that probably include it. The wider the given range, the more certain we are that it includes the true value. Thus, the specified accuracy of a measurement may express a range of values together with a measure of our confidence that the range really does include the true value.

The range of values thought to contain the true value is the **uncertainty interval** of a measurement. The uncertainty interval might be a percentage of the given value, a range of possible values, or a percentage of the measurement range. For example, we might specify a weight as:

- 70 kg ± 2 %
- 70 kg ± 1.4 kg
- 70 kg ± 1.4 % of full scale (with a scale of 0–100 kg)
- 70 kg ± 2.8 % of range (with a range of 50–100 kg)

Each of these statements indicates the same uncertainty of measurement. They differ in their representation of the possible error. The first one implies that the magnitude of the error is relative to the measured value; in other words, it's mainly a span error. The second one simply states the probable error at this one measurement value. The last two say the error is constant at any measured value; in other words, it's mainly a zero error.

With process measuring instruments, it's common to express the uncertainty as a percentage, like the first example above:

- 70 kg ± 2 %

We often interpret this to mean *2 % accuracy*, but the format is inadequate. It wrongly conveys the impression that the stated percentage is the maximum possible error. This is untrue; due to the random nature of error, there will always be an occasional result that falls outside any specified uncertainty interval.

To solve this dilemma, we turn to the statistics of random processes, which shows that random errors follow the so-called Gaussian *bell curve*. From this delightful piece of math, we calculate a parameter called the standard deviation (σ) that indicates the spread of the results. A small value of σ indicates high precision.

Statistical theory shows that about 99 % of all results will fall within ±2.6 σ of the average value. Therefore, if we specify an uncertainty interval equal to ±2.6 σ, we can be 99 % confident that it includes the elusive true value. Of course, there is always the 1 % chance that it doesn't.

From statistics, then, we might express the above measurement as:

- 70 kg ± 1.4 kg at the 99 % confidence level

This is a correct way to state a measurement because it contains three essential attributes: a quantity, a unit of measure, and an expression of uncertainty.

Refer to the SCI-FILE: *On Analytic Units* in Chapter 7 for a detailed discussion on the **unit of measure**. A process gas chromatograph injects a volumetric sample, so it fundamentally measures volume percent. Calibration in other units may be possible under certain limiting conditions, as detailed in the SCI-FILE.

The **confidence level** tells us how much we can trust the result. It also lets us estimate the standard deviation of the measurement. In this example, the confidence level is 99 %, so ±1.4 kg must be equal to ±2.6 σ. Thus: $\sigma = 0.54$ kg.

Unfortunately, though, not everyone uses the 99 % confidence level. You should be wary of analyzer specifications that claim ±1 % accuracy without stating a confidence level. Oftentimes, that specification is based on 95 % confidence ($\pm 2\,\sigma$), 68 % confidence ($\pm\,\sigma$), or no confidence at all.

Ideally, the stated uncertainty interval should include an allowance for systematic error, but it's difficult to quantify a calibration bias, so often without justification it's assumed to be negligible. To make that true requires an accurate calibration.

Accurate calibration

The accuracy of a measurement depends entirely on calibration. There are too many uncertainties to calculate an absolute calibration for a gas chromatograph, so our calibrations are invariably relative. The basic technique is to analyze a calibration sample and then adjust the instrument responses to match the known analyte concentrations.

It follows that a gas chromatograph is a comparative device. It compares the size of a peak on a process chromatogram with the size of that peak on a previous calibration chromatogram. Therefore, even with a perfect calibration procedure, the final accuracy of a measurement depends on the accuracy of the calibration standard itself. If a standard sample is inaccurate (and occasionally they are), all PGCs that use it for calibration will generate inaccurate process measurements.

It can be very difficult to tell whether a standard is accurate or not, because even a certified standard must be to some extent inaccurate. Often, the certification is based on analysis by a gas chromatograph, which itself needs calibration! One approach is to use the recently calibrated PGC to measure a second calibration sample, preferably from a different supplier. Then, having gained confidence in the calibration, always set up a validation program (described below) to track analyzer performance and recalibrate only when statistically justified.

When the PGC calibration drifts continually in one direction, don't recalibrate the PGC without first checking the validity of the calibration sample. It's a good plan to keep a spare calibration standard in abeyance and use it to double-check the PGC calibration from time to time. Immediately the spare standard is in regular service, buy another one.

Calibration and validation

The original calibration of a PGC establishes the relationship between a peak area (or height) and the concentration of that analyte in the analyzed sample. Then, all further calibration techniques aim to confirm or, as necessary, fine tune that original calibration.

The frequent and unnecessary adjustment of an instrument's calibration tends to reduce its overall accuracy. This happens because each analysis – even the analysis of the calibration sample – suffers a small random error. Adjusting the calibration to correct that small error affects all subsequent analyses, slightly biasing the results high or low. Frequent calibration adjustments jerk this bias up and down, reducing the overall precision of measuring the process sample.

It's possible to reduce calibration error by measuring the standard sample several times and calibrating to the average value, but most plants don't do that because each analysis takes time and consumes costly calibration fluid.

A better procedure is **validation**. After initial calibration, the PGC automatically measures a validation sample at regular intervals, perhaps daily. The procedure is the same as used for calibration, but it doesn't update the response factors. The processor stores the validation results for each analyte and typically calculates the standard deviation of at least 20 of the most recent values.

Figure 15.1 is a typical PGC display showing a **control chart** for one analyte. The control chart plots the validation data as deviations from the **mean** (the average value), and you can clearly see the variation of results, above and below the mean.

According to statistical theory, about 95.4 % of results should lie between $\pm 2\ \sigma$ of the mean (indicated by the yellow warning lines), and about 99.7 % should lie between $\pm 3\ \sigma$ of the mean (indicated by the red action lines). That means:

- There's about 1-in-20 chance that a normal result will go outside the yellow warning lines. A more frequent transgression is suspicious.
- There's almost no chance that a normal result would cross the red action lines, so any that do will require investigation.

If a validation result is inside the yellow warning limits, maintenance intervention is unnecessary – there's no need to adjust the calibration. But a repeated result falling outside the warning limits or a single result beyond the action limits reveals a non-random fault that demands maintenance attention. It's likely due to a specific cause that you should identify and eliminate.

Even when the validation results are safely inside the warning and action limits, a processor can perform statistical tests on the data to decide when recalibration is appropriate. It can even issue a warning alarm when an analyzer is beginning to fail. See ASTM 3764 (2015) for details.

For effective validation, the concentration of analytes in the validation sample must remain constant, but the actual values need not be accurately known, since the calibrated analyzer will measure them. A large cylinder of the process fluid will often suffice, thus saving the expense of blended samples.

Validation must be frequent enough to collect a statistical data set, so it's best automated. The process chromatograph should treat the validation sample as just another process stream. Daily validation is usually adequate. Less frequent validation risks the delayed detection of malfunction and could let the analyzer function in an out-of-control condition for an extended time.

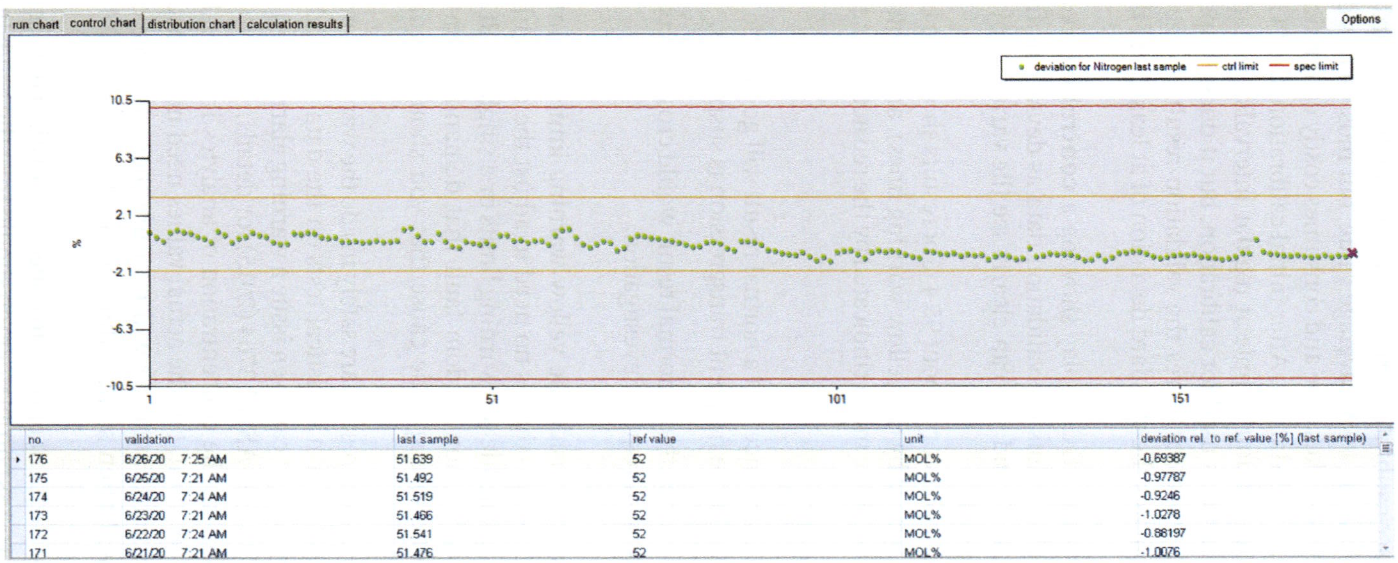

The PGC measures a constant validation sample at frequent intervals and plots the data as percent deviation from the mean value. This display shows validation results all within the orange warning lines, so no maintenance action is necessary.

Figure 15.1 A Validation Control Chart.

Source: Siemens Analytical Products and Solutions. Reproduced with permission.

Calibration methods

External standard method

The common way to calibrate a process gas chromatograph is to inject a sample fluid of known composition, known variously as the **calibration sample** or the **standard sample**.

> *If your calibration fluid is a gas, it's obviously okay to call it a* ***cal gas****. It's equally obvious that a liquid sample is* not *a* ***cal gas****, so don't make that common mistake.*

In this common procedure, the PGC injects and measures a standard sample by the same analytical method that it uses for the process sample. Most measured components have a linear detector response, so the analyzer needs only a single factor to convert peak area to concentration and a single point calibration will suffice.

As defined in laboratory practice, the **response factor** is simply the slope of a linear calibration curve, obtained during calibration by dividing the analyte peak area by its known concentration. Then, to calculate analyte concentrations in the analyzed sample, the analyst *divides* each peak area by its response factor.

However, to facilitate calculation, most automatic data processing systems store a **calibration factor** that is the reciprocal of the response factor. Then, to calculate analyte concentrations in the analyzed sample, the processor *multiplies* each peak area by its calibration factor (Miller 2005, 302–303).

> *It's important to remember the relationship between response factor and calibration factor. Sad to say, the calibration factors used by PGC processors are also called response factors, creating some confusion.*

Before updating the calibration factors in memory, the processor typically checks to see if the new calibration factors exceed a previously entered *deviation limit*. If they don't, the processor updates its stored factors with the new ones. If any factor should exceed its preset limit, the processor would ignore the calibration, keep the old factor, and issue a *calibration fail* notification.

Each analyte has a different calibration factor that also depends on the sample volume and detector response.

The PGC processor uses the certified concentrations of analytes in the calibration fluid to calculate calibration factors. When using a new calibration fluid, don't forget to manually enter the new values:

> *When you install a new container of calibration fluid don't forget to update the analyzer with the concentration of each analyte.*

Given several calibration fluids with different analyte concentrations, many PGC processors can compute a non-linear calibration. They often use a quadratic expression to relate peak area to concentration, although more complex relations are possible.

In some applications, the analytes can't all coexist together in one cylinder, so several calibration samples are necessary. Usually, the analysis method treats each calibration sample as a separate stream. The method specifies the analytes measured in each stream and their known concentrations.

The **external standard** technique is simple and convenient and doesn't require a full analysis of the calibration sample. It's the most common procedure in use, particularly when analytes have low concentrations.

Normalization

Normalization is a mathematical procedure that forces the measured concentrations of all components in a mixture to sum to exactly 100.00 %. To accomplish that, it's necessary to measure all significant components in the process sample.

Before normalization, it's necessary to accurately calibrate all components in the sample using the external standard technique. Then, the measured concentrations of the process samples should sum to exactly 100.00 %. If they don't, the PGC calculates the normalized concentration of each analyte:

$$\textit{normalized conc} = \textit{measured conc} \times 100\ \%/\textit{total}\ \%$$

In principle, normalization will compensate for any random or systematic error that affects one complete analysis, such as a small difference in sample volume or a slight change in detector sensitivity. It's very effective for a percentage-level application that analyzes all the components in the process stream, delivering results that are far more precise than would otherwise be possible with a PGC.

Before you choose to normalize results, take care to understand the limitations of the technique.

Normalization fails in trace-level analysis. The errors occurring when measuring very small peaks are different in kind than those experienced when measuring large peaks. Anyway, the miniscule part-per-million values are insignificant in the totalization. Therefore, the procedure doesn't compensate for errors inherent in the ppm measurements.

Normalization may not be appropriate when using separate analytical trains to measure a single process stream. Usually, you can't normalize results from trains that use different injector valves or different detectors. However, there is a way. Normalization becomes possible if the PGC measures a **bridge component** in both trains. The selected bridge component must be well-separated on both trains, without interference from other components. Then, its measured value should be the same in each case. If not, the processor uses the ratio of the bridge component measurements to adjust the measurement values from one train, thereby aligning them with the other train. Then, normalization of the aligned values yields an accurate analysis.

Be aware that normalization can mask PGC faults that need attention. When an analyte peak is partially backflushed or incorrectly integrated, the error would normally show as a low concentration measurement. When normalized, though, all the results conveniently sum to 100 %, and the error might escape notice. To guard against hiding a serious error, it's best to limit the automatic adjustment and issue an alarm if a normalization adjustment exceeds that limit.

Another concern with normalization is the treatment of **heavies**, the later peaks on the chromatogram. Typically, the later peaks are backflushed into a detector and regrouped as a single heavies peak. A perfect regrouping is unnecessary, since the area of the composite backflush peak is easily determined.

So far, this is good news; we are measuring all the components.

But the next step is not so nice. Whether normalized or not, it's always difficult to calibrate a **composite peak** because the molecules making up the peak are not all the same and they may have different response factors.

Calibrating a composite peak

Composite peaks are quite common in process chromatography. In addition to the backflushed peak mentioned above, a composite peak may comprise a few components that the column cannot easily separate. Sometimes the user is happy to get a measurement of the total concentration of those peaks.

When a composite peak comprises a few peaks that are difficult to separate, a reasonably accurate calibration may be possible. Since the individual peaks have almost the same retention times, it's likely that their molecules have almost the same response factors. Thus, an external calibration using one or two of those peaks will deliver acceptable accuracy.

When a composite peak contains radically different molecules, as is always true of a peak formed by backflushing, calibration is more difficult. Each component in the regrouped peak may have a different detector sensitivity and would have a different response factor if calibrated separately. For a backflushed peak, we don't even know what components are present, so perfect calibration is virtually impossible. The best we can do is to find an average calibration factor for all the regrouped peaks. Yet even the average value is unreliable because it requires the concentrations of all those peaks to miraculously stay in synch with each other.

In practice, we calibrate most backflushed peaks with a single analyte. Thus, it's common to calibrate a regrouped C_5+ peak "*as pentane,*" meaning that the calibration gas contained a known concentration of pentane, but no other heavies. Clearly, this is an approximation, but it's a reasonable assumption because the concentrations of heavier molecules in process samples are vanishingly small.

The rapid reduction in later heavies content occurs because the concentration of each heavy molecule in a process gas is often proportional to its vapor pressure. At a given process temperature, a heavier molecule has a markedly lower vapor pressure than a lighter molecule, so its concentration in the vapor phase is much lower. Many process liquids come from condensing a vapor, so the liquid concentrations follow suit. An example is the condensed overhead from a distillation tower.

Natural gas streams pose another good example. Without need of miracles, the concentrations of the heavies really do stay in synch with each other. Therefore, we can predict the average combined response factor for a regrouped C_6+ peak based on a solo hexane calibration. This is good, because the accurate analysis of the heavies content is important; it has a large effect on the calculated heating value of the gas.

Grab sample calibration

A common way to check the calibration of a process gas chromatograph is to take a **grab sample** (also known as a **spot sample**) from the process and analyze the sample in the laboratory. Spot samples are most often used for liquid process streams as it can be difficult to introduce a liquid standard to the PGC.

Unfortunately, the grab-sample method is the least-preferred way to calibrate a process gas chromatograph. The method has more potential sources of error than any other calibration technique. Yet it may be the only method available for calibrating toxic or reactive analytes not amenable to the external or internal standard procedures.

A main concern with grab-sample calibration is the additional sources of random and systemic error inherent with manual sampling and manual analysis.

For success with calibration by grab sample, the time of sampling must be accurately known, and the calibration results closely matched to the PGC analysis at that time. The calculation must also consider the time delays inherent in both grab and analyzer sampling systems, including any difference in process lag to each process tapping point. In addition, the extracted sample must remain stable long enough to perform the laboratory analysis, and there must be trust in the accuracy of the laboratory method and its associated standard.

It's found that the laboratory analysis often doesn't agree with the values given by the process analyzer. Some plants choose to adjust PGC calibration to comply with the laboratory result, but it's better to seek out the root cause for the difference. In practice, we often discover that disagreement is due to a design error or malfunction in the grab or analyzer sampling systems. The troubleshooting procedure is quite complex and is beyond the scope of the present book.

Internal standard method

The **internal standard** method may be employed in any PGC, but it's most useful when one or more of the analytes cannot be mixed into a calibration sample due to the high toxicity of the analyte, or because of its tendency to adsorb onto or permeate through the walls of the container or to react with other substances in the calibration mixture.

In the laboratory, calibration by internal standard usually involves spiking the measured sample with a known concentration of a reference component that will form a separate peak on the chromatogram. Some PGCs have attempted to emulate that technique by separately injecting a small constant volume of a pure standard substance. To accomplish this, they use dual injector valves or a single valve having two sample volumes. The technique has not found wide acceptance.

In a popular variation of the method, it's possible to choose a single analyte peak that is well-separated from other peaks and always present in the process sample and to nominate that peak as the internal standard. The chosen reference peak is the only peak calibrated by the external standard method. When the PGC analyzes a process sample, it accurately measures that peak and uses the measured value to calculate the concentration of one or more of the other analytes.

To calculate the concentration of the other analytes, the PGC processor needs to know the **relative response factor** (RRF) of the designated internal standard and other measured substances. These factors indicate the detector response of a component relative to the response of a standard substance such as benzene.

It's possible to determine the relative response factor of an analyte by experiment, but for most analytes the RRF values on different detectors are well documented and may be retrieved from the literature. Enter the RRFs into the analysis method.

During process analysis, the processor divides each analyte peak area by its RRF to allow for differences in detector response. It then calculates the concentration of each analyte by multiplying the known concentration of the reference peak by the ratio of analyte and reference peak areas.

The SCI-FILE *On Response Factors* has further details of the calculations.

Area percentage method

If given the above-mentioned relative response factors, a PGC can calculate the analyte concentrations directly from chromatogram peak areas.

The procedure is specific to peak area measurements and doesn't apply to peak height measurements.

The procedure relies on normalization, so the PGC must measure all components of the sample. A calibration sample is unnecessary, but you'll need to enter the RRF for every component in the sample. Alternatively, it might be possible to determine the factors by running a one-time calibration.

During a process analysis, the processor divides the measured area of each peak by its RRF to find the relative abundance of that component in the sample. For normalization, it's essential to measure every component. Then, the normalized measurement values are the true concentration of each component.

SCI-FILE: *On Response Factors*

External standard

When measuring a calibration sample using a linear detector, the area (A_i) of an analyte peak is directly proportional to its known concentration (χ_i). From this, we define a response factor (f_i) for that analyte as the slope of its linear calibration curve. Thus:

$$\chi_i = f_i \cdot A_i \tag{15.1}$$

To obtain a *response factor* for each analyte, divide its peak area by its known concentration:

$$f_i = \frac{A_i}{\chi_i} \tag{15.2}$$

Then, when analyzing an unknown sample, divide the area of each peak by the stored response factor:

$$\chi_i = \frac{A_i}{f_i} \tag{15.3}$$

Sounds good, but most process chromatographs don't work that way.

In practice, a processor finds it easier to multiply than to divide, so it typically stores a *calibration factor* (k_i) that is the reciprocal of the response factor:

$$k_i = \frac{\chi_i}{A_i} \tag{15.4}$$

Then, during process analysis, the processor *multiplies* each analyte peak area by its stored calibration factor to get the concentration:

$$\chi_i = k_i \cdot A_i \tag{15.5}$$

There's nothing wrong with this logic, except the stored calibration factors are often called "*response factors*," a widespread misnomer that causes much confusion. The subsequent PGC calculations are fully consistent, so the outcome is correct (Guiochon and Guillemin 1988, 571–572).

If the detector response is linear, the calibration factors may be determined by a single calibration, although the average of several calibrations will provide a more accurate instrument.

When the detector response is nonlinear – like a flame photometric detector measuring sulfur compounds – the analyzer must measure several different calibration samples that cover the expected range of analyte concentrations. From these measurements, a newer PGC can calculate a nonlinear calibration curve that best fits the data. Older instruments in the field may not support an automatic nonlinear calibration.

Normalization

To employ normalization, the analyzer must measure all significant components of the sample. The analysis may omit components at concentrations of less than 100 ppm as their contribution to the total composition is insignificant.

If feasible, normalization will improve the precision of measurement by forcing the sum of the component concentrations ($\sum \chi_i$) to equal 100 %.

Equation (15.6) calculates the normalized concentration of an analyte (χ_{ni}):

$$\chi_{ni} = \frac{100\ \chi_i}{\sum_{i=1}^{n} \chi_i} \tag{15.6}$$

Where n is the number of components.

Internal standard

The internal standard method may be useful when the calibration sample cannot include one or more of the analytes. The procedure uses a single peak of known concentration as a reference value.

Choose a reference peak that is always present in the process sample, easy to calibrate, and well-resolved from other components.

Set up the processor to measure the chosen reference peak concentration (χ_r) in the normal way. Then, use a certified external standard to calibrate that reference peak. There is no need to calibrate the other peaks.

To calculate the concentration of the other analytes, enter the relative response factors of the reference component (rf_r) and the other analytes (rf_i). Then, the processor calculates each analyte concentration by comparing its peak area (A_i) with the area of the reference peak (A_r):

$$\chi_i = \chi_r \cdot \frac{A_i}{A_r} \cdot \frac{rf_r}{rf_i} \tag{15.7}$$

Note that the relative response factors must be based on the same mass, mole, or volume units used to enter the concentration of the reference component, and that the calculated concentrations will have the same units.

Area percentage

This procedure also requires no calibration, but it relies on normalization, so the analyzer must measure all the significant components in the process sample.

The method relies on a relative response factor (rf_i) for each analyte, held in the processor memory.

During process analysis, the processor measures each analyte peak area in the normal way and then divides the measured area by its relative response factor to get the corrected peak area (A'_i):

$$A'_i = \frac{A_i}{rf_i} \tag{15.8}$$

The corrected peak areas are proportional to analyte concentrations, allowing the processor to normalize the analyte measurements using Equation (15.6).

As before, note that the relative response factors can be on a mass, mole, or volume basis and the calculated concentrations will have the same basis.

Knowledge Gained

- *A count is the quantity of a discontinuous variable and if done properly is exact.*
- *A measurement is an estimate of the quantity of a continuous variable and cannot be exact.*
- *The measurand is the measured variable and is the analyte when measuring concentration.*
- *A repeatable measurement suffers little random error and is precise, but it may be wrong.*
- *A precise measurement becomes accurate when calibrated to minimize systematic bias.*
- *The standard deviation of a measurement indicates the expected spread of results due to random error.*
- *The uncertainty of a measurement is expressed by a range of possible values and a confidence level.*
- *The confidence level expresses the probability that the given range of values contains the true value.*
- *A fully defined measurement has a value, a unit of measure, and a statement of confidence.*
- *Frequent calibration adjustments increase random error and reduce the precision of measurement.*
- *A validation procedure uses a reference fluid to check calibration but makes no adjustment.*
- *Statistical tests on validation results can predict the need for calibration or maintenance attention.*
- *An external standard sample is a calibration fluid with known analyte concentrations.*

- *The definition of response factor is the slope of a linear calibration curve.*
- *PGCs use a calibration factor that is the reciprocal of the response factor.*
- *The calculation of analyte response factors and concentrations is a simple job for a microprocessor.*
- *PGC processors can derive a nonlinear calibration equation from multiple calibration samples.*
- *When a PGC measures all sample components, the normalization procedure forces a total of 100 %.*
- *When feasible, normalization compensates for small changes in sample volume or detector sensitivity.*
- *The need to calibrate a composite peak with a single calibration factor leads to limited accuracy.*
- *It's possible to calibrate a process analyzer to match the laboratory analysis of a grab sample.*
- *Grab sample analysis is subject to many errors and is the least preferred method of PGC calibration.*
- *An internal standard may be appropriate when an analyte is too unstable to mix into a cylinder.*
- *A PGC can calculate analyte concentrations from a known reference peak and relative response factors.*
- *The reference peak must always be present in the process sample, fully separated from other peaks.*
- *The area percentage method also uses relative response factors but normalizes the results.*

Did you get it?

Self-assessment quiz: SAQ 15

Q1. In the archery analogy used in the text, which feature of the target represents random error and which feature represents systematic error? How can the archer reduce these two errors?

Q2. Why is it impossible for a measurement to be completely accurate? What is an uncertainty interval and a confidence level?

Q3. In what ways is validation different than calibration?

Q4. What is a control chart used for?

Q5. Explain why it's possible to calibrate the area of a composite peak, but not its height.

Q6. How does a PGC implement the internal standard method of calibration?

Q7. How can a PGC calibrate all its measurements without using a calibration standard?

Check your SAQ answers with those given at the end of the book.

Student evaluation test: SET 15

S1. For an analytical measurement made by a PGC, which one of the following statements is false?

Select the one false statement:

A. Careful calibration can reduce the random error of measurement.

B. Careful calibration can reduce the systematic error of measurement.

C. A perfectly accurate calibration is not possible because the calibration procedure is also subject to random and systematic error.

D. Trueness is the inverse of systematic error; perfect trueness would require the complete elimination of bias.

S2. When measuring a constant sample composition, which one of the statements below is true?

Select the one true statement:

A. A very precise measurement is certain to be accurate.

B. Averaging many results will reduce the systematic error and improve the accuracy of a measurement.

C. A precision of ± 1 % means the measured values found from repeated measurements of the constant variable will all lie in a 2 % band centered around the mean.

D. A confidence level of 95 % means there is 5 % chance that the stated uncertainty interval doesn't include the true value of the variable.

S3. When calibrating and using a process gas chromatograph to measure the concentration of selected analytes in a sample of the process fluid, which one of the below statements is true?

Select the one true statement:

A. When future PGCs accomplish the perfect measurement of analyte peak areas, calibration will no longer be required.

B. A certified standard sample has an exactly-known composition, specified on the certificate.

C. An analyzer properly calibrated against a certified standard always gives the correct measurement.

D. When an analyzer gives a different measurement value than the process laboratory, it must be recalibrated.

E. None of the above statements is true.

S4. When calibrating a process gas chromatograph using an external standard, which of the statements below are true?

To receive a score, you must select all the true statements and none of the false statements:

A. You must use a single calibration fluid containing known quantities of all the measured components.

B. To calibrate an analyte, you must know its concentration in the calibration fluid to a high degree of accuracy.

C. The processor will calculate a single calibration factor applicable to all the analytes.

D. It's best to adjust the calibration after every calibration check, even if only a small change occurs.

S5. When using peak area to measure concentration, why might it be difficult to calibrate a composite peak?

To receive a score, you must select all the true statements and none of the false statements:

A. Within the composite peak, each component peak may have a different width.

B. The different molecules in the composite peak may have different detector sensitivities and therefore different response factors.

C. The relative concentrations of the components comprising the composite peak might change, thereby changing the overall response factor for the group.
D. The calibration can be adjusted by dividing the area of each peak within the composite peak by its response factor.

S6. When a PGC analyzes a process sample, which of the below statements about normalization are true?
To receive a score, you must select all the true statements and none of the false statements:
A. Normalization is used to calculate the concentration of an unmeasured component.
B. Normalization can be used only when the heavies peaks are backflushed to vent.
C. When applied correctly, normalization improves the precision of measurement.
D. In some applications, normalization can be used together with relative response factors to obtain adequate measurement accuracy without using a calibration standard.

S7. Consider a measurement of carbon dioxide concentration reported as 10.4 mol% ±0.08 mol% at the 95 % confidence level.
What is the standard deviation of the measurement?
Express the standard deviation as a percentage of the measurement value and round your answer to the nearest value with two significant figures.

S8. Consider a validation control chart for methane measurement like the one shown in Figure 15.1.
When the analyzer is reliably measuring a constant sample, what percentage of the readings should fall between the lower yellow warning line and the lower red action line?
Use only the data about control charts found in this chapter. Work your calculations in three significant figures and express your final answer as a percentage also in three significant figures.

S9. When done correctly, validation is a useful procedure as it provides vital maintenance information that confirms nominal performance or warns of impending failure. When validating a process gas chromatograph, which of the statements below are false?
To receive a score, you must select all the false statements and none of the true statements:
A. The validation sample must contain only one analyte.
B. The validation sample composition must be accurately known.
C. The validation sample composition must remain constant during enough validation runs to establish statistical data such as the mean and standard deviation of the validation measurements.
D. If validation reveals only a minor calibration error, it's best to update the calibration before returning the PGC to analyzing process samples.

References

Cited

ASTM D3764 (2015). *Standard Practice for Validation of the Performance of Process Stream Analyzer Systems*. West Conshohocken, PA: ASTM International. Refer to latest edition: http://www.astm.org/cgi-bin/resolver.cgi?D3764

Miller, J.M. (2005). *Chromatography Concepts and Contrasts*, second edition. Hoboken, NJ: Wiley.

Guiochon, G. and Guillemin, C.L. (1988). *Quantitative Gas Chromatography for Laboratory Analyses and On-line Process Control*. Amsterdam, Netherlands: Elsevier Science Publishers.

ISO 5725-1 (1994). *Accuracy (Trueness and Precision) of Measurement Methods and Results – Part 1: General Principles and Definitions*. Geneva, Switzerland: International Organization for Standardization.

Figure

Equations

15.1	$\chi_i = f_i \cdot A_i$	Definition of response factor
15.2	$f_i = \dfrac{A_i}{\chi_i}$	Calculation of response factor
15.3	$\chi_i = \dfrac{A_i}{f_i}$	Calculation of concentration
15.4	$k_i = \dfrac{\chi_i}{A_i}$	PGC calculation of calibration factor
15.5	$\chi_i = k_i \cdot A_i$	PGC calculation of concentration
15.6	$\chi_{ni} = \dfrac{100\,\chi_i}{\sum_{i=1}^{n} \chi_i}$	Normalized concentration
15.7	$\chi_i = \chi_r \cdot \dfrac{A_i}{A_r} \cdot \dfrac{rf_r}{rf_i}$	Concentration by internal standard
15.8	$A'_i = \dfrac{A_i}{rf_i}$	Corrected peak area for normalization

Symbols

A_i	area of analyte peak
A'_i	analyte peak area corrected for relative response
A_r	area of reference peak
f_i	response factor for analyte
kg	a kilogram, the SI base unit of mass
k_i	PGC calibration factor for analyte
n	number of components in sample
rf_i	relative response factor of analyte
rf_r	relative response factor of reference peak
χ_i	concentration of analyte
χ_r	concentration of reference peak
χ_{ni}	normalized concentration of analyte

New technical terms

When first introduced, these new words and phrases are in bold type. For further information, refer to the Glossary at the end of the book.

bridge component
cal gas
calibration factor
calibration sample
composite peak
confidence level
continuous variable
control chart
count
external standard
grab sample
heavies
internal standard
mean
measurand
normalization
precision
random error
relative response factor
response factor
spot sample
standard sample
systematic error
trueness
uncertainty interval
unit of measure
validation
variable

Consult the Glossary for information on chemical names.

Answers to self-assessment questions

SAQ-01

1Q1. B. In gas-liquid chromatography, the liquid phase is stationary while the carrier gas moves down the column.

1Q2. A. Oxygen is not an appropriate carrier gas because it might react with and damage the liquid phase inside the column.

1Q3. D. For a gas-liquid chromatographic column only a non-volatile liquid is suitable as a stationary phase.

1Q4. C. While all parts of the chromatographic system are necessary, it's the contact with the stationary phase that causes the separation.

1Q5. B. It's only necessary to separate the designated analytes, not the unmeasured components.

1Q6. B. The detector generates a continuous signal proportional to the instantaneous number of component molecules leaving the column. The PGC processes this chromatogram signal to measure the concentrations of the analytes.

1Q7. D. The chromatogram shows the baseline, peak shape, and peak separation.

SAQ-02

2Q1. When the stationary phase is a liquid, the sample molecules partially dissolve in the liquid. The technical term is the *distribution* of component molecules between the gas and liquid phases. Each component has its own unique solubility in the liquid phase.

2Q2. When the stationary phase is a solid, the sample molecules adsorb onto the large surface area of the solid.

2Q3. The solubility of a gas in a liquid depends upon the temperature and pressure, the type of liquid, and the type of gas.

2Q4. At equilibrium, component molecules are leaving the liquid phase and entering the gas phase at the same rate that component molecules are leaving the gas phase and entering the liquid phase. In the absence of an outside influence, there is no net change, and the equilibrium is interminable.

2Q5. Peaks get wider because some of their molecules stay in the liquid phase longer than other molecules do. A more complete theory recognizes that a small part of the peak width is due to molecular diffusion in the gas phase.

2Q6. The component molecules dissolve less at higher temperature (remember the air bubbles in hot water?), so they spend less time in the liquid phase. Therefore, peaks take less time to get through a column when the column is at a higher temperature.

2Q7. The component molecules dissolve more at higher pressure (remember the carbon dioxide bubbles in champagne?), so they spend more time in the liquid phase. Therefore, peaks take more time to get through a column when the column is at a higher pressure.

Process Gas Chromatographs: Fundamentals, Design and Implementation, First Edition. Tony Waters.

SAQ-03

3Q1. At carrier gas speed. Component molecules can move along the column only when they are in the gas phase. Then, they move through the column at full carrier gas velocity.

3Q2. The liquid phase is the stationary phase. While it's true that dissolved molecules can randomly move by diffusion in all directions, their average rate of movement along the column is zero.

3Q3. Differences in solubility cause some component molecules to remain stationary in the liquid phase longer than others do. Separation occurs because some peaks stop longer than other peaks.

3Q4. The unretained peak stays in the gas phase and therefore travels through the column at the same speed as the carrier gas. Every component peak must spend this same time traveling. To calculate the average velocity of the carrier gas, divide the known length of the column by the observed holdup time.

3Q5. With no liquid to dissolve in, all the component peaks would arrive at the detector at the same time as the air peak. There would be no separation.

3Q6. For each component peak, the time after the air peak is the average time those molecules spent stationary in the liquid phase.

3Q7. Triangulation allows you to measure the base width of a peak. The intersection of the triangulating lines also marks the peak apex for measuring the peak retention time. An alternative is to measure the peak width at half of its height.

SAQ-04

4Q1. A spatial separation is a separation of component peaks by their distance along the column, typically measured in millimeters. It occurs in the column, before the peaks reach the detector.

4Q2. A temporal separation is a separation of component peaks by their time in the column, typically measured in seconds. It's the signal from the detector as recorded on the chromatogram.

4Q3. While in the column, the faster component peaks are wider than the slower ones. When on the chromatogram, though, the faster peaks are always narrower than the slower ones.

4Q4. The adjusted retention time (time peak spends in the liquid phase) is doubling with each increment to the number of carbon atoms in the component molecules.

4Q5. Answer: 18 minutes. From Figure 4.1 the holdup time (air peak) is 2 min, and the retention time of the n-butane peak is 10 min. Therefore, the adjusted retention time of n-butane is 8 min. Invoking the doubling rule for n-paraffins, the adjusted retention time of n-pentane is 16 min, and its retention time is therefore $2 + 16 = 18$ mins.

4Q6. Temperature programming is a controlled linear increase of column temperature for part or all the separation. It negates the doubling rule, allowing component peaks to elute in linear boiling point spacing, thus reducing the analysis time for strongly retained components.

4Q7. A resolution of 1.5 is adequate. Computer software techniques can measure peaks at lower resolution, but those measurements may suffer increased error.

SAQ-05

5Q1. Both. In conventional multistream systems, measuring fewer streams (ideally just one) always improves reliability. When a PGC measures process streams concurrently using separate sample injector valves for each stream, there are no shared sample pathways, reducing the impact of having several streams. However, it is still true that a single sample valve failure may take a whole analyzer out of service for repair, thereby reducing the availability of all streams.

Ideally, a single analyte would provide the highest reliability. But there may be additional analytes that are easy to separate and pose no threat. Reliability is likely to suffer only when the separation of additional components requires a more complex column system. So, reducing the number of analytes will improve reliability only when it enables a simpler column system, one that is more robust and easier to maintain.

5Q2. In 1966, the transmission of field measurements to the control room by serial data link.

5Q3. The measurement of simple analytes, typically those whose molecules contain just two elements like carbon dioxide does, can often be done continuously and more economically by photometric analyzers.

5Q4. Not Option A: Only a photometer can use a direct-insertion measurement probe.
Not Option B: If the PGC is well-designed, it will separate the analytes and measure them without interference from the other components. A photometer is more likely to suffer interference.
Yes on Option C: It's difficult for a photometer to measure several different components in a fluid stream. It may experience reduced accuracy due to their overlapping absorption wavelengths. A PGC is likely to be more accurate.
Not Option D: Photometers are usually much faster than PGCs and may require less maintenance.

5Q5. Because the PGC separates each analyte from all the other components before measuring it.

5Q6. Some spectrophotometers can process data from several sensor probes concurrently. In a similar way, some PGCs can now process data from several detectors concurrently, which substantially reduces PGC analysis time.

5Q7. Chemometrics is a mathematical procedure that discovers a set of absorption wavelengths that are predictive of the desired physical property. It does that by comparing the spectra of dozens of known training samples. If the training samples are representative of the likely process variation, the instrument can then predict the desired physical property of a process stream with adequate accuracy – and can do so very much faster than conventional methods.

SAQ-06

6Q1. Use the data in Table 6.1.

For hydrogen:	$rtc = 186.6/26 = 7.18$	(about 7)
For helium:	$rtc = 155.7/26 = 5.99$	(about 6)

You can see why these are the preferred gases for the TCD. The thermal conductivity of most other analytes is close to nitrogen.

6Q2. Hydrogen pros:

Most sensitive for TCD.
Low viscosity; gives faster analyses.
Plentiful and inexpensive.
May already be in use for an FID or FPD.

Hydrogen cons:

Flammable and potentially explosive.

6Q3. Helium pros:

Good sensitivity for TCD.
Totally inert, non-reactive, and safe in classified process areas.

Helium cons:

Slightly higher viscosity than nitrogen, no speed advantage.
Scarce, sometimes in short supply, more expensive.

6Q4. Nitrogen pros:

Plentiful and inexpensive.
Good for FID applications.
Adequate sensitivity for some TCD applications.
Non-flammable and safe in classified process areas.

Nitrogen cons:

Slower than hydrogen.
Low sensitivity for most analytes on TCD.

6Q5. Most PGCs use column switching. When columns switch, they may need a surge of carrier gas to quickly build up the pressure in the new column. A flow controller wouldn't provide the additional flow. A pressure regulator does.

6Q6. Option C is the preferred answer:

Not Option A. A negative concentration is impossible!
Not Option B. It's possibly true, but the question doesn't provide enough data to support that conclusion.
Yes on Option C. A valid conclusion, and general enough to be compatible with the question data.
Not Option D. An invalid explanation. This condition would generate a positive peak, somewhat smaller than expected.

6Q7. Option C is the preferred answer.

Not Option A. Too impure for PGC.

Not Option B. May be usable for percent-level analyses, but not for ppm measurements. This gas contains 100 ppm of impurities and would need high-capacity gas traps to remove moisture, oxygen, and hydrocarbons.
Yes on Option C. Good for ppm analyses as mentioned in the text. This gas contains 10 ppm of impurities and may need high-efficiency gas traps to remove moisture, oxygen, and hydrocarbons.
Not Option D. Expensive. Used only for extreme sensitivity applications.

SAQ 07

7Q1. A sample injector valve that uses an external sample loop would be a gas injector valve. The external sample loop would have too much volume for a liquid sample injection.

7Q2. The symbol "%" is a constant having the defined value 0.01.

7Q3. The sample temperature and pressure must be constant, particularly for a gas sample.

7Q4. For a weight percent calibration, the sample density must be constant.

7Q5. A different flow rate might cause the sample temperature or pressure to change. If the flow rate is too high, the gas may not reach the oven temperature and a cold gas would have more molecules in the sample volume. A high flow rate might also build up some extra pressure in the sample volume, with the same effect.

7Q6. If possible, inject a wide-boiling-range sample as a liquid, but to do so may require a high sample pressure to avoid fractionation.

7Q7. For this units conversion, assume that gas volume percent is equal to mole percent (it is). Then:

A. Calculate the molar mass of each component (g/mol), based on hydrogen = 1.0 and carbon = 12.
B. Multiply the mole percent of each component by its molar mass to get its weight (mass) in one mole of mixture.
C. Sum the g/mol of each component to get the weight (mass) of one mole of the mixture (= 37.0 g/mol).
D. Express each component weight as a percentage of the total weight to get its concentration in weight percent.

Conversion:

		g/mol	mol. %	g/mol	wt. %
Methane	CH_4	16	25 %	4.0	10.8 %
Ethane	C_2H_6	30	25 %	7.5	20.3 %
Propane	C_3H_8	44	25 %	11.0	29.7 %
n-Butane	C_4H_{10}	58	25 %	14.5	39.2 %
TOTAL:			100 %	37.0	100.0 %

SAQ 08

8Q1. The list doesn't include the sliding-plate valve.

8Q2. The plunger valve.

8Q3. The spool valve. You could also mention solenoid valves, but they did not work well. Soon, other valves replaced them.

8Q4. The sliders with holes in them have polished surfaces and glands on both sides of the slider. For sample injection, this allows the precision alignment of the hole in the gland and the hole in the slider (see Figure 8.3). These sliders may also use slotted glands that allow the more complex connections needed for gas injection and column switching (as in Figure 8.4). Alternatively, the slider may have only one polished and slotted face, so all the connections are on one side of the slider.

8Q5. The standard symmetrical valve rotates 60°. Yokogawa has a special 6-port rotary valve for liquid sample injection that rotates 90°, but that's an unusual arrangement. The extra two ports allow a purge fluid to enter and flush liquid out of the trapped slot.

8Q6. Traditional diaphragm valves block the flow paths *between* alternate adjacent ports, thereby providing the same flow patterns as a rotary valve does. Some modern valves apply pressure directly over each selected port, independently blocking the flow to or from that port.

8Q7. Two concentric groove rings encircle the ports to prevent leakage to or from the atmosphere. Additional grooves may run between ports to intercept cross-port leakage. Carrier gas purges all grooves.

SAQ 09

9Q1. The general elution problem applies to the analysis of several different substances on one isothermal column. It states that a good resolution of the early peaks will always result in a long analysis time for the later peaks. Temperature ramping and column switching are the two techniques that can overcome the problem, and in practice they tend to be mutually exclusive.

9Q2. The distribution column system is not on the list.

9Q3. Almost all PGCs need the backflush housekeeping function, and a 10-port valve can often inject the sample and backflush the first column. Most other column systems require two expensive valves to do that. The single 10-port valve can be a rotary, diaphragm, slide valve, or even a spool valve – or a valve of equivalent function like the Yokogawa double 6-port rotary valve.

9Q4. All injected components have the same amount of time to migrate along the first column regardless of their solubility. Therefore, when the flow reverses direction, they will all take the same time to migrate back to their starting point and exit the column. Due to peak broadening, though, only half of the peak will exit the column at this time – so a little extra time (or extra flow) is necessary to get the other half of the peak out of the column.

Note: This is true only if the substances enter the first column in the vapor state.

9Q5. With one detector, it can be difficult to ensure that the peaks from Column 2 and Column 3 don't overlap in time. With two detectors it doesn't matter.

9Q6. No; heartcut is often unnecessary because the tailing effect occurs after the major component, not before it. However, if a heartcut system is necessary to measure an additional small peak that comes out after the major component, the system must also cut the early peak into Column 2, so it reaches the detector.

9Q7. The restrictor gives the same flow resistance as Column 2, so the carrier flow rate into the detector doesn't change when the valve operates. If the column valve had only four ports, a flow change would be unavoidable, and this might upset the chromatogram baseline when the valve switches.

SAQ 10

10Q1. A concentration-sensing detector measures analyte concentration in the carrier gas with a peak height that is unaffected by flow rate, and a peak area that is inversely proportional to flow rate. A rate-sensing detector measures analyte rate of arrival at the detector with a peak height that is proportional to flow rate, and a peak area that is unaffected by flow rate.

10Q2. Thermistors can be intrinsically safe. They are more sensitive than hot-wires but lose that advantage at oven temperatures higher than about 120 °C. Thermistors may suffer reduced life in hydrogen carrier gas or when sensing corrosive gases.

10Q3. Constant voltage suffers from loss of power during the sensing of a peak and may cause element burnout if the carrier stops flowing. Constant current improves detector response, but is even more likely to suffer element burnout. Constant resistance maintains elements at a stable temperature, improves element life and speed of response, and protects against burnout.

10Q4. The FID flame ionizes organic analytes, creating positive carbon-containing ions and free electrons. The 300 V dc between collector electrode (−) and jet (+) accelerates electrons towards the jet, where they flow through the external circuit back to the collector electrode. Meanwhile positive carbon ions migrate to the negative collector, where the arriving electrons neutralize them.

10Q5. The FID responds to the rate of arrival of analyte molecules: adding a makeup gas would not change that. The TCD responds to analyte concentration: adding a makeup gas would dilute the analyte concentration and reduce the response.

10Q6. The FPD operates with a low-energy hydrogen-rich flame that promotes sulfur and phosphorus fluorescence but minimizes fluorescence from other elements. The optical filter passes the sulfur wavelengths and blocks the phosphorus fluorescence.

10Q7. The ECD outputs a constant baseline current of electrons released during the ionization of nitrogen carrier or makeup gas. When sensing a peak, the analytes capture some of those electrons and reduce the detector output current. The peak then appears as a dip in detector output current.

SAQ 11

11Q1. Any of the listed symptoms could be due to temperature variation:

- **A.** A cyclic baseline could be due to the direct effect of a cyclic oven temperature on a thermal conductivity detector. With a flame ionization detector, a cyclic baseline could be due to variations in column bleed caused by the oven heating cycle.
- **B.** The same effects as in (A) can cause a baseline to drift up or down in response to an unstable oven temperature. This happens during the warmup period, for instance.
- **C.** Sample temperature changes in the injector valve result in a variable number of molecules injected (particularly with a gas sample), and thus cause variable peak heights.
- **D.** The retention time of a peak is very dependent upon column temperature.

11Q2. Heating method most likely to:

- **A.** Achieve ultra-fast analysis: Direct column heating.
- **B.** Require instrument air: Airbath oven.
- **C.** Have reduced flexibility: Airless oven.

11Q3. Heating method often associated with:

- **A.** Explosionproof: Airless oven.
- **B.** Air purged: Airbath oven.
- **C.** Small and light: Direct column heating.

11Q4. Heating technique that:

- **A.** Analyzes a wide-boiling mixture: Direct column heating.
- **B.** Takes most time for heating and cooling: Airless oven.
- **C.** Accepts multiple valves and columns: Airbath oven.

11Q5. The two heating techniques preferred for:

- **A.** Analysis of natural gas: Airless oven or direct column heating.
- **B.** Temperature programming: Airbath or direct column heating.
- **C.** Most installed PGCs: Airbath oven and airless oven.

11Q6. Air-baths are less popular because of their high cost of operation. The ongoing cost is due to the large volume of instrument air they consume, and the power required to heat that air to oven temperature.

11Q7. The three methods are:

- **A.** Run a heating wire in direct contact with the column.
- **B.** Fit the column into a compact heated cartridge.
- **C.** Pass an electric current through the column itself.

SAQ 12

12Q1. The first PGC timers used a synchronous motor driving a cam, belt, or disc that tripped a microswitch at the designated times. Later models used optical sensing of marks on a film or disc, but a synchronous motor was still the basic driver. The next generation of programmer used digital counters, but the microprocessor soon superseded all other methods of event scheduling.

12Q2. The ten-port valve has only two positions, so only two events are necessary, one to turn it on and one to turn it off. This valve configuration injects a sample and backflushes the first column, so there is no need to actuate the valve more than once in an analysis cycle.

12Q3. PGC would have two independent applets, each with its own injector valve, column system, and detector. In effect, it would be two single-stream analyzers.

12Q4. The peak gate is set up by two events, a gate-on time and gate-off time that bracket the expected arrival time of an analyte peak. The processor assumes that any peak appearing in the gate is the nominated analyte.

12Q5. The analysis method (or program) is a collection of instructions that specifies all the desired event times. In a microprocessor system, it may also provide calibration data and procedures, and any necessary calculations.

12Q6. A processor identifies peaks solely by their retention times, so great care is necessary when specifying an acceptable time for each analyte peak.

12Q7. Some PGCs can shift the analyte peak gates to track small variations in retention time. A processor monitors the adjusted retention time of a reference peak or the actual retention time of an analyte peak, and then makes appropriate adjustments to the gate times. Special correlation software can also align the peaks to reduce the chance of misidentification.

SAQ 13

13Q1. Chromatogram, bargraph, trend record, and numerical printout.

13Q2. The chromatogram is necessary for checking the positions of peaks and setting peak gate times and for troubleshooting chromatographic faults, which always show symptoms on the chromatogram.

13Q3. The detector signal is proportional to the instantaneous number of analyte molecules in the carrier gas. The measurement of peak area adds all those “instants” for the entire width of the peak, effectively counting the number of analyte molecules that have passed through the detector. Therefore, the integrated area is the best measure of analyte concentration.

13Q4. For a linear detector, the height of a symmetrical peak is a good measure of its area. Most real peaks are somewhat asymmetric, but peak height may still be a more accurate measurement when the analyte peak partially overlaps another peak.

13Q5. Most analog systems measure peak height. An analog peak picker retains the instantaneous value of the peak. Then, the short-term memorized signal transfers to a memory unit dedicated to that analyte in that stream. The memory unit continuously transmits the measurement signal until a new one is available.

13Q6. When the common device fails it causes the loss of several measurements. Thus, one fault results in many failures.

13Q7. The data highway is a low-cost way to deliver lots of information to the process control system. It also allows a central maintenance system to communicate with any of the PGCs for maintenance overview and settings. Although they are also a common pathway, data highways incorporate redundancy and are almost perfectly reliable.

SAQ 14

14Q1. The smoothing effect of averaging 16 raw data points reduces the signal noise by $\sqrt{16}$, i.e. by a factor of 4. This reduces the noise level to 25 % of its former value.

14Q2. With a peak width parameter set to 8 s, the processor will average each set of eight raw data points. The measurement frequency is then $40/8 = 5$ Hz. Since the peak integration time is 10 s, the number of data points used is $10\text{ s} \times 5\text{ s} = 50$ points.

14Q3. At 40 Hz, the processor reads the detector signal every 25 ms. For each smooth data point acquired, the processor averages the last nine readings. But this average value occurred at the central time of the nine averaged points, which is four sampling periods (4 ✘ 25 ms) back in the past. Events appearing on the reconstructed chromatogram therefore occurred 100 ms ago. The chromatogram is 100 ms later than real time.

14Q4. A falsely low value for the measured peak area occurs when the slope detection threshold is set to high. This setting may cause the integration to start too late on the peak rise or end too early on the peak tail. Another reason might be detector noise that triggers the slope-detection algorithm and causes an earlier end-detection of a peak. When the integration zero-line between the two slope detections is higher than the true baseline, the integration misses a large part of the peak.

14Q5. The correct settings are:

- **A.** Integrate from baseline to baseline and allocate the shared area by perpendicular drop from valley point.
- **B.** Don't try to measure a shoulder peak; get a better separation.
- **C.** Since the baseline after the analyte peak continues to fall, it's best to use tangent skim for measuring the small peak.

14Q6. Valley-to-valley is appropriate only if there is a repeatable bump on the baseline underneath the peaks. In this situation valley-to-valley might be the best estimate of the baseline, but it would be better to improve the chromatography and remove the bump.

14Q7. A tangent skim always starts at a valley point. The integration zero-line starts at the valley and ends at a tangent to the baseline.

SAQ 15

15Q1. In the archery analogy, the closeness of grouping of the arrows in the target represents precision. To improve precision, an archer might need to practice their technique to make their shooting more repeatable.
The offset between the group of arrows in the target and the bull's eye represents systematic error. To improve trueness, the archer might need to adjust the bow sight or deliberately shoot off-center.

15Q2. A measurement has an infinite number of possible values; so to express it accurately would require an infinite number of decimal points. Therefore, real measurements have some uncertainty. The uncertainty interval is a range of possible values and has a stated confidence level indicating the probability that the true value is within that uncertainty interval. The wider the interval, the more confident we are that it contains the true value.

15Q3. During calibration, the PGC stores response factors to match the known concentrations of the analytes. During validation, the PGC subjects the stored analyte measurements to a statistical analysis that reveals when recalibration is beneficial and may even predict certain types of incipient failure.

15Q4. A control chart is a simple way of tracking validation results. It plots the data as deviation from the mean and may have warning and action lines marking the limits of acceptable variation.

15Q5. To reliably measure peak height of a composite peak would require that all individual peaks have the same retention time, which is quite impossible to do. Area measurement doesn't require that alignment because area is additive. All peaks within the composite are integrated; the calibration is never perfect, but often adequate.

15Q6. The PGC uses a peak that is always present in the process sample as a reference peak and calibrates that peak by the external standard method. During a process analysis, the processor measures the concentration of that reference peak and uses it to calculate the concentration of all other analyte peaks by reference to relative response factors stored in memory.

15Q7. The PGC processor divides each peak area by its relative response factor to get a relative concentration and then normalizes all results to sum to 100.0 %. The total must include all significant peaks.

Bibliography

Suggestions for further reading

Annino, R. and Villalobos, R. (1992). *Process Gas Chromatography: Fundamentals and Applications*. Research Triangle Park, NC: Instrument Society of America. ISBN 1-55617-272-9

Clemons, J. (2016). Chromatography in process analysis. In: *Encyclopedia of Analytical Chemistry* (ed. R.A. Meyers). Chichester, UK: Wiley. doi:10.1002/9780470027318.a2107.pub3

Lipták, B.G. and Venczel, K. (eds.) (2016). *Instrument and Automation Engineers' Handbook – Process Measurement and Analysis, Fifth Edition, Vol. 2: Analysis and Analyzers*. Boca Raton, FL: CRC Press. ISBN 9781466559325

McNair, H.M. and Miller, J.M. (2008). *Basic Gas Chromatography, Second Edition*. Hoboken, NJ: Wiley. doi:10.1002/9780470480106

Miller, J.M. (2005). *Chromatography: Concepts and Contrasts*. Hoboken, NJ: Wiley. ISBN 0-471-47207-7

Waters, T. (2017). The fine art of chromatogram reading. In: Proceedings of the 62nd ISA Analysis Division Symposium, *Pasadena, California*. Research Triangle Park, NC: International Society of Automation.

Waters, T. (2013). *Industrial Sampling Systems*. Solon, OH: Swagelok Company. ISBN: 978-0-615-73250-3

Process Gas Chromatographs: Fundamentals, Design and Implementation, First Edition. Tony Waters.

Glossary

This Glossary defines the new technical terms listed at the end of each chapter and some general terms used in gas chromatography. In addition, the list includes the names and formulae of components that often appear on chromatograms, including all the C_1–C_4 hydrocarbons and the chemical elements involved. The cryptic notation shown for some of these is a shorthand version of their name often used by chromatographers. The included atomic mass or molar mass (M) of each substance may be useful when converting measurement units.

Cross-references within the Glossary are in bold font.

A

Absorption photometer	An analytical instrument that determines the concentration of an analyte by measuring the absorption of electromagnetic radiation at certain wavelengths characteristic of that analyte.
Accuracy	See **trueness**.
Acetone	See **2-propanone**.
Acetylene	See **ethyne**.
Acetylenes C_nH_{2n-2}	The **alkyne** family of hydrocarbons.
Active solid	A granular solid that has a large surface area capable of adsorbing gas molecules. Used as a stationary phase in early PGCs but now mostly replaced by synthetic column packings. Examples include molecular sieves, silica gel, charcoal, and alumina.
Adjusted retention time (t'_R)	Peak retention time as measured from the air peak, equal to the time component has spent in the stationary phase.
Adsorbent solid	A granular solid used as a stationary phase in gas-solid chromatography.
Adsorption	The entrapment of gas molecules onto the surface of an active solid due to physical or electronic attraction.
AID	See **argon ionization detector**.
Air-bath oven	A chromatograph oven that uses compressed air for heating.
Airless oven	A chromatograph oven that doesn't use compressed air for heating.
Air peak	An unretained peak. Any component that is insoluble in the liquid phase and therefore travels through the column at the same speed as the carrier gas.

Process Gas Chromatographs: Fundamentals, Design and Implementation, First Edition. Tony Waters.

Aldehyde [–CHO]	A member of the aldehyde family of organic chemical compounds.
Alkane C_nH_{2n+2}	The generic name for a member of the **paraffin** family; i.e., non-cyclic hydrocarbons without double or triple bonds.
Alkene C_nH_{2n}	The generic name for a member of the **olefin** family: i.e., non-cyclic hydrocarbons with at least one double bond.
Alkyne C_nH_{2n-2}	The generic name for a member of the **acetylene** family: i.e., non-cyclic hydrocarbons with at least one triple bond.
Allene	See **propadiene**.
Allotrope	A different physical form of the same chemical element: e.g., diamond and graphite are allotropes of carbon.
Alumina Al_2O_3 $M = 101.960$ g/mol	An active solid capable of adsorbing gas molecules, previously used as a solid stationary phase.
Amount of substance	The number of molecules present. A more fundamental measure than the weight or volume of the molecules. The actual number is enormous, so chemists often count them in multiples of a **mole**.
Ampere (A)	The SI base unit of electric current (I).
Analysis time	The time from sample injection to final presentation of results. May be less than **cycle time**.
Analyte	A substance or a group of substances whose concentration in the sample fluid is the target of an analytical measurement. A more general term applicable to any measured quantity is the **measurand**.
Analytical measurement	A measurement of composition.
Analytical units	The fraction of molecules in a sample that conform to a desired identity or quality, usually expressed as a percentage by volume, mole, or weight.
Analyzer management system	A central analyzer maintenance station allowing digital access to any PGC on the network. Allows maintenance people to perform many maintenance tasks without entering the plant environment.
Anion	An atom or a fragment of a molecule that has gained an electron, becoming a negatively charged ion. Attracted to an anode.
Anode	A positively charged electrode that attracts anions.
Applet	In PGC, a simple valve and column arrangement that injects, separates, and measures some of the desired analytes in a sample, but not all. To perform the complete analysis, two or more independent applets are necessary.
Application engineering	The art of configuring a PGC to perform a desired analysis. Central to the art, is the design of a **column system** and the specification of the individual columns.
Argon Ar $M = 39.95$ g/mol	The chemical element with atomic number (Z) = 18. A monatomic gas, the third of six noble gases. An extremely inert gas sometimes used as a carrier gas. As an analyte, argon is difficult to separate from oxygen.

Argon ionization detector (AID) — A GC detector that is highly-sensitive to most organic compounds; uses a radionuclide to ionize argon gas, which in turn ionizes the analyte molecules. Now obsolete in PGCs, mostly superseded by the flame ionization detector.

Aromatics — A group of aromatic compounds.

Aromatic compound — Any organic compound that contains a structure comprising six carbon atoms in a ring, apparently with alternating single and double bonds. **Benzene** [C_6H_6] is the founding member of the category.

Asymmetry ratio — The ratio of peak fall time to peak rise time. For symmetrical peaks the ratio would be unity, but the asymmetry ratio for most chromatographic peaks is higher. An asymmetry ratio up to about 1.2 is normal.

Atmospheric pressure — The pressure of the atmosphere at the location of the measurement, which varies by elevation and local weather. For reporting data, the international standard reference pressure is 1.0 bara (about 14.5 psig). In the United States, the traditional reference pressure is 14.7 psig (101.325 kPa).

Atmospheric referencing — Allowing a gas sample to equilibrate to atmospheric pressure immediately before sample injection.

Atmospheric referencing valve (ARV) — A device to adjust the pressure of a gas sample volume to atmospheric pressure before sample injection. One or two valves stop the sample flow and connect the sample loop to atmosphere for a few seconds. Must vent to outside air, not to flare.

Atomic mass — The mass of one atom of a chemical element, measured in daltons or amu.

Atomic mass unit (amu) — One twelfth of the mass of the carbon-12 atom. The **dalton** (Da) is the name now given to the amu.

Automatic peak detection — When selected, the processor recognizes the onset or end of a peak when the rate of change of slope of the detector signal (second differential) exceeds a preset threshold value.

Autosampler — A laboratory device that sequentially draws a sample from the next vial in a moving line of vials and injects that sample into a laboratory chromatograph.

Autozero — A timed event that resets the detector **baseline** signal to zero. The chromatogram baseline is often set slightly higher than zero to make negative excursions visible.

Availability — A measure of reliability. Availability is the fraction of time that an analyzer is properly calibrated and available to deliver the measurements, expressed as a percentage of the time the plant is running. Most PGCs achieve 95– 98% availability.

B

Backflush — A method of removing **heavies** from a column by reversing the direction of carrier flow. Used in most PGCs to remove all undesired components from the first column before injecting another sample.

Backflush column	An initial column in a column system, designed to separate **heavies** from the analytes so the heavies can be backflushed.
Backflushed peak	A peak that has been backflushed from a column into a detector and may comprise several components fully or partially regrouped together by the reversed backflush flow.
Backflush-to-detector	A backflush column system that includes a detector to measure the backflushed peak.
Backflush-to-vent	A backflush column system that does not include a detector to measure the backflushed peak.
Backflush vent	An unrestricted outlet to atmosphere, not connected to a flare. Emits a small flow of carrier gas plus a recurrent but miniscule amount of **heavies**.
bar	An international unit of pressure defined as 100 kPa and adopted as the standard reference pressure for measurements that depend on pressure. 1 bar is approximately equal to 14.5 psig.
bara	Absolute pressure in bar, equal to the gauge pressure plus the local barometric pressure.
barg	Gauge pressure in bar.
Bargraph	An obsolete method of displaying concentration data by the length of vertical bars on a strip-chart recorder.
Base width (w_b)	The width of a chromatogram peak in distance or time units, measured between the intersections of the triangulated peak sides with the extended baseline.
Baseline	The detector signal as displayed on a screen or printed on a chart at times when no peak is passing through the detector. The PGC computes peak heights or peak areas relative to the baseline.
Baseline drift	Gradual change in the baseline signal due to detector temperature variation, detector and electronics aging, atmospheric pressure variation, impurities in the carrier gas, elution of highly retained peaks, or **column bleed**.
Benzene C_6H_6 ○ $M = 78.114$ g/mol	A colorless, flammable, liquid hydrocarbon. The benzene molecule has six carbon atoms arranged in a ring structure. Simply stated, the bonding comprises alternate double and single bonds between carbon atoms. Benzene is the founding member of all aromatic compounds and an important intermediate in the synthesis of many useful chemicals. Therefore, it often occurs on chromatograms.
Boiling point	The temperature at which the vapor pressure of a liquid becomes equal to the pressure on its surface, allowing the free escape of liquid molecules into the vapor phase.
Bridge component	A component measured by two or more parallel applets to link their measurement sensitivities and thus enable full sample computations such as **normalization**.

1,2-Butadiene
C_4H_6 1,2-$C_4''\,''$
$M = 54.092$ g/mol
: A hydrocarbon having four carbon atoms and two double bonds, with the carbon skeleton C=C=C–C. Often measured in the analysis of 1,3-butadiene streams and in other light olefin samples.

1,3-Butadiene
C_4H_6 1,3-$C_4''\,''$
$M = 54.092$ g/mol
: A hydrocarbon having four carbon atoms and two double bonds, with the carbon skeleton: C=C–C=C. 1,3-butadiene is an important monomer used in the production of synthetic rubber, notably for vehicle tires. It tends to polymerize in sample systems to form the notorious **green oil**, particularly at temperatures above 60 °C.

1,3-Butadiyne
C_4H_2 $C_4'''\,'''$
$M = 50.060$ g/mol
: Also called diacetylene. A hydrocarbon having four carbon atoms and two triple bonds, with the carbon skeleton: C≡C–C≡C. Often an analyte in light hydrocarbon samples.

Butane
C_4H_{10} nC_4'
$M = 58.124$ g/mol
: Also called n-butane. A straight-chain **paraffin** having four carbon atoms that commonly occurs in light hydrocarbon gas streams.

1-Butene
C_4H_8 $1C_4''$
$M = 56.108$ g/mol
: Also called 1-butylene. One of the four butene isomers. A straight-chain **olefin** having four carbon atoms with the carbon skeleton: C=C–C–C. Commonly occurs in light olefin gases such as ethylene, propylene, or butadiene plant streams.

cis-2-Butene
C_4H_8 cC_4''
$M = 56.106$ g/mol
: One of the four butene isomers. A straight-chain **olefin** having four carbon atoms with the carbon skeleton: C–C=C–C. The "cis" tag indicates that the terminal methyl groups are adjacent, on the same side of the double bond. This makes the molecule more compact than the "trans" version, so this peak elutes first on a chromatogram. Commonly occurs in light olefin gases such as ethylene, propylene, or butadiene plant streams.

trans-2-Butene
C_4H_8 tC_4''
$M = 56.106$ g/mol
: One of the four butene isomers. A straight-chain **olefin** having four carbon atoms with the carbon skeleton: C–C=C–C. The "trans" tag indicates that the terminal methyl groups are on opposite sides of the double bond. This makes the molecule less compact than the "cis" version, so this peak elutes after that component on a chromatogram. Commonly occurs in light olefin gases such as ethylene, propylene, or butadiene plant streams.

Butenyne
C_4H_4 $C_4''\,'''$
$M = 52.075$ g/mol
: Also called vinyl acetylene. A straight-chain hydrocarbon having one double bond and one triple bond with the carbon skeleton: C=C–C≡C. Commonly occurs in light olefin gases such as ethylene, propylene, or butadiene plant streams.

1-Butyne
C_4H_6 1-C_4'''
$M = 54.091$ g/mol
: Also called ethyl acetylene. A straight-chain hydrocarbon having an initial triple bond with the carbon skeleton: C≡C–C–C. Commonly occurs in light olefin gases such as ethylene, propylene, or butadiene plant streams.

2-Butyne
C_4H_6 2-C_4'''
$M = 54.091$ g/mol
: Also called dimethyl acetylene. A straight-chain hydrocarbon having a central triple bond with the carbon skeleton: C–C≡C–C. Commonly occurs in light olefin gases such as ethylene, propylene, or butadiene plant streams.

C

Cal gas	A colloquial contraction of calibration gas. See **calibration sample**.
Calibration	The procedure for ensuring that a PGC will accurately measure the concentration of the selected analytes in a process sample.
Calibration factor	The mathematical relationship between measured peak area (or height) and the known concentration of an analyte in the injected calibration sample. Typically, the calibration factor stored by a PGC is the reciprocal of the **response factor**.
Calibration fluid	A gas or liquid calibration sample.
Calibration sample	A contained gas or liquid mixture having specified analyte concentrations accepted as correct. Often this acceptance will be contingent upon a certified chemical analysis by a trusted laboratory.
Candela (cd)	The SI base unit of luminous intensity (I_v).
Capillary column	General term for any narrow-bore open-tubular column, including **WCOT**, **PLOT**, or **SCOT** versions
Carbon C $M = 12.011$ g/mol	The chemical element with atomic number $(Z) = 6$. The sixth element, having six electrons and four chemical bonds. The foundation of all life on Earth and present in every organic molecule. Exists in several allotropic forms: charcoal, graphite, and diamond.
Carbon dioxide CO_2 $M = 44.01$ g/mol	A colorless, asphyxiant gas. Usually separated by a porous polymer column and measured by a thermal conductivity detector. For low ppm measurement, a PGC may catalytically convert it to methane for measurement by a flame ionization detector.
Carbon disulfide CS_2 $M = 76.139$ g/mol	A highly volatile and flammable liquid that usually has a pungent odor due to sulfurous impurities. A flame photometric detector can measure low-ppm concentrations of carbon disulfide.
Carbon monoxide CO $M = 28.01$ g/mol	A toxic and flammable gas. Usually separated by a molecular sieve column and measured by a thermal conductivity detector. For low ppm measurement, a PGC may catalytically convert it to methane for measurement by a flame ionization detector.
Carbonyl sulfide COS $M = 60.07$ g/mol	A toxic and flammable gas with an unpleasant odor. Has a linear molecule with the structure: O=C=S. A flame photometric detector can measure low-ppm concentrations of carbonyl sulfide.
Carrier gas	The mobile phase of a gas chromatograph that carries the component molecules through the column. A PGC maintains constant carrier gas pressure and flow, except in the rare applications of pressure and flow programming. Common carrier gases include hydrogen, helium, nitrogen, and argon.
Carrier gas velocity	See **velocity**.
Cathode	A negatively charged electrode that attracts positive ions.

Cation	An atom or a fragment of a molecule that has lost an electron, becoming a positively charged ion and attracted to a cathode.
Cavity ring-down spectroscopy	A powerful photometric technique capable of measuring parts-per-trillion of certain analytes in gas samples.
Celite	A granular solid with high surface area often used as the inert support in packed columns. Comprises the fossilized remains of diatoms – a common type of algae found in natural waters.
Charcoal	An amorphous granular form of carbon that is capable of adsorbing gas molecules; previously used as a solid stationary phase.
Chemical reaction	Many measurements in analytical chemistry involve a chemical reaction with the analyte. For process analysis, though, we prefer instrumental methods like PGC where no chemical reactions occur.
Chemometrics	Chemometrics is a mathematical procedure that discovers a set of characteristics (such as photon absorption wavelengths) that can predict a physical property of a fluid. It does that by comparing the spectra of dozens of known training samples. If the training samples are representative of the likely process variation, the instrument can predict the desired physical property or analyte concentration of a process stream with adequate accuracy – and can do so very much faster than conventional methods.
Chlorine Cl $M = 35.45$ g/mol	The chemical element with atomic number $(Z) = 17$. Usually encountered in PGC as the chlorinated derivatives of hydrocarbons, the lightest of which is chloromethane.
Chlorine gas Cl_2 $M = 70.90$ g/mol	Chlorine is a yellow-green, highly-poisonous gas at room temperature and extremely reactive.
Chloromethane CH_3Cl $M = 50.49$ g/mol	Also called methyl chloride. A colorless, odorless, and flammable gas widely used in industrial chemistry. Has the structure of methane in which a chlorine atom replaces one hydrogen atom.
Chromatogram	A chart or display showing the variation in a GC detector output signal over time. Generally, the chromatogram shows a predicable number of peaks on a mostly flat baseline. The chromatogram is an essential aid for setting up or troubleshooting a PGC.
Chromatogram signal	The detector output signal. It may be a raw analog signal direct from the detector, a live digitized detector signal, or a representation of an analog chromatogram reconstructed from digital memory.
Chromatograph	An analytical instrument using the principles of chromatography.
Chromatography	The science and art of separating substances by a mobile phase moving across or through a stationary phase. For analysis, quantitative detection follows the chromatographic separation.
Colorimetry	In this very sensitive wet-chemistry method of laboratory and process analysis, the analyte reacts with a chemical to produces an intense color. The instrument then measures the color density using a photometer tuned to respond to that color.

Column	The heart of any gas chromatograph. A long tube that allows the carrier gas to transport the injected sample across a large surface area of solid or liquid stationary phase.
Column bleed	The presence of heavy molecules in the column effluent due to gradual evaporation of the liquid phase or elution of residual solvents. The detector may respond, giving an elevated baseline or may become coated with a solid deposit; like silica, for instance.
Column length (L)	The length of a column in meters.
Column oven	The column oven maintains a very constant column temperature or precisely varies the temperature during analysis to follow a predefined pattern.
Column switching	The technique of allowing selected components to pass through different columns, thus improving analyte resolution and analysis time.
Column system	Two or more columns that work together to separate analytes from each other and from other components of the sample.
Column train	An ABB term for an **applet**.
Column valve	A mechanical device to switch carrier gas flow and selected component peaks into different columns. Ideally it should have zero volume and no unpurged pockets that would entrap and delay component molecules.
Column vent	A carrier gas outlet that must flow to atmosphere, not to a recovery or flare line.
Come read	An instruction transmitted to a computer to inform that new data is available. The computer then reads the new value.
Component	Any unique chemical substance present in the sample fluid. A measured component is an analyte.
Composite peak	A peak or group of partially separated peaks containing two or more components. Either the column doesn't separate those components, or the column system has intentionally regrouped them together. Since perfect synchronicity of peaks is unlikely, the composite peak may have an irregular shape.
Concentration	See **mole fraction**.
Concentration detector	A gas chromatograph detector that responds to the concentration of the analyte in the column eluent.
Concurrent separations	The concurrent analysis of different components of the same process stream by two or more independent applets housed in a single process gas chromatograph, each applet having its own sample injector, columns, and detector.
Concurrent streams	The concurrent analysis of two or more process streams using independent applets housed in a single process gas chromatograph, each applet having its own sample injector, columns, and detector.

Confidence level	The estimated probability that the true value of the measurand falls within the **uncertainty interval**.
Continuous analyzer controller (CAC)	A device previously used to interface the continuous analog signal from external process analyzers to the digital PGC network. Most continuous analyzers now provide digital measurement signals.
Continuous variable	A variable having an infinite number of possible values. It would require an infinite number of digits to exactly represent the value of a continuous variable, so a measurement can only estimate its approximate value.
Control chart	A plot of sequential measurements, represented as deviations from the mean value, that also exhibits warning and action limit lines. These limits are usually at ± 2 and ± 3 standard deviations from the mean, respectively. Typically, maintenance action is necessary when two successive values cross the warning line, or when one value crosses the action line.
Control unit	An electronic device capable of controlling the time-dependent events in a process gas chromatograph and processing a chromatogram signal to produce useful information. All modern instruments use an integral microcomputer for these functions.
Count	An exact value having zero uncertainty, but not necessarily an integer. Any defined quantity is a count, not a measurement. For instance, pi (π) is a count.
Cutter column	See **heartcut**.
Cycle time	Usually the time elapsed between sample injections. May also refer to the time needed to complete the whole cycle of a multistream sequence.
Cyclobutane C_4H_8 cyclo-C_4 $M = 56.107$ g/mol	An isomer of butene. A hydrocarbon gas with a molecular structure comprising four carbon atoms connected in a circular formation by single bonds. Sometimes appears as a peak on chromatograms of light hydrocarbon samples.
Cyclohexane C_6H_{12} cyclo-C_6 $M = 84.16$ g/mol	A hydrocarbon liquid with a molecular structure comprising six carbon atoms connected in a circular formation by single bonds. Cyclohexane is a powerful industrial solvent. It often occurs on chromatograms separating aromatic compounds.
Cycloparaffins C_nH_{2n}	The family of cyclic hydrocarbons having only single bonds between carbon atoms. No double or triple bonds. Cyclopropane is the smallest member.
Cyclopentane C_5H_{10} cyclo-C_5 $M = 70.1$ g/mol	An isomer of pentene. A hydrocarbon gas with a molecular structure comprising five carbon atoms connected in a circular formation by single bonds. Rarely appears as a peak on chromatograms of light hydrocarbon samples.
Cyclopropane C_3H_6 cyclo-C_3 $M = 42.08$ g/mol	An isomer of propene. A hydrocarbon gas with a molecular structure comprising three carbon atoms connected in a circular formation by single bonds. Often appears as a peak on chromatograms of light hydrocarbon samples.

D

dalton (Da)	The SI unit of atomic mass, defined as one twelfth of the mass in grams of a carbon-12 atom. Thus, one **mole** of daltons is a mass equal to one gram.
Dead leg	An unpurged cavity, recess, or tube that can retain prior concentrations of sample molecules and allow them to diffuse into the current analyzed sample.
Density	The weight (or mass) of a unit volume of a gas or liquid.
Detector	A device that generates a signal, usually electronic, proportional to the instantaneous number of component molecules in the effluent from a column. All PGC detectors are differential, responding to the difference between the baseline condition and the presence of a peak.
Detector vent	The outlet for gases discharged from a detector. It's important neither to restrict the detector vent nor to subject it to pressure pulses.
Diaphragm valve	A valve that applies pressure to a diaphragm in a manner that blocks one or more flow paths while opening others.
Dielectric constant	Traditionally, the ratio of the capacitance of a capacitor filled with the measured material to the capacitance of an identical capacitor in a vacuum. Now called relative permittivity.
Diffusion	A movement of molecules that gradually ensures uniform composition in an enclosed space. For a gas, the driving force of diffusion is a difference in its partial pressure at separate locations in the enclosed space. The rate of diffusion is proportional to the square root of the gas density (Graham's Law).
1,2-Dimethylbenzene C_8H_{10} o-Xy $M = 106.16$ g/mol	Also called ortho-xylene. Frequently measured by PGCs in an aromatics plant. The three xylene isomers are difficult to separate with liquid stationary phases.
1,3-Dimethylbenzene C_8H_{10} m-Xy $M = 106.16$ g/mol	Also called meta-xylene. Frequently measured by PGCs in an aromatics plant. The three xylene isomers are difficult to separate with liquid stationary phases.
1,4-Dimethylbenzene C_8H_{10} p-Xy $M = 106.16$ g/mol	Also called para-xylene. Frequently measured by PGCs in an aromatics plant. The three xylene isomers are difficult to separate with liquid stationary phases.
Dimethylether C_2H_6O DME $M = 46.07$ g/mol	Also called ether. The smallest member of the ether family having the structure CH_3–CO–CH_3. A volatile and highly flammable liquid.
2,2-Dimethylpropane C_5H_{12} neo-C_5 $M = 72.15$ g/mol	Also called neopentane. An isomer of pentane with a central carbon atom connected to four other carbon atoms. Its compact shape gives neopentane a low boiling point and low retention time on liquid columns, being the first C_5 peak to appear on the chromatogram.
Diolefin C_nH_{2n-2}	A hydrocarbon having two double bonds between carbon atoms. The smallest diolefin is **propadiene**.
Directly-heated column	See **resistive heating**.

Distribution	In gas chromatography, distribution is the process by which solute molecules distribute themselves between a gas phase and a liquid phase to form a steady state equilibrium.
Distribution column system	One that partially separates the analytes into two or more groups, then routes each group to a different secondary column optimized for the final separation of that group.
Distribution constant (K_D)	The distribution constant of a component is the ratio between its liquid-phase concentration and its gas-phase concentration, at equilibrium.
Disulfide	A chemical compound whose molecule contains two sulfur atoms. For organic disulfides the two sulfur atoms bond together and have carbon chains attached at each end. The simplest example is dimethyl disulfide: $H_3C–S–S–CH_3$
Doubling rule	The adjusted retention time approximately doubles for each larger member of a homologous series.
Dual column	A term that is not well defined but may indicate a **distribution column system**.
Dynamic equilibrium	A stable state where two opposing processes are occurring at the same rate, so the outcome is unchanging.

E

ECD	See **electron capture detector**.
Electrochemical	A chemical reaction that produces or consumes electrons. Electrochemical devices often measure contaminants in water – such as dissolved oxygen.
Electrode potential	A voltage that forms on a wire immersed in an aqueous solution due to metal ions entering the solution. It may also occur between two solutions due to a difference in their concentrations. Electrode potentials can measure oxygen, pH value, and selective ion concentrations, among others.
Electron capture detector (ECD)	A detector used in chromatography to measure low concentrations of gases with high electron affinity, like chlorine. It uses a radionuclide to ionize nitrogen carrier gas, releasing low-energy electrons. Two polarized electrodes then generate a constant baseline current that decreases when an analyte captures some electrons.
Electronic pressure controller (EPC)	An electronic device that actively controls gas pressure and may have a set point under the control of the PGC microprocessor.
Electronic signal (4-20 mA)	The standard electronic analog data-transmission signal for process control.
Elute	To carry a component peak out of a column. All PGCs use elution chromatography, where a carrier gas elutes a band of component molecules out of the column and into the detector, thereby forming a peak on the chromatogram.

Ethane C_2H_6 C_2' $M = 30.07\,g/mol$	A hydrocarbon gas, the second member of the paraffin homologous series. Occurs on chromatograms of petroleum gases, including natural gas.
Ethene C_2H_4 C_2'' $M = 28.05\,g/mol$	Also called ethylene. A hydrocarbon gas, the first member of the olefin homologous series. Commonly occurs on chromatograms of samples from petroleum and petrochemical plants.
Ether [C–O–C]	A series of organic chemicals containing an oxygen atom between two carbon atoms. Colloquially, *ether* means the simplest member of the series: **dimethylether**.
Ethyl group $[-C_2H_5]$	A common feature of a branched-chain hydrocarbon molecule. Not a separate chemical compound.
Ethylene	See **ethene**.
Ethylene splitter	Also called a C2 Splitter. A process distillation tower that separates **ethene** in the overhead from **ethane** in the bottoms.
Ethyne C_2H_2 C_2''' $M = 26.04\,g/mol$	Also called acetylene. A hydrocarbon gas, the first member of the acetylene homologous series. Occurs on chromatograms of light hydrocarbon gases, particularly in olefin process streams.
Explosion-proof enclosure	See **flameproof enclosure**.
Exponential	A curved line that gradually approaches but never quite achieves a constant slope.
External standard	A calibration gas or liquid in a container, external to the process analyzer.

F

FID	See **flame ionization detector**.
Film thickness	The thickness of a liquid phase film on the walls of a capillary column.
Flame arrestor	A safety device to prevent a flame from propagating into a flammable process environment. Comprises a set of narrow metal pathways that cool the flame below the gas ignition temperature.
Flame ionization detector (FID)	A popular PGC detector that uses a hydrogen flame to ionize hydrocarbon molecules. High-voltage electrodes then collect the ions and released electrons. The resulting electron current provides an extremely sensitive and linear measure of carbon content that follows the profile of each eluting peak. The detector responds only to organic compounds.
Flame photometric detector (FPD)	A PGC detector that uses a hydrogen flame to excite the fluorescence of infrared radiation by sulfur or phosphorus atoms. An optical filter selects the fluorescence due to sulfur, which then impinges on a photomultiplier tube. This device is very sensitive to light and outputs a signal that follows the profile of each eluting peak that contains sulfur.

Flame photometry — A method for the chemical analysis of certain metal elements that examines the wavelengths of light emitted by an analyte sprayed into a flame.

Flameproof enclosure — Also called an explosion-proof enclosure. An enclosure that complies with applicable codes for installation in a specified hazardous area. Quenches a flame by cooling it below the ignition temperature of the gas in question.

Flash point — The lowest temperature at which a flammable fluid will burn in air when exposed to an ignition source.

Fluoresce — To emit electromagnetic radiation.

Fluorescence — The emission of photons due to electrons returning to their ground state after being excited by heat, radiation, or chemical reaction. The emitted fluorescence may be in the infrared, visible, ultraviolet, or x-ray regions of the spectrum.

Fluorine
F
M = 18.998 g/mol — The chemical element with atomic number (Z) = 9. The lightest of the halogen group of elements.

Fluorine gas
F_2
M = 37.996 g/mol — A yellow-green gas that is extremely reactive.

Force (F) — The physical influence that accelerates the motion of an object, defined as mass multiplied by acceleration.
The SI unit of force is the newton (N) where:

$$N = kg \cdot m/s^2$$

Forced integration — A method of chromatogram peak area integration. The integration proceeds continuously from the start command until the stop command. The accumulated integral includes all deviations of the baseline during that period.

Formaldehyde — See **methanal**.

Fourier transform — A mathematical procedure that converts a signal occurring in time to an equivalent set of sinusoidal waveforms that are easier to analyze.

FPD — See **flame photometric detector**.

Fractionation — The conversion of a single-phase gas or liquid sample to a two-phase mixture of vapor and liquid. Usually, this results in a composition change that invalidates the sample for analysis.

Freon — One member of a family of fluorocarbon refrigerant gases.

Fused peaks — Two or more partially separated chromatogram peaks, particularly those that exhibit no valley between them.

G

Galvanic current	An electric current produced solely by chemical reaction as in a dry cell battery. In process analyzers, the most common application for a galvanic cell is to measure oxygen at sub-ppm concentrations in a gas stream.
Gas chromatograph	The instrument.
Gas chromatography (GC)	The technique of separating analytes prior to measuring them by injecting a vapor sample into a gas mobile phase that moves in contact with a liquid or solid stationary phase.
Gas-liquid chromatography (GLC)	Gas chromatography using a liquid stationary phase.
Gas-solid chromatography (GSC)	Gas chromatography using a solid stationary phase.
Gate	See **peak gate**.
Gaussian	The curve shape of a normal statistical distribution, also known as the standard curve of error. When inside the column, peaks closely approximate to this shape. On the chromatogram, the peaks appear slightly skewed because the back of the peak stays in the column longer than the front, thus becoming a little wider.
General elution problem	When analyzing a sample that contains a wide range of components, a single chromatographic column that separates the early peaks will always take an excessive time to elute the later peaks.
Grab Sample	Also called a spot sample. A discrete sample of a process fluid contained in a balloon, cylinder, or bottle. Most grab sampling is manual and is subject to uncontrolled variations in procedure.
Gram mole	The mass in grams of one mole of molecules, numerically equal to their molar mass in atomic mass units.
Green oil	A polymer oil containing the dimer and tetramer of butadiene or other unsaturated C_4 hydrocarbons. Commonly occurs in gas samples containing C_4 olefins and acetylenes, particularly if they flow in heated lines. Tends to form more rapidly at temperatures above 60 °C.

H

Halogenated solvent	Excellent solvents capable of dissolving many compounds. They have a basic hydrocarbon structure but contain one or more halogen atoms in the place of hydrogen atoms.
Halogens	A group of similar elements that includes fluorine, bromine, chlorine, and iodine.
Hastelloy™	A nickel-molybdenum-chromium-tungsten alloy with excellent corrosion resistance in a wide range of severe environments.

Hazardous area	An area, or more correctly a volume, of space in a processing plant where a flammable atmosphere might exist.
Heartcut	A column switching technique that enables the measurement of ppm-level analyte peaks following a large major component peak of high percentage concentration. Essentially, the technique concentrates the analytes by separating and removing most of the major component.
Heartcut column system	Uses an initial column to achieve a partial separation of the analytes while venting the bulk of the major component. A column valve then operates to allow the analyte peak(s) and a small amount of the major component to enter a second column for final separation.
Heating value	A measure of the energy available from a fuel and often the basis of the contract value of a natural gas shipment. Some PGCs have software to calculate the heating value of natural gas from a chromatographic analysis.
Heavies	Component peaks that elute from a column later than the main analyte peaks. Typically, heavies are backflushed to vent or regrouped into a single peak for measurement.
Helium He $M = 4.0026$ g/mol	The chemical element with atomic number (Z) = 2. Also, a monatomic gas; the lightest of the six noble gases and extremely inert. Because of its safety and high thermal conductivity, it was the favorite carrier gas for PGCs in North America but is becoming scarce and expensive. PGCs are now more likely to use hydrogen carrier gas.
Helium ionization detector (HID)	A highly-sensitive detector that can measure almost any analyte. A radionuclide ionizes the helium carrier gas, which then ionizes the analytes. Now obsolete in PGCs due to its instability and the onerous regulations for operating a radioactive source.
Henry's law	At equilibrium, the amount of dissolved gas in a liquid is directly proportional to its partial pressure above the liquid.
HETP	Height equivalent to a theoretical plate. See **plate height**.
Hexane C_6H_{14} nC_6' $M = 86.18$ g/mol	Also called n-hexane. A normal paraffin often measured in liquid process streams and in natural gas. The column may group the hexane with its many isomers to give a composite C_6 measurement.
HID	See **helium ionization detector**.
Holdup time (t_M)	The time taken by the carrier gas to pass through a column, conveniently measured by the retention time of an **air peak**.
Holistic property	A measured quality of a whole fluid such as density, flash point, or color.
Homologous series	A set of carbon compounds having identical molecular structure except for the length of the main carbon chain. A classic example is the paraffin series, whose members differ only by the incremental addition of a **methylene** group: $-CH_2-$

Host computer	A computer, typically a process-control system, which receives analytical measurements from a PGC and may exercise some control over its operation.
Hot-wire detector	A thermal conductivity detector that uses heated wire elements, rather than thermistors.
Housekeeping column	A column designed to remove all heavies from the column system before injecting another sample.
Housekeeping rule	A column system must remove all components of an injected sample from the columns before it injects another sample.
Hydrocarbon C_xH_y	A chemical compound containing only carbon and hydrogen atoms.
Hydrogen H $M = 1.00794$ g/mol	The chemical element with atomic number $(Z) = 1$. The first element.
Hydrogen gas H_2 $M = 2.01588$ g/mol	The lightest gas; transparent, odorless, and flammable. As a carrier gas, hydrogen is less expensive than helium. It generates a somewhat lower **plate number** than helium or nitrogen carrier gas but provides much faster separations.
Hydrogen chloride HCl $M = 36.46$ g/mol	A toxic gas that readily dissolves in water to form hydrochloric acid. When wet, it is highly corrosive. The molecule is covalent but highly polarized, so it tends to adsorb on surfaces.
Hydrogen cyanide HCN $M = 27.0253$ g/mol	A colorless and highly toxic liquid or gas with an almond odor. The liquid boils at 26 °C. The molecule is covalent and has the structure: N≡C–H
Hydrogen fluoride HF $M = 20.006$ g/mol	A colorless and highly toxic liquid or gas. The liquid boils at 19.5 °C and is highly acidic. Dissolves in water to form the corrosive hydrofluoric acid.
Hydrogen sulfide H_2S $M = 34.08$ g/mol	A colorless, flammable, and highly toxic gas with an odor of rotten eggs. Commonly occurs in samples of industrial gases and when at low ppm levels is often a designated analyte for PGC measurement.

I

Inert support	An inert solid like **celite** that can hold a liquid phase immobile in a packed column. The liquid coats the large surface area of the support.
Infrared (IR)	The portion of the electromagnetic spectrum with wavelengths between 700 nm and 1 mm.
Infrared absorption	The absorption by sample molecules of infrared energy at certain wavelengths that are characteristic of the desired analyte. The transmitted light thus experiences a loss of energy at those same wavelengths proportional to the number of analyte molecules in the light beam.
Inorganic compound	Generally, a chemical compound that contains no hydrogen-carbon bonds.

Integration	The measurement of peak area.
Intercolumn detector	A low-volume thermal conductivity detector placed between columns to allow visual indication of peaks transiting from one column into another and thereby to facilitate the setting of column switching times.
Internal standard	A selected component that is always present in the analysis. The calibration procedure uses an external standard to calibrate only that peak. The PGC then calculates the concentration of other peaks based on their response factors relative to the internal standard.
Intrinsically safe	An electrical device or circuit that is incapable of releasing enough energy to ignite an explosive atmosphere, under normal or abnormal conditions.
Ion	An atom or group of atoms that has gained or lost one or more electrons, becoming a negatively or positively charged entity.
Ionization	The process of becoming an ion.
Isobutane	See **2-methylpropane**.
Isobutene	Also called isobutylene. See **2-methylpropene**.
Isomers	Two or more compounds with the same chemical formula and the same molar mass, but having a different structure, leading to different physical properties and different chemical reactions. Hydrocarbon isomers having branched carbon chains tend to have a more compact molecule than their straight-chain homologs and therefore elute first on a gas-liquid column.
Isoparaffins	Branched chain isomers of the normal paraffins, of which there are many. When used to indicate a homologous series, typically refers to the 2-methyl isomers of the n-paraffins. For example, see **2-methylbutane**.
Isopentane	See **2-methylbutane**.
Isothermal analysis	A chromatographic separation performed at constant column temperature, followed by detection and measurement of the analyte concentrations.

J

joule (J)	The SI derived unit of energy (*E*). An electrical definition: one joule is the energy dissipated as heat when an electric current of one ampere passes through a resistance of one ohm for one second.

K

Katharometer	Prior name for a thermal conductivity detector.
Kelvin (K)	The SI base unit of temperature (*T*).
Ketone [–CO–]	A member of the ketone family of organic compounds.
Kilogram (kg)	The SI base unit of mass (*m*).

Krypton Kr $M = 83.798$ g/mol	The chemical element with atomic number $(Z) = 36$. A monatomic gas, the fourth of six noble gases. Not used in PGC.

L

Liquid chromatography (LC)	The technique of separating analytes prior to measuring them by injecting a sample into a liquid mobile phase that flows over or through a solid stationary phase. A process liquid chromatograph was once available for sale but was not commercially successful. The technique is common in the laboratory for the analysis of food, drugs, and other complex mixtures, but is rarely used online in the industrial process industries.
Liquid loading	The percentage by weight of liquid phase deposited on a solid support. In effect, a measure of liquid film thickness.
Liquid-phase column	A column having a liquid as the stationary phase.
Liter	Also called a litre. A unit of volume equal to one thousandth of a cubic meter.
Live switching	A method of column switching that uses gas pressure to control flow direction rather than mechanical valves. Used mainly with capillary columns to minimize intercolumn volume.
Low-pass filter	An electronic or digital filter that passes low frequencies without attenuation but attenuates or blocks higher frequencies.

M

Major component	The main constituent of an analyzed sample. Usually refers to a very large peak that tends to interfere with the analysis of ppm concentrations of other components in the sample.
Mass-flow detector	A rate-sensitive detector that responds to the instantaneous population of analyte molecules present, thus effectively measuring their mass flow.
Mass percent	See **weight percent**.
Mass spectrometry (MS)	An analytical technique that ionizes the sample gas under vacuum and then accelerates the ions through a magnetic or electronic field, which defects their motion in proportion to their mass and electronic charge. Alternatively, the analyzer separates the ions according to their time of flight. The analyzer may detect specific ions by their mass-to-charge ratio or output a complete mass spectrum for further analysis.
Mean	The arithmetic average of a set of results.
Measurand	A general term used in metrology to identify the “quantity intended to be measured.”
Memory unit	An electronic device previously used to provide an analog **trend record** signal. Stores and retransmits a continuous analog signal representing the value of a single PGC measurement,

MEMS	Micro-electromechanical systems refers to tiny integrated devices or systems that combine mechanical and electrical components. They are fabricated using integrated circuit batch processing techniques and can range in size from a few micrometers to millimeters.
Mercaptan [–SH]	A hydrocarbon derivative containing the –SH functional group. Used to odorize natural gas.
Meter (m)	Also called a metre. The SI base unit of length (*L*).
Methanal CH_2O M = 30.026 g/mol	Also called formaldehyde. The smallest aldehyde; a colorless, toxic gas with a pungent odor. Not detected by an FID, unless converted to methane by a methanator.
Methanator	A device that converts ppm amounts of carbon monoxide, carbon dioxide, or formaldehyde to methane, so a flame ionization detector can measure them. The column effluent mixed with hydrogen passes through a heated catalyst tube to convert the peaks to methane.
Methane CH_4 C_1 $M = 16.04$ g/mol	A hydrocarbon gas; initial member of the paraffin homologous series. Occurs on chromatograms of petroleum gases and comprises about 85 % of natural gas.
Methanol CH_3OH MeOH $M = 32.04$ g/mol	Also called methyl alcohol. The smallest member of the alcohol family of organic chemical compounds.
Method	A tabulation of instructions to a PGC specifying the timed events and calculations necessary to perform an analysis of one or more process streams.
Methyl acetylene	See **propyne**.
Methylbenzene C_7H_8 $M = 92.141$ g/mol	Also called toluene. An aromatic compound having a single methyl group attached to a benzene ring. Often designated as an analyte for PGCs in aromatics plants.
2-Methylbutane C_5H_{12} iC_5' $M = 58.12$ g/mol	Also called isopentane. A paraffin having five carbon atoms with the second carbon atom connected to three other carbon atoms. An isomer of n-pentane.
Methyl chloride	See **chloromethane**.
Methylcyclopropane C_4H_8 $M = 56.108$ g/mol	An isomer of cyclobutane that often occurs in light hydrocarbon streams.
Methylene group [$-CH_2-$]	The repeating chain component in **paraffin** molecules and many polymers.
Methyl group [$-CH_3$]	A common feature of branched-chain hydrocarbon molecules. Not a separate chemical compound.
2-Methylpropane C_4H_{10} iC_4' $M = 58.12$ g/mol	Also called isobutane. The simplest branched chain hydrocarbon. Due to its more compact molecule, isobutane elutes before n-butane on all liquid columns. Frequently occurs on chromatograms of light hydrocarbon gases.

2-Methylpropene C_4H_8 iC_4'' $M = 56.106$ g/mol — Also called isobutene or isobutylene. The simplest branched chain **olefin**, and one of the four isomers of butene. Frequently occurs on chromatograms together with other light olefins.

Microliter (μL) — One millionth of a liter (=1 mm^3).

Migration rate — The net rate of progress of a whole peak through the column. On a given column, the migration rate of a peak is inversely proportional to its retention factor.

Milliliter (mL) — One thousandth of a liter (=1 cm^3). Often colloquially called a cc.

Mobile phase — For a PGC, this is the carrier gas. Common choices are hydrogen, helium, or nitrogen, but others are possible. The gas must be pure and dry.

Molar mass (M) — The quantity of material in one molecule of a substance, measured in daltons. Equal to the sum of the atomic masses of the individual atoms contained within the molecule, as designated by the formula of that molecule.

Molar weight — The force of gravity acting on one molar mass.
The SI base unit of quantity (n).

Mole (mol) — A way to count molecules or other small entities. One mole is about 6.022×10^{23} of them.

Molecular sieves — A series of solid stationary phases capable of separating oxygen and nitrogen as well as other light gases such as hydrogen, methane, and carbon monoxide. Is prepared from a mineral known as a zeolite, which can selectively sort molecules based on their size, due to its very regular pore structure. The maximum size of molecule that can enter the pores of a zeolite depends upon the dimensions of the pores.

Molecular weight — See **molar mass**.

Molecule — The smallest amount of a chemical substance that can independently exist, and still retain the properties of that substance.

Mole fraction — The fraction of the molecules in a sample that conform to a specified identity or type.

Mole percent — The mole fraction expressed as a percentage. For a gas sample at low pressure, the mole percent and the volume percent are approximately equal.

Monel™ — A group of nickel alloys, primarily composed of nickel (from 52 to 67 %) and copper, with small amounts of iron, manganese, carbon, and silicon. Monel alloys are resistant to corrosion by many agents, including rapidly flowing seawater.

Moving average — An averaging technique used to reduce the noise content of a signal. The processor averages an odd number of measurements equally spaced in time and accepts the average as the correct value at the time of the center measurement. Each time the processor reads a new data value it drops the oldest value and recalculates the average, thus moving the average.

Multidimensional chromatography
A fancy name for gas-chromatographic separations that use serial columns containing different stationary phases.

N

Naphthalene
$C_{10}H_8$
$M = 128.17$ g/mol
A dual-ring aromatic hydrocarbon with a strong mothball odor.

Naphthenes
An obsolete and confusing term for cyclic hydrocarbons, not related to the chemical substance naphthalene.

Near infrared (NIR)
The portion of the electromagnetic spectrum having wavelengths between 780 nm and 2500 nm.

Neon
Ne
$M = 20.1797$ g/mol
The chemical element with atomic number $(Z) = 10$. The second noble gas, after helium. Extremely inert monatomic gas. Might make an interesting carrier gas, but too expensive for PGC.

Neopentane
See **2,2-dimethylpropane**.

Newton (N)
$N = kg \cdot \frac{m}{s^2}$
The SI unit of **force**: the force required to give a mass of one kilogram an acceleration of one meter-per-second-per-second.

Nitrogen
N
$M = 14.007$ g/mol
The chemical element with atomic number $(Z) = 7$, and a valency of 3.

Nitrogen gas
N_2
$M = 28.014$ g/mol
The familiar, colorless, and odorless gas. Mostly inert. Sometimes used as a PGC carrier gas.

Noble gas
Generic name given to the elemental gases helium, neon, argon, krypton, xenon, and radon. All are chemically inert, occur in trace amounts in the atmosphere, and are often measured by gas chromatography.

Normalization
A mathematical procedure to proportionally adjust all measured concentrations so they total 100 %. The technique is valid for percent-level measurements but may not work for ppm-level measurements. The PGC must measure all components of the sample that have a significant effect on the total – typically all those greater than 0.1 % by volume.

O

Octane
C_8H_{18} nC_8
$M = 114.232$
Also called n-octane. A straight-chain paraffin having eight carbon atoms in each molecule.

Octane number
A measure of the quality of gasoline related to its tendency to pre-ignite under compression and "knock."
On the arbitrary octane number scale, 2,2,4-trimethylpentane has an octane number of zero and n-octane has an octane number of 100. Other values of octane number derive from simple binary mixtures of these standard liquids.

Olefin
[>C=C<]
C_nH_{2n}

A member of the **alkene** family of hydrocarbons having one or more double bonds between carbon atoms.

Open tubular column

A column that has no packing so there's an open path for gas flow. Such columns have little flow resistance, so they can be long and narrow, and thereby very powerful. The stationary phase is a thin film of liquid phase or a thin layer of solid particles attached to the wall.

Optical filter

A device that allows a selected band of photon frequencies to pass through but blocks any higher or lower frequencies.

Order of magnitude

To differ by a factor of ten; e.g., a difference of three orders of magnitude is one thousandfold.

Organic compound

Any chemical compound containing carbon, except the carbon oxides and sulfides.

Outlier

A measurement result that doesn't fit the normal distribution of results and can be statistically shown to be highly unlikely. The term is used in chemometrics to indicate a sample concentration that is so outside the scope of the calibration model that the value cannot be trusted.

Oven

A temperature-controlled enclosure.

Oxygen
O
$M = 15.999$ g/mol

The chemical element with atomic number $(Z) = 8$. Present in the molecular structure of many chemical compounds analyzed by PGCs.

Oxygen gas
O_2
$M = 31.998$ g/mol

The familiar gas in the atmosphere and present in many gases analyzed by PGCs. Difficult to separate from argon by gas chromatography.

P

Packed column

A tube packed with small particles that may or may not have a thin film of liquid phase deposited on them. The original PGCs used packed columns, and they are still popular today. Tube sizes have gradually changed from ¼-inch o.d. to 1/16-inch o.d. The most common size found in a PGC is ⅛-inch o.d.

Paraffin
[–C–]
C_nH_{2n+2}

A member of the **alkane** family of hydrocarbons having only single bonds between carbon atoms.

Parallel chromatography

Two or more independent chromatographic separations occurring concurrently, each using a separate sample injector, column system, and detector.

Paramagnetism

A form of magnetism induced in some materials by an external magnetic field. In analysis, the paramagnetism of oxygen is a prime example. The movement of oxygen when exposed to a magnetic field is the principle employed by many process oxygen analyzers.

Partial pressure	The portion of the absolute pressure of a gas that is due to a specific compound, often expressed in kilopascals (kPa). The total pressure of a mixture of gases is the sum of the partial pressures of its constituents.
Partition	An obsolete technical term now superseded by **distribution** in gas chromatography but retained to describe the similar process in liquid chromatography.
Pascal (Pa)	The SI derived unit of **pressure** (P). A force of one newton per square meter.
Pascal-second (Pa · s)	The SI derived unit of dynamic viscosity (η or μ).
PDD	See **pulsed discharge detector**.
Peak	Literally, the shape of the detector response to an eluting component, when recorded on the chromatogram. Colloquially, a compact group of component molecules migrating through the column.
Peak area	During a peak, the integrated detector response above the chromatogram **zero line**: i.e., the best estimate of the baseline signal that would be present if the peak was not there. Measured in millivolt-seconds or microampere-seconds, depending upon the detector in use.
Peak gate	The time between two preset scheduled events bracketing the expected retention time of a single named peak or a group of peaks.
Peak height	During a peak, the maximum detector signal relative to the chromatogram **zero line**: i.e., the best estimate of the baseline signal that would be present if the peak was not there. Measured in millivolts or microamperes, depending upon the detector in use.
Peak picker	An analog pneumatic or electronic device used in early PGCs to capture the height of a peak so a computer could read its value. Alternatively, the peak picker would transfer the held value to a long-term memory to output a continuous trend signal.
Peak tail	A trail of component molecules following a peak that distorts the peak and may form an extended baseline offset declining exponentially. May be due to slow sample injection, non-linearities in the phase equilibria, adsorption of component molecules on surfaces, or entrapment of molecules at unpurged cavities in the walls of the flow path.
Peak width at base (w_b)	The width of a chromatogram peak in distance or time units, measured between the intersections of the triangulated peak sides with the extended baseline.
Peak width at half height ($w_{0.5}$)	The width of a chromatogram peak in distance or time units, measured horizontally at half the peak height.
Peak width parameter (s)	A method table parameter used by some PGCs during data smoothing. The processor divides its scan rate (Hz) by the stored peak width parameter (s) to determine the number of raw data points to average.
PEEK	Polyether ether ketone. A thermoplastic polymer having good mechanical properties and chemical resistance at high temperatures.

Pentane
C_5H_{12} nC_5'
$M = 72.15$ g/mol

Also called n-pentane. A straight-chain paraffin having five carbon atoms in its molecule. Often seen on chromatograms of light hydrocarbons.

Percent (%)

A concentration of one part in one hundred. The percent sign is numerically equal to 0.01.

Percent-off

The percentage amount of a liquid mixture that evaporates into the vapor phase at a specified temperature and pressure.

Perfect resolution

Defined as a resolution of 1.5, which is just adequate for measuring equal symmetric peaks. More resolution is necessary for adjacent peaks of different size, particularly when the smaller peak elutes after the larger one.

Permeation

The act of gas molecules passing through an apparently solid material like the walls of polymer tubing. Don't use polymer tubing for ppm gas samples. May also refer to gas leaks at threaded or compression joints in stainless steel tube assemblies.
As in **diffusion**, the rate of permeation is proportional to the partial pressure differential. Increasing the sample pressure will not stop its contamination by permeation of gases from the atmosphere or flare line.

Perpendicular drop

A method of allocating total peak area between two overlapping peaks. In effect, the processor separates the two areas by a vertical line dropped from the valley point between the peaks.

Phosphorus
P
$M = 30.974$ g/mol

The chemical element with atomic number $(Z) = 15$. As a pure solid, it exists as two allotropic forms: red and white. Phosphorus is a highly reactive element and a very common component of minerals and living things, but not often encountered in PGC.

Photoionization detector (PID)

A photoionization detector employs ultraviolet photons from a discharge lamp to ionize the analyte molecules. The ions and electrons formed then migrate to the charged electrodes, as they do in an FID.

Photometer

A generic term for any analytical instrument that uses electromagnetic radiation to measure analyte concentration. Measurement techniques may employ any spectral wavelength and work by absorption, fluorescence, or refraction of the light beam.

Photometry

The science and art of measuring analytes by their absorption or emission of electromagnetic radiation.

Photomultiplier

A highly-sensitive electronic device to detect photons of light and generate a proportional electric current.

Photon absorption

Most chemical compounds absorb photons of certain characteristic wavelengths, allowing measurement of their concentration by detectors tuned to those wavelengths.

Photon emission

Some chemical compounds absorb photon or heat energy and then emit photons of certain characteristic wavelengths, allowing measurement of their concentration by detectors tuned to those wavelengths.

Physical property	A unique property of a molecule that can indicate its concentration in the sample, such as the paramagnetism of oxygen or the thermal conductivity of hydrogen.
Physical property analyzer	An instrument that measures a quality of the process sample other than its chemical composition. PGCs can calculate a few physical property measurements such as heating value or boiling point.
Physical separation	The separation of different types of molecule from each other by non-chemical means.
PID	See **photoionization detector**.
Piston Valve	See **spool valve**.
Planar chromatography	A liquid-chromatographic technique that uses a two-dimensional substrate to separate the analytes. Examples include paper chromatography and thin-layer chromatography.
Plate	A theoretical concept equivalent to the physical space within the column needed to create one equilibrium between the component molecules in the gas phase and those in the liquid phase.
Plate height (H)	The length of column required to generate one plate (see Equation 3.5).
Plate number (N)	The effective number of equilibria exhibited by a component peak (see Equations 3.3 and 3.4).
PLOT column	The stationary phase in a porous-layer open-tubular column is a very thin layer of solid material coated on the inside wall of the tube to selectively adsorb sample molecules from the sample gas.
Plug injection	A rapid sample injection to ensure that all the sample molecules enter the column in a compact group, thereby minimizing the widths of peaks and maximizing their resolution.
Plunger Valve	A valve designed to inject liquid samples. The sample volume is a hole or circumferential groove in a rod. Liquid sample fills the volume in a cold part of the valve. Actuation drives the rod into a hot zone typically within the column oven where the sample rapidly vaporizes.
Pneumatic detector	A PGC detector that uses a flow orifice and capillary tube in series to generate a pressure signal. The pressure signal is due to changes in gas viscosity and density when a component peak enters the detector.
Pneumatic signal (3-15 psig)	The standard pneumatic analog data-transmission signal for process control.
Polyimide	A lightweight, heat-resistant, chemical-resistant, and flexible plastic used for electrical and thermal insulation and for the manufacture of molded parts.
Polymer	A substance having long-chain molecules of indeterminate length comprising repetitive structural elements due to the repeated attachment of monomer segments. The simplest polymer is polyethylene with the chain structure: $[-CH_2-CH_2-]_n$

Porous polymer	A synthetic solid stationary phase. The original porous polymers developed by Hollis were small polystyrene beads with varying degrees of polymer crosslinking. Hollis, O.L. (1966). Separation of Gaseous Mixtures Using Porous Polyaromatic Polymer Beads. *Analytical Chemistry* 1966382309-316 (February 1, 1966) doi.org/10.1021/ac60234a038
Pound mass (lb)	The American unit of mass or weight, now defined as exactly 0.453 592 37 kilograms.
Poundal (pdl)	An American unit of **force**: the force required to give a mass of one pound an acceleration of one foot-per-second-per-second.
Pound force (lbf)	An American unit of **force**: the gravitational force exerted on a mass of one pound on the surface of Earth.
ppb	A concentration of one part in a billion (1×10^{-9}).
ppm	A concentration of one part in a million (1×10^{-6}).
ppt	A concentration of one part in a trillion (1×10^{-12}).
Precision	Also called repeatability. See **uncertainty interval**.
Pressure (P): $P = \frac{F}{A}$	Defined as the **force** (F) exerted by a fluid on a unit area (A) of the walls of its container, in newtons-per-square-meter. The **SI unit** of pressure is the pascal (Pa), where: $\text{Pa} = \text{N/m}^2$. The American unit of pressure is the pound-per-square-inch (psi).
Pressure, absolute	The total pressure exerted by a fluid.
Pressure, gauge	The fluid pressure in excess of atmospheric pressure at the location and time of measurement.
Prime rule	See **housekeeping rule**.
Priority interrupt	An older data-acquisition protocol that sends a signal to instruct a host computer to interrupt routine processing and read new data available at the PGC.
Process analyzer	Colloquially, any quality-measuring instrument (QMI).
Program	A sequence of timed events and calculations that controls the operation of a process gas chromatograph.
Programmer	A device, usually electromechanical, that controlled the sequence of operations in early process gas chromatographs. Superseded by the microprocessor-based Control Unit.
Propadiene C_3H_4 C_3'''' $M = 40.065$ g/mol	Also called allene. The smallest **diene** with the carbon skeleton: C=C=C. Occurs on chromatograms of light olefin process streams.
Propane C_3H_8 C_3' $M = 44.097$ g/mol	A hydrocarbon gas. The third member of the paraffins homologous series, with the straight carbon skeleton: C–C–C. Frequently present in petroleum gases.

2-Propanone
CH_3–CO–CH_3
$M = 58.08$ g/mol
: Also called acetone or methyl-methyl-ketone. Smallest molecule in the ketone family. A flammable, volatile solvent used to clean tubing.

Propene
C_3H_6 C_3''
$M = 42.081$ g/mol
: Also called propylene. A hydrocarbon gas, the second member of the olefins homologous series with the straight carbon skeleton: C=C–C. An important monomer for producing polypropylene.

Propylene
: See **propene**.

Propyne
C_3H_4 C_3'''
$M = 40.064$ g/mol
: Also called methyl acetylene. A reactive hydrocarbon gas, the second member of the acetylenes homologous series with the straight carbon skeleton: C≡C–C. Often a desired analyte in light olefin gases.

Pulsed discharge detector (PDD)
: The PDD uses a pulsed electric discharge in helium, or helium doped with another noble gas, to raise helium atoms to an energetic state while emitting ultraviolet photons. The detector is popular in PGCs as it can mimic other ionization detectors without using a radionuclide.

Q

Quality-measuring instrument (QMI)
: An instrument to measure a quality of a process fluid as distinct from its current condition. The category includes analytical instruments measuring composition and non-analytical instruments measuring a physical property of the fluid, but both types are "process analyzers."

Quantum cascade laser (QCL)
: A wide-range tunable laser that produces a high optical power output at ambient temperature.

R

Raman spectroscopy
: A type of spectroscopy used in chemistry to identify molecules and study chemical bonding. The vibrational frequencies of a molecule are specific to its chemical structure and bonding. Industrial Raman spectrometers can identify and quantify the various types of molecules present in a sample.

Random error
: Unpredictable variation in a measurement due to uncontrollable variations in environment or procedure. Limits the **precision** of a measurement.

Rate-of-arrival detector
: A detector that outputs a signal proportional to the arrival rate of analyte molecules into the sensor. Often called a **mass-flow detector**.

Raw chromatogram
: A stored digital chromatogram before applying any signal-processing techniques such as noise suppression.

Raw data
: Individual samples of the chromatogram signal before any data processing occurs.

Reference peak
: A selected peak that is always present on the chromatogram and used to predict the location of other peaks. The processor predicts the position of another analyte by using the fixed ratio between the retention times of analyte peak and reference peak.

Relative response factor (RRF)	The detector response factor of an analyte relative to a standard substance like benzene, measured under the same operating conditions. Retrieve RRFs from the literature. The ratio of two RRFs gives their relative sensitivity on the specified detector and may allow the calibration of an analyte without using an external standard.
Relative retention	The ratio of the adjusted retention times of two peaks on a chromatogram. Formally called their **separation factor**.
Repeatability	See **precision**.
Resistive heating	Term used for the direct heating of a column by a heating wire in contact with the column tubing or by passing an electric current through the tubing itself.
Resolution (R_S)	The degree to which the peak areas of two adjacent peaks are separate from each other, defined as the separation between the peaks divided by their average base width.
Response factor	A factor obtained during calibration by dividing the analyte peak area by its known concentration. To calculate analyte concentration the analyst divides each peak area by its response factor. See **calibration factor**.
Response time	The total response time of a PGC is the sum of the delays in the sampling system plus the analysis time.
Retention factor (k)	The retention factor of a peak is the ratio of its adjusted retention time to the holdup time of the column.
Retention time (t_R)	The elapsed time to a component peak apex, measured from the instant of sample injection, being the average time that those component molecules spend in the column. Also, equal to the sum of the holdup time and the adjusted retention time of that component.
Rider peak	A small peak that appears after a very large peak on a chromatogram. Since large peaks always form an exponentially decaying tail that is visible on a highly sensitive detector, the small peak appears to "ride" of the tail of the large peak.
Rotary valve	A chromatographic valve, typically having two positions. When actuated, a rotor turns on a stator to make the desired connections.

S

Sample	A small aliquot of the material to be analyzed. In a process gas chromatograph, it's vitally important that the injected sample volume is highly repeatable. Also (colloquially) the flow of process fluid to and through the gas chromatograph from which the aliquot is taken. See **sample flow**.
Sample conditioning	The process of adjusting the condition of the incoming sample to comply with analyzer requirements. The reliability of a PGC is heavily dependent on the performance of its sample conditioning system, so the highest quality of design and maintenance is mandatory.

Sample conditioning system	An assembly of devices to adjust the condition of the sample to match analyzer requirements. May include flow control, temperature or pressure adjustment, filtration, phase separation, or phase change.
Sample flow	As applied to a PGC, the flow of sample fluid passing through the sample inject valve. Liquid flow is commonly set at 30 mL/min but might be as low as 10 mL/min. Gas flow is much higher, often 100–150 ml/min.
Sample injection valve	See **sample injector**.
Sample injector	A means to inject reproducible samples of the process fluid into the carrier gas stream.
Sample loop	An external loop of tubing on a gas sample injector valve that provides the desired sample volume.
Sample splitter	A device for injecting extremely small samples into capillary columns. The carrier gas containing the injected sample is split into two flows, one entering the column and the other going to vent.
Sample volume	The volume of sample gas or liquid injected into a chromatographic column.
Sampling rate	The number of times per second that a processor measures the analog chromatogram signal, commonly expressed as a frequency in hertz, such as 40 Hz.
SCOT column	The stationary phase in a support-coated open-tubular column is a liquid coated on very fine support particles in a uniform layer on the inner wall of the tube.
second (s)	The SI base unit of time (t).
Second derivative	The rate of change of the slope of a line. For the chromatogram baseline it increases very rapidly at the onset of a peak giving a good indication that the peak has started. It's also used to sense the apex of a peak and the valley point between peaks.
Semi-diffusion	An arrangement of passages in a thermal conductivity detector that has the thermal elements partly in and partly out of the direct flow path of the carrier gas.
Sensitivity	The sensitivity of a **concentration detector** is the change in signal output per unit change in the *concentration* of analyte in the detector. The sensitivity of a **rate of arrival detector** is the change in signal output per unit change in the *rate of arrival* of analyte into the detector.
Separation (S)	The distance or time between the apexes of two adjacent peaks on a chromatogram. Separation does not indicate the degree of overlap between peaks as it doesn't account for peak width. See **resolution**.
Separation factor (α)	The ratio of adjusted retention times between a specified later peak and a specified earlier peak on an isothermal chromatogram. The ratio is usually constant on a given liquid phase and independent of column operating conditions, provided those conditions are constant during analysis.

Shoulder peak	A chromatogram peak that has insufficient resolution from another peak to form a valley between them, so the smaller peak appears as a lump on the side of the larger peak.
Signal noise	Minor random variation in detector output signal caused by a myriad of small electronic, fluidic, or environmental instabilities that are uncontrollable.
Signal smoothing	Usually refers to the **moving average** method of signal noise reduction.
Silica gel SiO_2	An active solid capable of adsorbing gas molecules, previously used as a solid stationary phase.
Silicon Si $M = 28.085$ g/mol	The chemical element with atomic number $(Z) = 14$. A nonmetal with chemical properties similar to carbon.
SimDis	Simulated distillation: a chromatographic method of estimating the **percent-off** at any specified boiling temperature.
SI unit	An international standard unit of measure approved and certified by the Système international d'unités in France. All SI units are based upon seven invariant base units and are directly related to other SI units by the equations of physics. Therefore, they need no conversion factors.
Slide valve	A two-position chromatographic valve that uses a sliding plate to switch the connections between ports. Slide valves may use an electric or pneumatic actuator and are functionally equivalent to rotary or diaphragm valves.
Slope detection	A software method of determining when a peak is about to emerge from a column by measuring the rate of change of a chromatogram signal. Typically, the processor recognizes the start of a peak when the rate of change exceeds a preset threshold value. A similar procedure determines when the peak has finished.
Software autozero (SAZ)	Not a true **autozero** because the detector signal remains unchanged. The processor simply measures the baseline signal and compensates for that value when quantifying the peaks.
Solenoid valve	A two-position valve operated electromagnetically by an electric current in a wire coil. Solenoid valves for ac power have a copper ring that may be in contact with the fluid. Copper in contact with acetylenes can form an explosive acetylide, so don't use ac solenoid valves with hydrocarbon samples.
Solubility	As used here, solubility refers to the percentage of component molecules dissolved in the liquid phase at equilibrium. This usage of solubility incorporates the volume ratio of the two phases, so a change in phase volume ratio would also change the observed solubility. A more technical approach using the concentration of molecules in the phases is independent of phase volume. See **distribution constant**.
Solute	The substance that dissolves in a solvent. Mostly synonymous with **component** in gas-liquid chromatography.

Solvent	The liquid substance that dissolves a solute. Not much used in gas chromatography, but used in **liquid chromatography**, for a liquid stationary phase.
Spatial separation	The distance-separation of components within a column, for which the doubling rule does not apply.
Specific heat (c)	Also called the specific heat capacity. Simply stated, the specific heat of a substance is the amount of energy required to raise the temperature of a unit mass of the substance by one degree. The SI unit for specific heat is the joule-per-kelvin-per-kilogram J/(kg· K). The specific heat of a substance may vary, sometimes substantially, depending on the starting temperature and pressure of the sample. For gases, the specific heat at constant pressure (c_P) is different from the specific heat at constant volume (c_V).
Spectrophotometer	An analytical instrument that splits the radiation absorbed or emitted by a sample into its spectral wavelengths and measures the intensity of certain characteristic wavelengths to determine the concentration of an analyte.
Spectroscopy	The science and art of measuring the intensity of emitted or absorbed radiation as a function of wavelength, particularly when applied to the identification and analytical measurement of chemical compounds.
Split ratio	When using a split injection technique, the split ratio is the amount of an injected sample that enters the column versus the amount vented. In practice, it's simply the ratio between the carrier gas flow rate into the column and the carrier gas flow rate exiting the splitter vent.
Splitter	See **sample splitter**.
Spool valve	An early PGC valve, now obsolete, used for gas sample injection and column switching. A rod carrying O-rings restrained in grooves slides in a tube to make the desired connections.
Spot sample	See **grab sample**.
Standard deviation (σ)	A measure of the random variation in a measurement equal to the square root of its **variance**.
Standard sample	A contained fluid with composition known to the desired level of uncertainty.
Start time	The preset time that a forced integration will start, sometimes called the on-gate.
Stationary phase	The solid or liquid material that remains stationary in a column as the carrier gas moves the sample molecules across its surface. It's the stationary phase that separates the analytes from each other and from the other components of the sample.
Step stream	An instruction in the program or method to stop the current sample flow and start sample flow from the next process stream scheduled for analysis.
Stop time	The preset time that a forced integration will stop, sometimes called the off-gate.

Stream switching	The manual or automatic practice of sequencing samples for analysis from two or more process streams.
Stripper column	See **backflush**.
Sulfur S $M = 32.06$ g/mol	Also called sulphur. The chemical element with atomic number (Z) = 16. A yellow solid that forms several stable allotropes.
Supercritical fluid	A substance at a temperature and pressure above its critical point, where distinct liquid and gas phases do not exist.
Systematic error	A constant bias from the correct measurement value due to a zero or span error, or because of a nonlinearity in the response curve.

T

Tangent skim	A method of measuring a **rider peak** by constructing an integration **zero line** that starts at the valley point before the rider peak and ends as a tangent to the receding baseline.
TCD	See **thermal conductivity detector**.
Temperature programming	Operating a column on a preprogrammed temperature schedule consisting of one or more periods of running at a constant set temperature, one or more periods of ramping at a set rate to a higher temperature, and a final period of rapid cooling.
Temporal separation	The time separation of components on a chromatogram and subject to the doubling rule.
Thermal conductivity	The ability of a material to transfer heat energy between two points when a temperature differential exists between them.
Thermal conductivity detector (TCD)	The original and still common detector in process gas chromatographs, capable of detecting any substance in the carrier gas. It detects the change in heat loss from a heated element when analyte molecules enter the detector.
Thermistor	A temperature-sensitive metal-oxide resistor, typically glass coated. Used as a detector element in some TCDs.
Theoretical plate	See **plate**.
Threshold value	A limiting value, below which no action occurs. PGCs use threshold values for automatic peak detection and for rejecting peaks that have insignificant height, width, or area.
Time constant	The time constant of a detector is the time it takes to output 63% of the value of a step change applied to its inlet port. After three time constants, the output will reach about 95% of the inlet value.
Timed event	An event such as a valve actuation or peak gate that must occur at a preset time after sample injection.
Timer	Any device that enables the occurrence of timed events.

Titanium Ti $M = 47.867$ g/mol	The chemical element with atomic number (Z) = 22. A lightweight, strong, and corrosion-resistant metal.
Titrator	A wet-chemical analyzer that measures the volume of a standard reagent that reacts with the analyte in a fixed volume sample.
Toluene	See **methylbenzene**.
Trap-and-hold	A PGC column system that traps selected peaks in a non-flowing column and releases them later for further separation.
Trend record	The historical record of a trend signal presented on a chart recorder trace.
Trend signal	A continuous measurement signal, that changes after each new analysis to represent the latest analyte value.
Triangulating	A graphical method of characterizing a peak by drawing straight lines through the inflexion points at each side of the peak and extending them to intersect the projected baseline.
1,1,2-Trichloroethene	A halogenated solvent.
Trichloroethylene	See **1,1,2-trichloroethene**
Trueness	The closeness of a measurement to the true value of the measurand. In standard terminology accuracy is equal to trueness plus precision.
Tunable diode laser (TDL)	A sensitive photometric gas analysis technique. It's high sensitivity comes from using a powerful diode laser source tuned to exactly match the wavelength most strongly absorbed by the analyte.

U

Ultraviolet	The portion of the electromagnetic spectrum with wavelength from 10 nm to 400 nm, shorter than that of visible light but longer than x-rays. Many process gas analyzers use ultraviolet absorption photometry, wherein the bonding electrons in analyte molecules strongly absorb ultraviolet energy at certain wavelengths that are characteristic of molecular structure.
Ultraviolet absorption	The absorption by sample molecules of ultraviolet energy at certain wavelengths that are characteristic of the desired analyte. The transmitted light thus experiences a loss of energy at those same wavelengths proportional to the number of analyte molecules interacting with the light beam.
Uncertainty interval	The range of measurement values believed to contain the true value of the measurand. The measurement must also state the **confidence level**.
Unit of measure	An accurately defined quantity of a measured variable that is used as a standard for the measurement of that variable. See **SI unit**.
Unretained peak	See **air peak**.

V

Vacancy peak	A negative peak appearing on the chromatogram at the retention time of a known component, due to the carrier gas containing more of that component than the injected sample does.
Valency	The number of chemical bonds exhibited by an element. For instance, carbon has a valency of 4.
Validation	A procedure for checking the calibration of an instrument by periodically measuring a sample of constant composition. Validation doesn't change the calibration of the instrument. The PGC compares the validation results statistically with a large number of previous validations to evaluate the need for recalibration or maintenance attention.
Validation fluid	The validation sample must be constant, so a large volume of fluid is necessary. However, the PGC doesn't need to know the concentration of the analytes because it will measure them. Sometimes the validation sample is simply a large cylinder filled with the process fluid.
Validation frequency	10–20 prior validations are necessary for statistical analysis, so the PGC should automatically validate at least once daily to build up a collection of historical values.
Valley point	The lowest point of the chromatogram signal between two overlapping peaks when their resolution is sufficient to show two apexes but not enough for a baseline separation.
Valve	Most process gas chromatographs use special chromatographic valves. They route the flow of carrier gas and component peaks into a selected column, detector, or vent.
Valve timing	The actuation and deactuation times of chromatographic valves are timed events specified in the analysis method. To achieve the desired resolution of analyte peaks, most column systems rely on highly repeatable valve timing.
Vapor pressure (of liquid)	The vapor pressure of a liquid is the equilibrium partial pressure of its vapor above the surface of that liquid. Liquids boil when their vapor pressure becomes equal to the surrounding pressure. Under that condition, the gas phase above the liquid is all vapor from the liquid.
Variable	A measurable property of an object or phenomenon that may change. Metrology is the science of evaluating a **continuous variable**.
Variance (σ^2)	A measure of the variability of a measurement due to random error. Estimated from a set of results by dividing the sum of the squares of the deviations from the mean value by the number of results minus one.
Velocity (u) of carrier gas (m/s)	More correctly, the average velocity of the carrier gas. Calculated by dividing the column length in meters by the **holdup time** in seconds.

Viscosity – dynamic (μ or η) (Pa·s)	A measure of a fluid's resistance to flow due to internal friction within the moving fluid. A high viscosity fluid resists motion because it has a lot of internal friction. The SI unit of viscosity is the pascal-second (Pa·s). The more common unit is the centipoise (cP), equal to the millipascal-second (mPa·s).
Viscosity – kinematic (ν) (m^2/s)	The ratio of the viscosity of a fluid to its density. The SI unit of kinematic viscosity is the square-meter-per-second (m^2/s) but rarely used. The common unit is the stokes (St) or centistokes (cSt) where: 1 St = 1 cm^2/s
Volatile liquid	A liquid that rapidly evaporates at the temperature of the sample injector.
Voltage-to-frequency converter	A device that converts an analog voltage signal to a proportional frequency signal. As used in chromatographs, the output frequency was a square wave so the peak integrator could count the pulses.
Volume (V)	The size of a three-dimensional space, measured in cubic-meters (m^3) or cubic-feet (ft^3).
Volume percent	A unit of concentration: the ratio of analyte volume to total volume of a sample, expressed as a percentage.

W

watt (W)	The SI derived unit of power (P). The definition of one watt is a power dissipation of one joule-per-second (J/s).
WCOT column	A wall-coated open-tubular column comprising a liquid stationary phase coated onto the walls of capillary tube typically made of fused silica. While highly effective, this kind of column is often too fragile for use in process gas chromatographs.
Weight percent	A unit of concentration: the ratio of analyte weight to total weight of a sample, expressed as a percentage. Effectively equal to mass percent.
Wet chemistry	Methods of chemical analysis involving chemical reactions in aqueous or non-aqueous solution.
Wheatstone bridge	An electrical circuit for detecting a small change of resistance. Has four resistors connected in a series-parallel arrangement. When the resistances are equal, the voltages at the junction of each series pair are equal giving a differential of zero millivolts. If one or more of the resistances changes, the differential voltage rapidly varies from zero giving a sensitive detection of resistance change.
Width at half height	See **peak width at half height**.
Width of peak	See **peak width at base** or **peak width at half height**.

X

Xenon Xe $M = 131.293$ g/mol	The chemical element with atomic number (Z) = 54. A heavy, colorless, odorless, and monatomic gas, the fifth of six noble gases. Not used in PGC.

X-ray fluorescence (XRF) Exposing an element to high-energy x-rays or gamma rays may cause the atoms to **fluoresce**, emitting x-rays at wavelengths that are characteristic of that atom. For each element measured, a dedicated process instrument would measure the incident fluorescent energy at one selected wavelength. However, a laboratory instrument might scan a range of x-ray wavelengths and output a spectrum of peaks representing the concentration of many different elements.

m-Xylene See **1,3-dimethylbenzene**.

o-Xylene See **1,2-dimethylbenzene**.

p-Xylene See **1,4-dimethylbenzene**.

Z

Zero line As used herein, the zero line is the baseline for integration. The integrated peak area comprises the total area above the zero line. Thus, setting up an appropriate zero line is an important prerequisite to accurate measurement.

Zeolite See **molecular sieves**.

Index

Process Gas Chromatographs: Fundamentals, Design and Implementation, First Edition. Tony Waters.

Printed and bound by CPI Group (UK) Ltd, Croydon, CR0 4YY

07/12/2025

14786064-0001